1st International Conference on Sustainable Energy and Future Electric Transportation (SeFet 2019)

E3S Web of Conferences Volume 87 (2019)

Hyderabad, India
14 - 16 February 2019

Editors:

Suresh Kumar Tummala
Phaneendra Babu Bobba

ISBN: 978-1-5108-8163-1

Printed from e-media with permission by:

Curran Associates, Inc.
57 Morehouse Lane
Red Hook, NY 12571

Some format issues inherent in the e-media version may also appear in this print version.

No changes have been made to the content of these proceedings. There may be changes to pagination and minor adjustments for aesthetics.

Printed by Curran Associates, Inc. (2019)

For additional information, please contact EDP Sciences – Web of Conferences at the address below.

EDP Sciences – Web of Conferences
17, Avenue du Hoggar
Parc d'Activité de Courtabœuf
BP 112
F-91944 Les Ulis Cedex A
France

Phone: +33 (0) 1 69 18 75 75
Fax: +33 (0) 1 69 28 84 91

contact-edps@webofconferences.org

Additional copies of this publication are available from:

Curran Associates, Inc.
57 Morehouse Lane
Red Hook, NY 12571 USA
Phone: 845-758-0400
Fax: 845-758-2633
Email: curran@proceedings.com
Web: www.proceedings.com

TABLE OF CONTENTS

Preface

1st International Conference on
Sustainable Energy and Future Electric Transportation

February 14-16, 2019, at GRIET, Hyderabad, INDIA

www.sefet.griet.ac.in

SeFet 2019 has organized at Gokaraju Rangaraju Institute of Engineering and Technology, Hyderabad, INDIA from 14-16 February 2019.

SeFet 2019 aims to provide a platform for Engineers, Researchers, Scientists & Academicians to present their work, to share experiences and ideas in the areas of Renewable Energy, Electric Transportation, Power Electronics, Electric Drives, Control Techniques, Smart Grids, Communication Protocols, Intelligent Charging Infrastructure and Standards. In view of changing scenario under sustainable energy & electric transportation, this conference aims to put together the experts from these areas to disseminate their knowledge and experience for working in years to come.

SeFet 2019 has been organized with the following Tracks:

- Electric Transportation and Vehicle System
- Power Electronics Systems and Electric Drives
- Renewable Energy – Solar, Wind Integration
- Machine Learning, IoT based Charge Stations and Storage
- Communication Protocols for Electric Vehicles
- FACTS, Smart Grids
- Control System and Automation

SeFet 2019 was organized in collaboration with Woosong University, South Korea and Central University of Karnataka, INDIA. Conference is Technically sponsored by Insta Sine Power Technologies, Mumbai, INDIA and Master PCB Ltd., Hyderabad, INDIA

SeFet 2019 followed double blind review with minimum of two reviewers for each paper submitted. Care has been taken for plagiarism of all the article. Licensed plagiarism tool (Turnitin and Plagiarism Checker X) have been used and ensured each article has less than 20% plagiarism

The SeFet 2019 peer-reviewed and accepted papers have been edited as conference proceedings to be published with E3S Web of Conferences (Open Access proceedings

https://doi.org/10.1051/e3sconf/20198700001

in Environment, Energy and Earth Sciences), which is indexed by Scopus, CPCI (Web of Science), Google Scholar, CAS, DOAJ, EBSCO, ProQuest and Ei Compendex.

As editors of this special issue of the SeFet 2019 proceedings, I wish to express my sincere gratitude to the plenary speakers, authors, participants, reviewers, program and technical committee members, and organizing members; who have contributed in organization of the conference and publication of this special issue. With high standard and high-quality submissions and presentations in the SeFet, I am very much sure, one day SeFet will become a leading conference in this specific academic area of 'Sustainable Energy and Future Electric Transportation'.

Dr. Suresh Kumar Tummala
Dr. Phaneendra Babu Bobba

Editors of the SeFet 2019 Proceedings
Department of Electrical & Electronics Engineering,
Gokaraju Rangaraju Institute of Engineering & Technology,
Hyderabad, INDIA 500090
www.eee.griet.ac.in
www.griet.ac.in

edp sciences

Web of Conferences

Statement of Peer review

In submitting conference proceedings to *Web of Conferences*, the editors of the proceedings certify to the Publisher that

1. They adhere to its **Policy on Publishing Integrity** in order to safeguard good scientific practice in publishing.
2. All articles have been subjected to peer review administered by the proceedings editors.
3. Reviews have been conducted by expert referees, who have been requested to provide unbiased and constructive comments aimed, whenever possible, at improving the work.
4. Proceedings editors have taken all reasonable steps to ensure the quality of the materials they publish and their decision to accept or reject a paper for publication has been based only on the merits of the work and the relevance to the journal.

Title, date and place of the conference

International Conference on Sustainable Energy and Future Electric Transportation
14-16 February 2019
Hyderabad, INDIA

Proceedings editor(s):

Dr. Suresh Kumar Tummala
Dr. Phaneendra Babu Bobba

Date and editor's signature

07.02.2019

17, avenue du Hoggar - PA de Courtabœuf – BP 112 - 91944 Les Ulis Cedex A (France)
Tél. : 33 (0)1 69 18 75 75 - Fax : 33(0)1 69 07 45 17 – www.edpsciences.org

An optimised system for energy monitoring and data acquisition in substations/domestic applications using IoT

V. Badri Rama Krishnan[1*], *Keerthana* Sandepudi[1], and *Shaista* Gazal[1]

[1]G.Narayanamma Institute Of Technology and Science, Electrical and Electronics Engineering Department, Hyderabad, Telangana, India

Abstract. The rising demand and tariff rates of electric power have derived the need to develop innovative methods to monitor, manage, economize and save energy. Energy monitoring and conservation are pre-eminent today because of the imbalance between power generation and demand. This paper proposes the outline and application of electrical energy measurement by an Arduino micro-controller that is used to calculate the power consumed by any individual electrical device. The main intention of the presented energy meter is to monitor, acquire and interpret the power consumption data online by integrating smart plugs, sensors at the device level and upload it to the server using IoT based communication protocol. Thus, this project enables the consumer to act accordingly to save power and to render a reliable power supply by making maximum use of energy sources in the cheapest ways.

1 Introduction

The data acquired by a system including software and hardware allows one to measure the energy consumed. Efficient usage of energy in smart homes reserves capital, sustainability and decreases carbon footprint at the whole. Necessarily, the need for smart energy management has ascended for the suburbans. However, the expense, complexity and high maintenance technology has somewhat limited a large-scale implementation of such systems. As IoT technologies are ubiquitous, they can provide a large scale computing platform for sensing, monitoring and controlling the household appliances energy consumption. This data can be collected using different wireless sensors which are installed in residential areas. Using predictive analysis and advanced methods, the data can be collected, analyzed and monitored in the form of reports, graphs and charts. Thus, this analyzed data can help common public, utilities, suppliers and utility operational stakeholders to gain significant intuitions on energy consumption of smart homes. The energy suppliers can use the power consumption data available to give an adaptable and on-demand supply with pertinent energy marketing plans [1]. Having a easy access to the electricity utilities and knowledge of their of consumption history, the users can adjust and optimize their power consumption and reduce their overall electricity bills. To have an functional cost saving system, it becomes necessary to monitor and control the operation of domestic applications depending on the collective power consumption over a time period, the maximum power consumption, the effect of weather/atmospheric conditions and consumption slab rates. This is where the combination of IoT technology, energy management solutions on a local and national level can be implemented using Big Data analytics and BI [2]. Moreover, the use of IoT also allows seamless remote control of home electronics permitting the customers to their ON/OFF usage pattern via a personal computer or a mobile phone.

The monitoring of power consumption and energy of residential loads using ZigBee as the communication protocol was first introduced. Although, the data that was collected and aggregated solely by the home server in this system could lead to data loss in case of a system shut-down. Additionally, a bridge between ZigBee and TCP/IP stack would be needed to connect it to a community of homes. It has also been extended to extensive domains in the IoT paradigm employing the GSM/GPRS networks to remotely control the end-devices [3]. Correspondingly, authors in [4] emphasize on IoT based DC powered homes to develop a DC distribution system circumscribing all residential DC-based loads that interact with each other. Nevertheless, the dearth of standardized protocols and regulations were the main challenges in considering smart DC powered homes as appropriate substitute to AC power systems. IoT provides an coherent interface for DC powered technologies in economic energy distribution to combat these problems.

Installing home EMS can help the public-owners to comprehend contribution of each device towards the electricity bill they receive every month. Furthermore, most of former work has basically prioritised individual smart homes and lack the energy management provisions for regional utility providers or national level utility centers. The technology to collect huge volumes of data from home sensor networks is available, however, managing the collected data efficiently and extracting deeper insights from it remains a challenge.

The existing models on EMS and cost saving models are implemented on discrete units while the proposed model can be built on top of the existing paradigms to provide a distributed EMS platform from consumer to community levels. It has allies found from a study found that home energy management system can reduces electricity consumption by 17–20% potentially with tolerable discomfort [5].

Smart energy meter[6] is developed using the Wifi module based on three main objectives. They are:-

* Corresponding author: badriramakrishnan@gmail.com

1. To provide automated load energy immediately.

2. To optimize the usage of electricity.

3. To reduce power wastage and make it more economic.

The proposed system design removes human inclusion in electricity maintenance and introduces cost-effectiveness. It allows the consumer to audit the energy consumption in watts from a webpage. The webpage evaluates and provides a detailed description and visualization of the energy usage statistics. Wi-Fi unit performs IoT operation by transferring the data of the load to Google sheets(docs). Hence, the consumer can manage power by realizing the energy usage time to time. This system utilizes an Arduino microcontroller. The units that are generated are displayed via the Wifi module. The IoT hypothesis gives the basic framework and opportunities to form a connection between the physical world and computer based systems. Hardware devices are connected with each other over the internet.

2 Components Required

1. SENSOR DATA: Smart plugs (Sensors MODULES)

2. GATEWAY (WIFI Modules)

3. GOOGLE SHEETS (that will help individual users to post and retrieve sensor data)

2.1 Block diagram

The smart electricity meter consists of an ARDUINO Uno Board, ESP8266 WiFi module, current and voltage sensors and a power supply.

The system is designed on an ARDUINO microcontroller. It's structure is differentiated into three sections viz.,the controller, sensors and a WiFi unit. The controller does the basic calculations and processes the information. The voltage and the currents sensors transfer the readings to ESP8266 with the help of the CP2102 USB to UART driver and upload to Google sheets using the Wifi module.The system sample the signals and converts the resulting samples into digital numeric values that can be manipulated by a computer. The components of data acquisition systems include:

- Sensors that convert the physical parameters to electrical signals.
- Signal conditioning circuitry that converts the sensor signals to be further converted into digital values.
- Analog-to- digital converters that convert conditioned sensor signals to digital values.

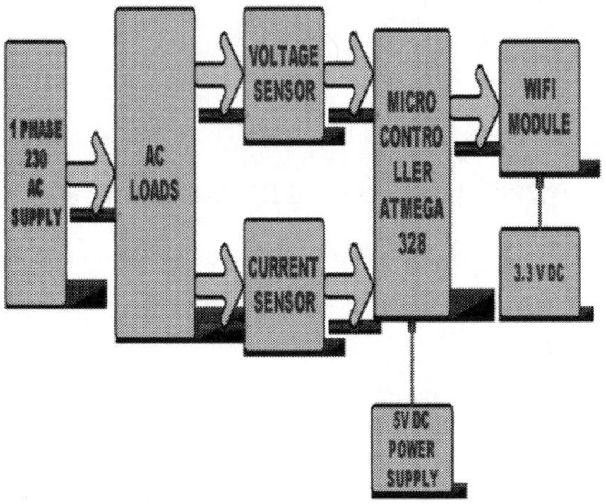

Fig. 1. Block diagram of the electricity meter

3 Hardware Implementations

3.1 Current Sensing

The measurement of AC current has always been laborious due to coupling of noise signals with its inappropriate isolation etc. But, with the help of ACS712 module devised by Allegro Microsystems, it has become a lot easier. According to the Hall-effect, when a current carrying conductor is placed in a magnetic field, an emf is generated across its edges perpendicular to the directions of both the current and the magnetic field that is measured by the hall sensor. This is called as the hall-voltage, measured in terms of millivolts. The measured hall-voltage is directly proportional to the current flowing through the conductor.

The major advantage of using ACS712 Current Sensor is that it can measure both AC and DC currents and it also provides isolation between the Load (AC/DC load) and the Measuring Unit (Microcontroller part). As shown in the picture we have three pins on the module which are Vcc, Vout and Ground respectively. The current from output voltage of ACS712 current sensor can be calculated as follows:

- When there is no current flowing through the sensor, output voltage will be Vcc / 2. Where Vcc is power supply voltage given to acs712 to current sensor.

- If the Vcc = 5 volt, then the output voltage of current sensor will be equal to 2.5 when there is no current passing through a sensor. 2.5 volt is the offset voltage or base voltage of the sensor which should be subtracted from the measured voltage.

- The output voltage decreases when current starts passing through the sensor.

- So we can calculate dc current by using following commands:

https://doi.org/10.1051/e3sconf/20198701001

adcvalue = analogRead(A0);

Voltage = (adcvalue / 1024.0) * 5000;

Current = ((Voltage- voltage_offset) / mVperAmp);

Fig. 2. ACS712 Current Transformer

3.2 Voltage Sensing

ARDUINO is generally operated and manufactured to work with voltage ratings less than or equal to 5V. So, it is not possible to directly measure AC voltage greater than 230V by giving high input voltages to microcontrollers which may cause temporary or permanent damage to microcontrollers. Hence, a simple transformer with less current rating which steps down 230V to 12V is used. The voltage divider decreases the voltage being measured from 12V to 5V. Thereby, ARDUINO analog inputs can be used to measure DC voltage between 0 and 5V (on 5V ARDUINOS such as the Arduino Uno when using the standard 5V analog reference voltage). Code in the ARDUINO sketch is then used to calculate the actual voltage being measured. This allows voltages greater than 5V to be measured.

Fig. 3. Potential Transformer

The input voltage value is given to pin A2 of the Arduino. It provides some simple filtering by adding up 10 analog values from pin A2 sampled at 10 ms intervals. After taking 10 samples, the voltage is calculated using the average of the 10 sampled values and is sent out of the serial port for display on the Arduino Serial Monitor window. Ten analog samples are taken using the following code 1. The sum or total of all 10 values added together are stored in sum. The variable sample count keeps track of the number of samples. The voltage is calculated using:

voltage = ((float)sum / (float)NUM_SAMPLES * 5.0) / 1024.0;

A calibrated value is used in place of 5.0 in the above sketch. The 10 samples (sum) is divided by 10 (NUM_SAMPLES) to get the average value read.5.0 is the 5V ADC reference voltage. 1024.0 is the maximum value that the ADC can have plus 1 (1023 + 1 or 2 to the power of 10 plus 1) 1023.0 can also be used here.

4 Operation

The supply voltage is connected to the AC loads with the current sensor in series and the transformer connected in parallel. The sensed values of current and voltage are given to the A1 and A2 pins of the Arduino respectively. This ARDUINO is based on ATmega328 microcontroller where the program to evaluate the sensed values is loaded into. Now, the IoT operation is performed by ESP8266 Wifi module by sending energy meter data to web-page which can be accessed through an IP address. The Tx, Rx of ESP8266 pins are connected to the pins 7 & 8 of the ARDUINO board. CP2102 is an USB2.0 to TTL UART Converter module based on CP2102 Bridge by SiLabs. It is used with Laptop's not having a standard serial port. This module creates a virtual COM port using USB on your computer which can support various standard Baud Rates for serial communication. The Rx and Tx pin can be connected directly to the MCUs pins (assuming 5v I/O). The suitable interfacing between Arduino with ACS712 and arduino with ESP8266 is done and the respective codes are dumped on to it using Arduino computer program interface.

When the supply is given and the Arduino code is executed, the sensed voltage and the current readings are transferred to ESP8266 with the help of the CP2102 USB to UART driver. By following the TCP/IP stack protocol, we send this data to the suitable webpage using a WIFI router. By using Google sheets, we list down the values of voltage, current and power being updated continuously over the required period of time. Using ESP8266 User Data Program (UDP) protocol data can be sent to local routers or intermediate stations for data continuity or redundancy. Hence provides extra support for data transfer during internet shutdowns too. Analysis of power quality will be much more user or customer understandable with the Google sheets data platform, As the data is free to be utilized with many simulation as well as real time data software's.

https://doi.org/10.1051/e3sconf/20198701001

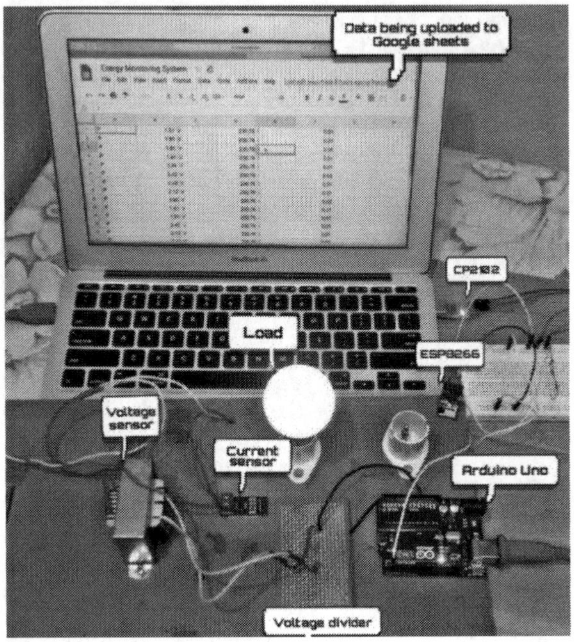

Fig. 4. Hardware Circuitry with output being displayed in the Google sheets

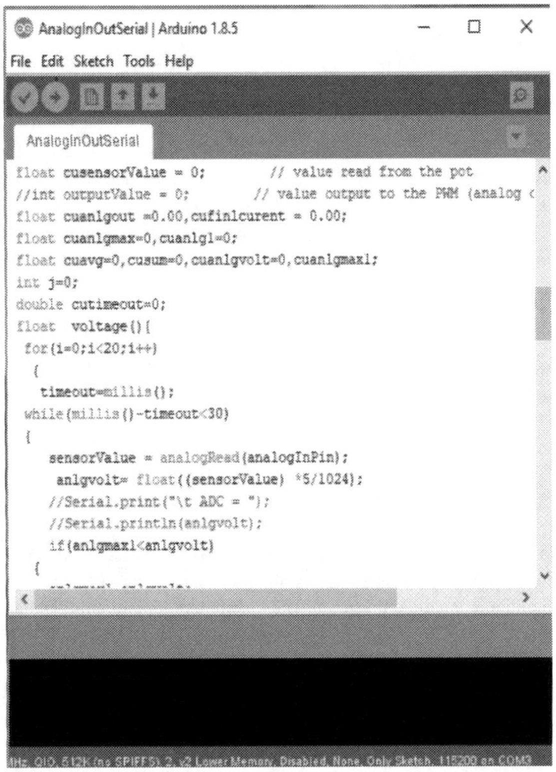

Fig. 5. Code for Arduino uno

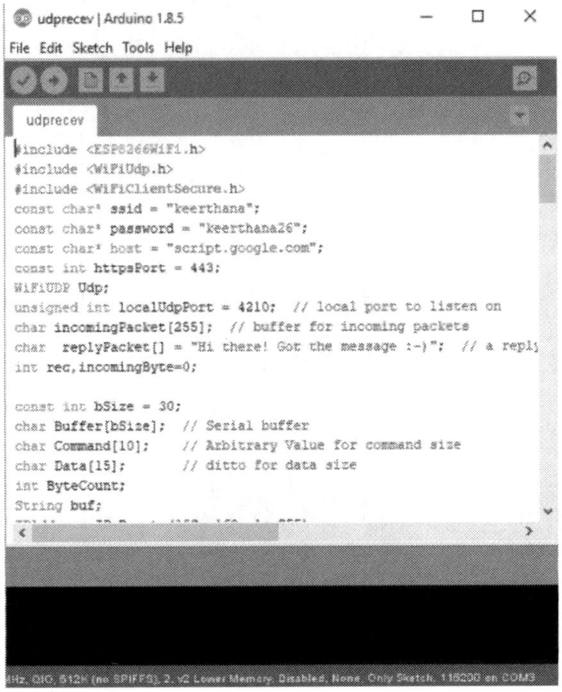

Fig. 6. Code for ESP8266 Wifi Module

5 Results

The power, voltage and current values sensed and measured are seen in the below picture in the serial monitor of the Arduino. The data Is further sent to the ESP8266. The measured data from the ESP8266 is transmitted to the Google sheets following the TCP/IP protocol as shown in fig. 7. and fig. 8.

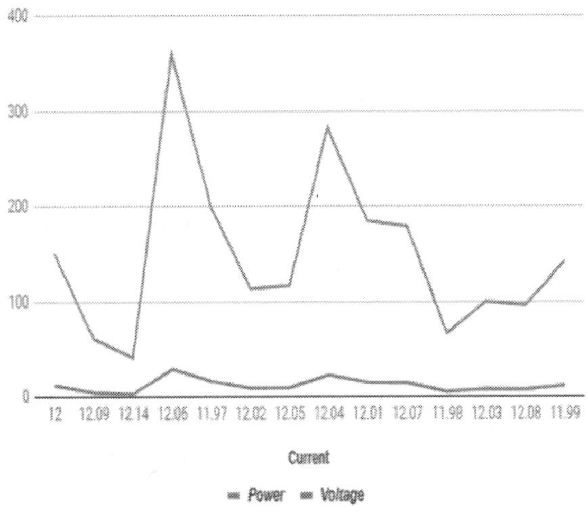

Fig. 7. Output power, voltage and current as displayed in Google sheets chart

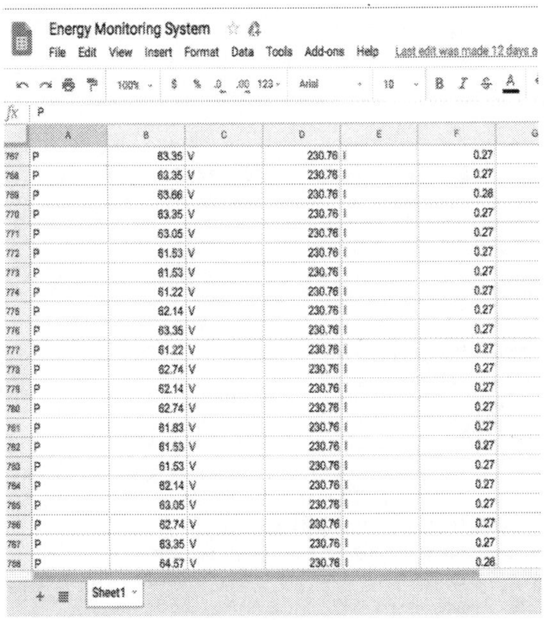
Energy Monitoring System
File Edit View Insert Format Data Tools Add-ons Help Last edit was made 12 days a

	A	B	C	D	E	F
767	P	83.35	V	230.76	I	0.27
768	P	63.35	V	230.76	I	0.27
769	P	63.66	V	230.76	I	0.28
770	P	63.35	V	230.76	I	0.27
771	P	63.05	V	230.76	I	0.27
772	P	61.53	V	230.76	I	0.27
773	P	61.53	V	230.76	I	0.27
774	P	61.22	V	230.76	I	0.27
775	P	62.14	V	230.76	I	0.27
776	P	63.35	V	230.76	I	0.27
777	P	61.22	V	230.76	I	0.27
778	P	62.74	V	230.76	I	0.27
779	P	62.14	V	230.76	I	0.27
780	P	62.74	V	230.76	I	0.27
781	P	61.83	V	230.76	I	0.27
782	P	61.53	V	230.76	I	0.27
783	P	61.53	V	230.76	I	0.27
784	P	62.14	V	230.76	I	0.27
785	P	63.05	V	230.76	I	0.27
786	P	62.74	V	230.76	I	0.27
787	P	63.35	V	230.76	I	0.27
788	P	64.57	V	230.76	I	0.28

Sheet1

Fig. 8. Output power, voltage and current as displayed in Google sheets

6 Conclusion

The proposed design is set to unfurl a new approach for smart energy management on the IoT platform. The system allows its users to monitor and control devices remotely via a user-friendly interface. The google sheets data provides a platform for many other vendors to use this data for further improvement of the power quality of the system. Data obtained can be sent to other cloud applications or Artificial Intelligence techniques can be enforced to predict further data analysis.

Tracking the performance of your electrical devices and system is also possible by power and energy management system. It provides real-time monitoring of the electrical parameters and makes you aware of any irregularities. That helps you to recognize and address any potential problems before the power supply completely goes off. The system also gives you data on how well your electrical system is supplying power to the devices and critical loads, helping to fortify optimal efficiency and safety for the electrical system and its various components.

References

1. International Research Journal of Engineering and Technology (IRJET), "IoT Based Smart Energy Monitoring", **e-ISSN: 2395-0056 Volume: 05 Issue: 03** | Mar-2018.
2. IEEE Transactions on Consumer Electronics, "A Smart Home Energy Management System Using IoT and Big Data Analytics Approach", **Vol. 63, No. 4** | November 2017.
3. G. Mingming, S. Liangshan, H. Xiaowei and S. Qingwei, "The System of Wireless Smart House Based on GSM and ZigBee", 2010 International Conference on Intelligent Computation Technology and Automation, Changsha | 2010, pp. **1017-1020**.
4. E. Rodriguez-Diaz, J. C. Vasquez and J. M. Guerrero, "Intelligent DC Homes in Future Sustainable Energy Systems: When Efficiency and Intelligence Works Together", IEEE Consumer Electron. Magazine, vol. 5, no. 1, pp. **74-80** | Jan. 2016.
5. K.Dittawit and F. A. Aagesen, "Home energy management system for electricity cost savings and comfort preservation", 2014 IEEE Fourth International Conference on Consumer Electronics Berlin (ICCE-Berlin), Berlin, 2014, pp. **309-313**.
6. Internet of Things (Iot) Based online Energy Monitoring System, Prof.M.Chakravarthy.
7. http://microcontrollerslab.com/acs712-current-sensor-interfacing-arduino/.
8. http://microcontrollerslab.com/data-receiving-webpage-arduino-esp8266/.
9. http://engineerexperiences.com/ac-voltagemeasurement.html.
10. https://www.silabs.com/products/interface/usb-bridges/classic-usb-bridges/device.cp2102.

Speed Control of an Induction Motor Fed by an Inverter Using dSPACE Controller

Srinivasrao Janiga[1], *Syed* Sarfaraz Nawaz[2], *Suresh Kumar* Tummala[3] and *Srinivasa Varma* Pinni[4]

[1]Research Scholar, EEE Department, KLEF & Assistant Professor, EEE Dept., Anurag Engineering College
[2]Associate Professor, EEE Department, GRIET, Hyderabad, India
[3]Professor, EEE Department, GRIET, Hyderabad, India
[4]Associate Professor, EEE Department, KLEF, India

Abstract. This paper presents the design and implementation of Inverter system for driving three phase Induction motor using DSPACE DS1104 controller with the controlling objective space vector pulse width modulation (SVPWM) technique. AC motor drives are commonly used over DC motor drives because of their more advantages. Induction motor is the most commonly used AC motor drive for various industrial and domestic applications. The project will be commenced by a basic understanding of SVPWM inverter, components used in the design and study the mathematical equations of the Induction motor. The performance of SVPWM based Induction motor (IM) in open loop is presented with simulation. Here the hardware implementation of the three phase inverter which is fed to a three phase induction motor driven by DSPACE CP1104 is been implemented.

1 Introduction

AC motor drives are commonly used over DC motor drives because of their more advantages. Induction motor is the most commonly used AC motor drive for various industrial and domestic applications. Generally Pulse Width Modulation (PWM) technique is used for driving a motor drive. In this paper design of variable speed induction motor drive using Space Vector Pulse Width Modulation (SVPWM) inverter is explained. PWM signals are generated using a DSPACE Controller.

1.1 Open loop Block diagram

Block diagram of single phase or three phase induction motor drive fed by an inverter driving in open loop is given below.

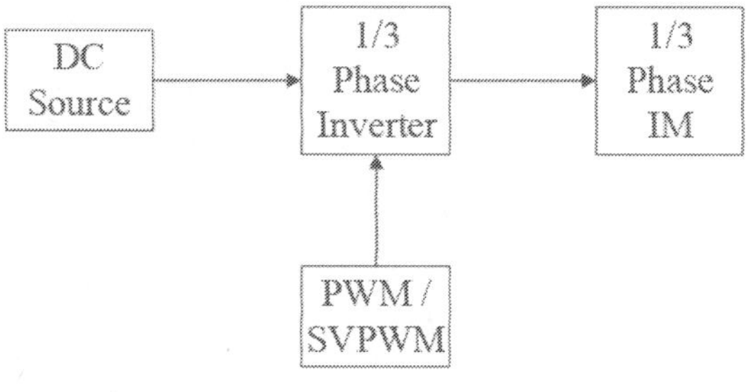

Fig. 1. Open Loop Block Diagram

2 Inverter

Depending on the supply inverters are classified in to two types. They are

- Single-phase Inverter
- Three-phase Inverter

2.1 Space vector pulse width modulation

Space vector pulse width modulation (SVPWM) is a pulse width modulation (PWM) technique which is used to apply to a three phase induction motor. The main objective of this technique is to use a dc voltage source and by means of six IGBT switches produce a three phase sine waveform in which frequency and amplitude is adjustable.

To implement space vector pulse width modulation, the voltages equations in abc reference frame can be transformed in to the stationary d-q reference frame that consists of horizontal and vertical axis

* Corresponding author: janigasrinivasrao@gmail.com

https://doi.org/10.1051/e3sconf/20198701002

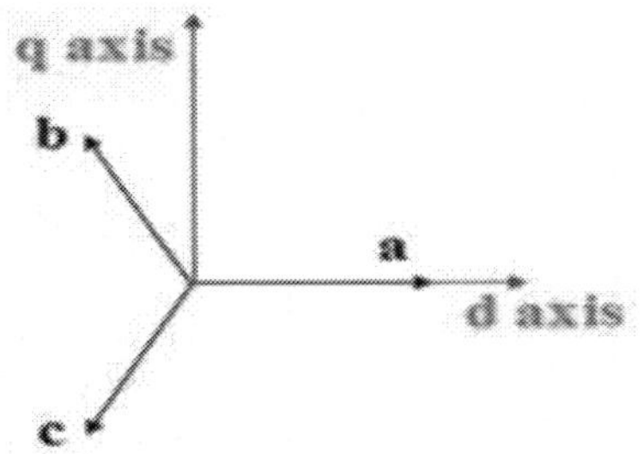

Fig. 2. Transformation of three axis to two axis reference frame

In space vector pulse width modulation six non-zero vectors (V1, V2, V3, V4, V5, V6) and two zero vectors (V0, V7) are formed. These are called the space vectors. The objective of space vector pulse width modulation technique is to approximate the reference voltage vector Vref using the eight switching patterns.

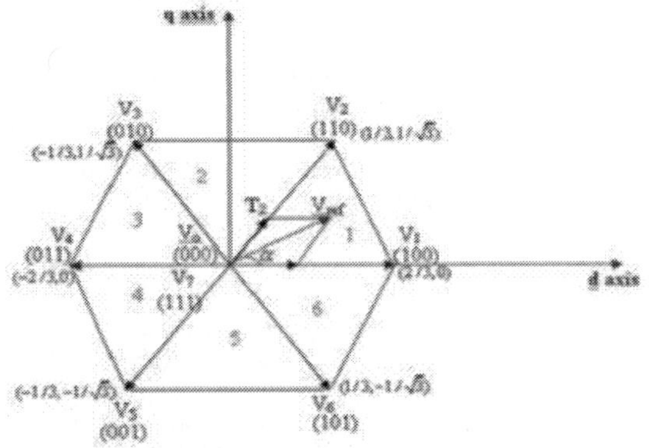

Fig.3. Basic switching vectors and sectors

Therefore, space vector pulse width modulation can be implemented by the following steps:

Step 1. Determine Vd, Vq, Vref, and angle (α)
Step 2. Determine time duration T1, T2, T0
Step 3. Determine the switching time of each transistor (S1 to S6)

Step 1: Determine the Vd, Vq, Vref, and an angle (α)

From figure 4, the Vd, Vq, Vref, and angle (α) can be determined as follows:

$$V_d = V_{an} - V_{bn}.\cos(60) - V_{cn}.\cos(60)$$

$$= V_{an} - \frac{1}{2}V_{bn} - \frac{1}{2}V_{cn}$$

$$V_q = 0 + V_{bn}.\cos(30) - V_{cn}.\cos(30)$$

$$= 0 - \frac{\sqrt{3}}{2}V_{bn} - \frac{\sqrt{3}}{2}V_{cn}$$

$$\begin{bmatrix} V_d \\ V_q \end{bmatrix} = \frac{2}{3}\begin{bmatrix} 1 & -\frac{1}{2} & -\frac{1}{2} \\ 0 & \frac{\sqrt{3}}{2} & -\frac{\sqrt{3}}{2} \end{bmatrix}\begin{bmatrix} V_{an} \\ V_{bn} \\ V_{cn} \end{bmatrix}$$

$$\left|V_{ref}\right| = \sqrt{V_d^{\,2} + V_q^{\,2}}$$

$$\alpha = \tan^{-1}\left(\frac{V_q}{V_d}\right) = \omega t = 2\pi f t$$

Where f= fundamental frequency

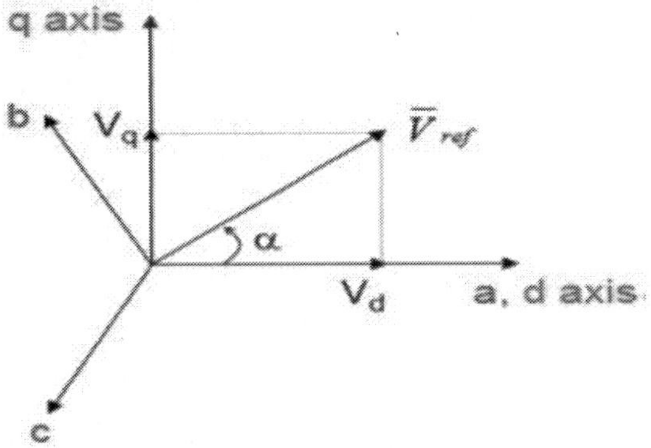

Fig. 4. Voltage Space Vectors and its components in d-q axis

Step 2: Determine time duration of T1, T2, T0
From figure5, the switching time duration can be calculated as follows:

Switching time duration at Sector 1

$$\int_0^{T_z} V_{ref} = \int_0^{T_1} V_1 dt + \int_{T_1}^{T_1+T_2} V_2 dt + \int_{T_1+T_2}^{T_z} V_0 dt$$

$$T_z V_{ref} = T_1 V_1 + T_2 V_2$$

$$T_z \left|V_{ref}\right| \begin{bmatrix} \cos\alpha \\ \sin\alpha \end{bmatrix} = T_1 . \frac{2}{3} . V_{dc} \begin{bmatrix} 1 \\ 0 \end{bmatrix} + T_2 . \frac{2}{3} . V_{dc} \begin{bmatrix} \cos(\frac{\pi}{3}) \\ \sin(\frac{\pi}{3}) \end{bmatrix}$$

(Where, $0 \le \alpha \le 60$)

$$T_1 = T_z . a . \frac{\sin(\frac{\pi}{3} - \alpha)}{\sin(\frac{\pi}{3})}$$

$$T_2 = T_z . a . \frac{\sin(\alpha)}{\sin(\frac{\pi}{3})}$$

$$T_0 = T_z - (T_1 + T_2)$$

$$(\text{where } T_z = \frac{1}{f_z} \text{ and } a = \frac{|V_{ref}|}{\frac{2}{3} V_{dc}})$$

Switching time duration at any Sector

$$T_1 = \frac{\sqrt{3}.T_z.|V_{ref}|}{V_{dc}} (\sin(\frac{\pi}{3} - \alpha + \frac{n-1}{3}\pi))$$

$$= \frac{\sqrt{3}.T_z|V_{ref}|}{V_{dc}} (\sin\frac{n\pi}{3}\cos\alpha - \cos\frac{n\pi}{3}\sin\alpha))$$

$$T_2 = \frac{\sqrt{3}.T_z.|V_{ref}|}{V_{dc}} (\sin(\alpha - \frac{n-1}{3}\pi))$$

$$= \frac{\sqrt{3}.T_z|V_{ref}|}{V_{dc}} (-\cos\alpha \sin\frac{n-1}{3}\pi + \sin\alpha\cos\frac{n-1}{3}\pi)$$

$$T_0 = T_z - (T_1 + T_2)$$

[where, n=1 to 6 (i.e, sector 1 to 6)]
($0 \leq \alpha \leq 60$)

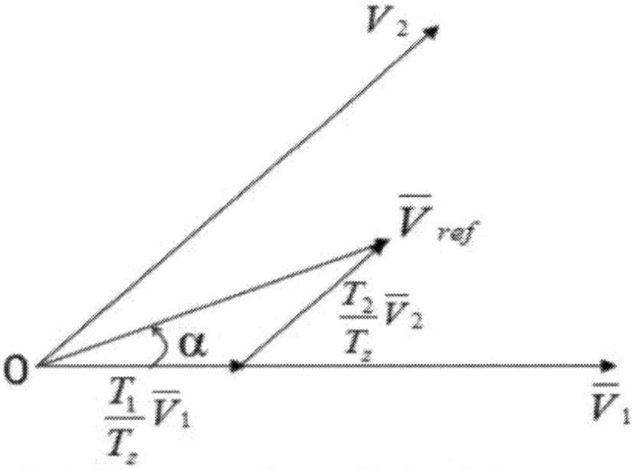

Fig.5. Reference vector from combination of adjacent vectors

Step 3: Determine the switching time of each transistor (S1 to S6)

Table 1. Switching time at each sector

Sector	Upper Switches (S1, S3, S5)	Lower Switches (S4, S6, S2)
1	S1=T1+T2+T0/2 S3=T2+T0/2 S5=T0/2	S4=T0/2 S6=T1+T0/2 S2=T1+T2+T0/2
2	S1=T1+T0/2 S3=T1+T2+T0/2 S5=T0/2	S4=T2+T0/2 S6=T0/2 S2=T1+T0/2
3	S1=T0/2 S3=T1+T2+T0/2 S5=T2+T0/2	S4=T1+T2+T0/2 S6=T0/2 S2=T1+T0/2
4	S1=T0/2 S3=T1+T0/2 S5=T1+T2+T0/2	S4=T1+T2+T0/2 S6=T2+T0/2 S2=T0/2
5	S1=T2+T0/2 S3=T0/2 S5=T1+T2+T0/2	S4=T1+T0/2 S6=T1+T2+T0/2 S2=T0/2
6	S1=T1+T2+T0/2 S3=T0/2 S5=T1+T0/2	S4=T0/2 S6=T1+T2+T0/2 S2=T2+T0/2

3 Induction motor

An Induction motor works on the principle of Electro Magnetic Induction (EMI). Stator and rotor are the two parts of induction motor which are separated by an air gap. When a three phase supply is given to an induction motor, it produces a constant amplitude and rotating magnetic field (RMF). This field cuts the rotor conductor and induces an Electro Magnetic Force (emf) which drives current in short circuited rotor conductor.

3.1 Dynamic Modelling of Induction Motor

Dynamic model of an induction motor can be obtained from two-axis theory of electrical machines. The following assumptions are made for dynamic model of induction motor.

1. Space harmonics are negligible.
2. The slotting in stator and rotor produces negligible variations in respective inductances.
3. Mutual Inductances are equal.
4. Voltage and Current harmonics are neglected.
5. Saturation, hysteresis and eddy current losses are negligible.

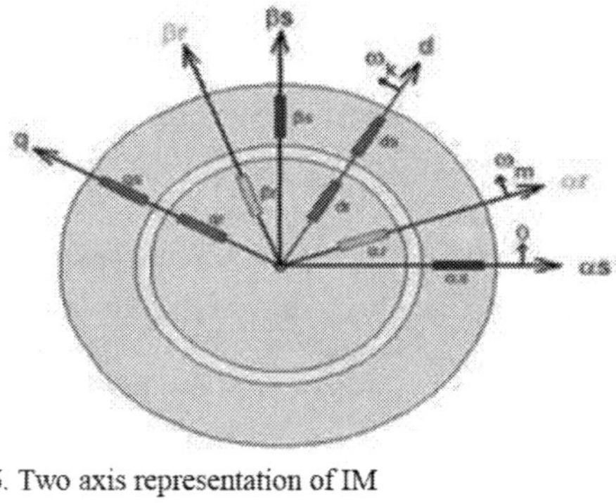

Fig.6. Two axis representation of IM

Equations corresponding to the two axis representation of Induction Machine are reduced through KVL as follows

$$V_{qs} = R_q i_{qs} + P(L_{qq} i_{qs}) + P(L_{qd} i_{ds}) + P(L_{q\alpha} i_d) + P(L_{q\beta} i_\beta) \text{----(1)}$$

https://doi.org/10.1051/e3sconf/20198701002

$$V_{ds} = P(L_{dq} i_{qs}) + R_d i_{ds} + P(L_{dd} i_{ds}) + P(L_{dd} i_d) + P(L_{d\beta} i_\beta) ----(2)$$

$$V_\alpha = P(L_{\alpha q} i_{qs}) + P(L_{\alpha d} i_{ds}) + R_\alpha i_\alpha \; P(L_{\alpha\alpha} i_\alpha) + P(L_{\alpha\beta} i_\beta) ----(3)$$

$$V_\beta = P(L_{pq} i_{qs}) + P(L_{pq} i_{ds}) + P(L_{\beta\alpha} i_\alpha) + R_\beta i_\beta + P(L_{\beta\beta} i_\beta) ----(4)$$

$$\begin{aligned} L_{\alpha\alpha} &= L_{\beta\beta} = L_{rr} \\ L_{dd} &= L_{qq} = L_{ss} \\ L_{\alpha\beta} &= L_{\beta\alpha} = 0 \\ L_{dq} &= L_{qd} = 0 \end{aligned}$$

$$\begin{aligned} L_{\alpha d} &= L_{d\alpha} = L_{sr}\cos\theta_r \\ L_{\beta d} &= L_{d\beta} = L_{sr}\sin\theta_r \\ L_{\alpha q} &= L_{q\alpha} = L_{sr}\sin\theta_r \\ L_{\beta q} &= L_{q\beta} = -L_{sr}\cos\theta_r \end{aligned}$$

With respect to the fictitious rotor

$$i_\alpha = i_{drr}\cos\theta_r + i_{qrr}\sin\theta_r$$
$$i_\beta = i_{drr}\sin\theta_r + i_{qrr}\cos\theta_r$$

with respect to the arbitrary reference frame

$$i_{ds} = i_{ds}^c\cos\theta_c - i_{qs}^c\sin\theta_c$$
$$i_{qs} = i_{ds}^c\cos\theta_c + i_{qs}^c\sin\theta_c$$

$$\begin{bmatrix} V_{qs}^c \\ V_{qr}^c \\ V_{ds}^c \\ V_{dr}^c \end{bmatrix} = \begin{bmatrix} R_s + L_s P & -w_c L_s & L_m P & w_c L_s \\ w_c L_s & R_s + L_s P & w_c L_s & L_m P \\ L_m P & -(w_c - w_r)L_m & R_s + L_s P & -(w_c - w_r)L_m \\ -(w_c - w_r)L_m & L_m P & -(w_c - w_r)L_m & R_s + L_s P \end{bmatrix} \begin{bmatrix} V_{qs}^c \\ V_{qr}^c \\ V_{ds}^c \\ V_{dr}^c \end{bmatrix}$$

Torque: $V = [R]i + [L]Pi + [G]w_r i + [F]w_c i$

$$I^t.V = i^t[R] = i^t[L]Pi + i^t[G]w_r i + i^t[F]w_c i$$

$i^t[L]Pi$ Rate of change of stored magnet energy

$i^t[G]w_r i$ Air gap power = Mech Rot speed x Air gap

$$w_n T_e = P_a = i^t[G]iw_r$$
$$T_e = \frac{P}{2} i^t[G]i$$

$$[G]i = i^t \begin{bmatrix} 0 & 0 & 0 & 0 \\ 0 & 0 & 0 & 0 \\ 0 & -l_m & 0 & -l_m \\ l_m & 0 & l_m & 0 \end{bmatrix} \begin{bmatrix} i_{qs}^c \\ i_{qr}^c \\ i_{ds}^c \\ i_{dr}^c \end{bmatrix}$$

$$I^t[G]i = L_m\left[i_{qs}^c i_{dr}^c - i_{ds}^c i_{qr}\right]$$

$$T_e = \frac{3}{2}\cdot\frac{P}{2} L_m (i_{qs}^c i_{dr}^c - i_{ds}^c i_{qr}^c)$$

$$T_e = \frac{3}{2}\cdot\frac{P}{2}\cdot\frac{1}{w_b}\left[\psi_{ds}^c\, i_{qs}^c - \psi_{qs}^c i_{ds}^c\right]$$

Through the above analysis speed, torque of an Induction Motor can be controlled by voltage current parameters.

4 Simulation

Matlab is an interactive software program for numerical computations and graphics. In this paper three phase SVPWM inverter is explained in open loop.

Open loop three phase inverter:

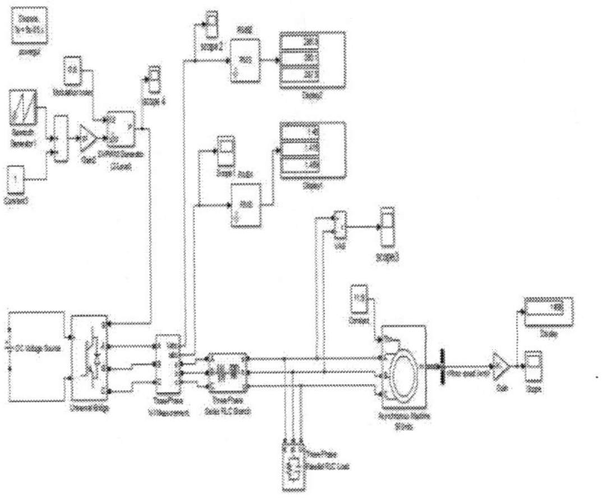

Fig.7. Simulink of three phase inverter driving IM in open loop

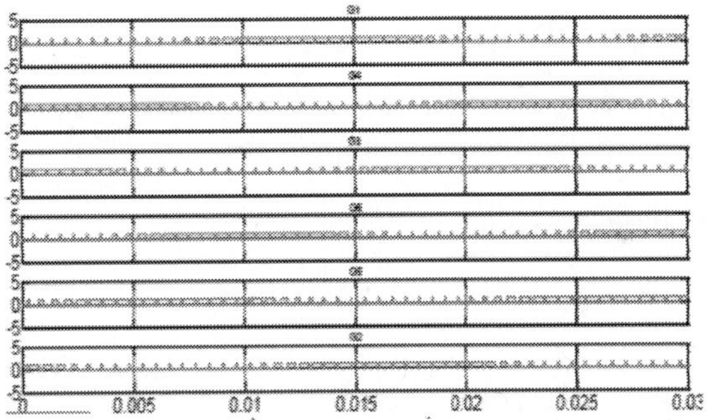

Fig.8. Gate pulses

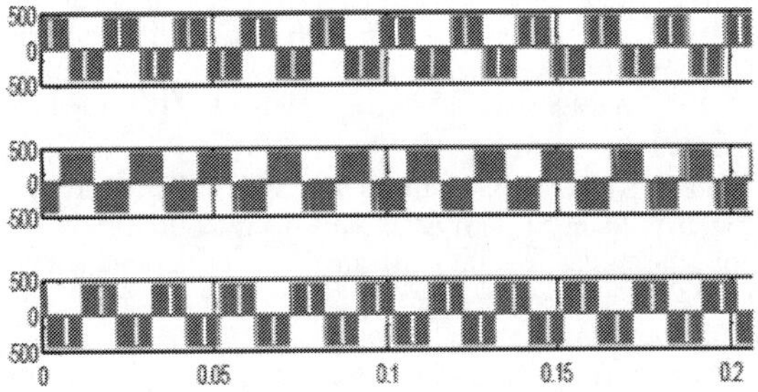

Fig.9. Output phase Voltages

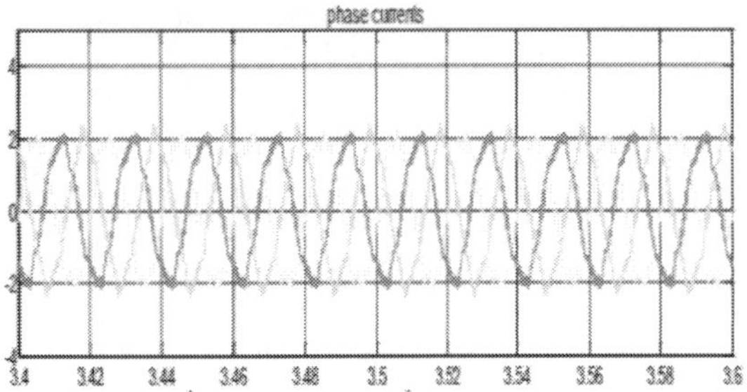

Fig.10. Output phase Currents

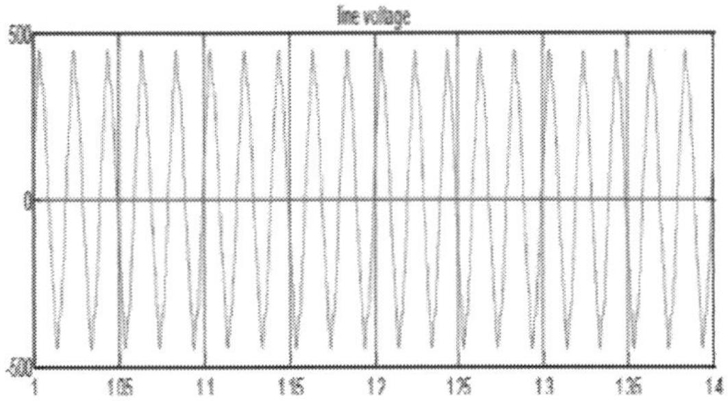

Fig.11. Output Line Voltage

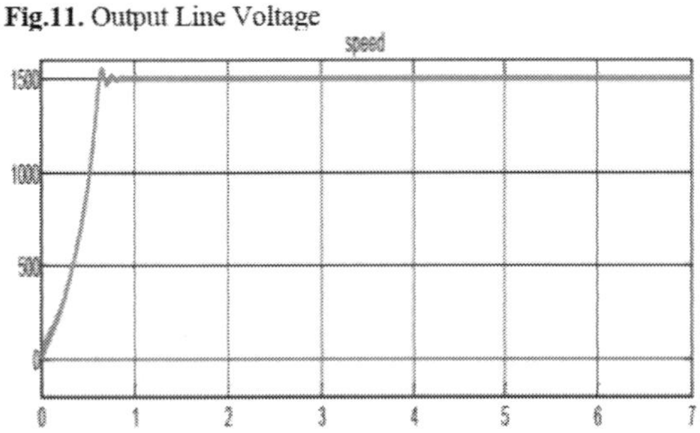

Fig.12. Response of Speed

5 DSPACE DS1104 Controller

The hardware unit comprises of the DSPACE unit installed to a desktop where in the MATLAB files can be interfaced through the DSPACE package software. Three phase Inverter is been designed with an isolation circuit to protect the controller board from the sudden voltage fluctuations feeding back to the controller board. Pulses from the controller board reaches the switches through the isolation circuit through opto-couplers EL817, pulses passes through the EL817 by light manner.

Inverter feeds the induction motor. And it is well suited for drive control. DSPACE unit operates for 250MHz and the pulse width generated is of 2k frequency. DSPACE unit comprises of 8 analog and 8 digital pins. 20 input and output signals can be interchanged randomly. Input to the Inverter unit is 440volts where the output of the Inverter received is 400 Volts 2 A

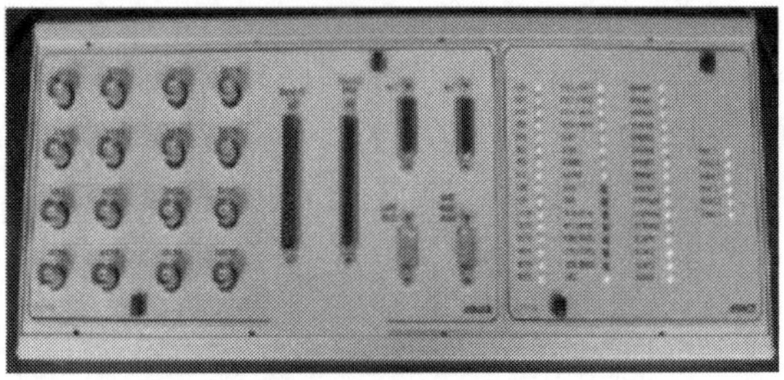

Fig.13. DSPACE Controller Board

6 Experimentation Results

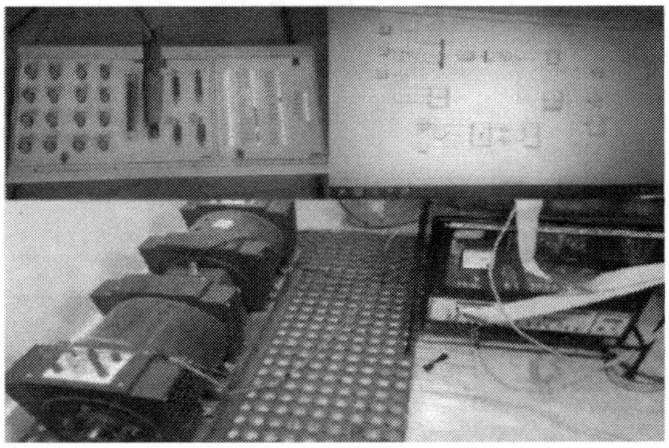

Fig.14. Hardware unit of Three phase Inverter driving IM using DSPACE

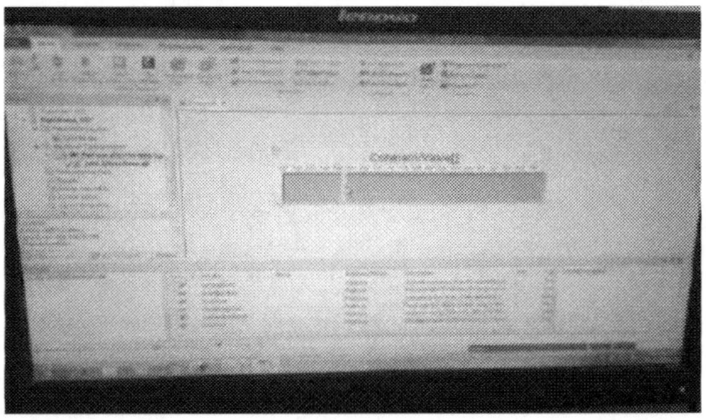

Fig.15. Varying Modulation index in DSPACE

Fig.16. Pulses generated through DSPACE

https://doi.org/10.1051/e3sconf/20198701002

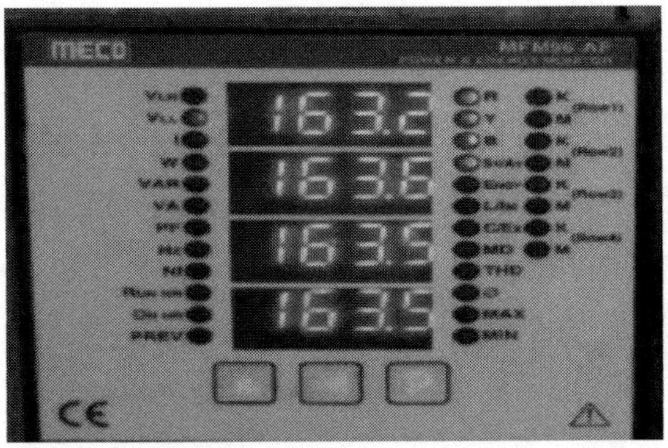

Fig.17. Output Voltage of Inverter

Fig.18. Speed of Induction motor

Practical Results:

Table 2. Output Voltage and Speed

S.NO	DC InputVoltage (V)	Modulation Index (M)	AC Output Voltage (V)	Speed (rpm)
1	330	0.8	164.6	1128
2	330	0.85	164.0	1189
3	330	0.9	163.5	1250
4	330	0.95	162.1	1309
5	330	1.0	162.0	1410
6	330	1.1	161.9	1427

By varying the width of the PWM pulse in slider control display of the DSPACE desk speed o induction motor is varying accordingly.

7 Conclusion

In this paper Induction motor speed drive is designed using DSPACE DS1104 Controller. Speed, output voltage and current of induction motor at different duty cycles are obtained. The pulse width modulation(PWM) technique used here is space vector pulse width modulation (SVPWM). Simulation of three phase inverter fed induction motor in open loop operation is explained and its respective output waveforms are shown

References

1. L. Vinod Kumar and Syed Sarfaraz Nawaz, "Design and Implementation of Inverter System to Drive Induction Motor using DSPACE" International Journal for Research in Applied Science & Engineering Technology (IJRASET) Volume **4** Issue XI, November (2016), ISSN: 2321-9653.
2. "DSPACE direct torque control Implementation for induction motor" by A.Abbou , Y Sayouti, H Mahmoudi M. Akherraz in 18th Mediterranean Conference on Control &Automation
3. M. Depenbrock, "Pulse width modulation control of a three phase inverter with non-sinusoidal phase voltages," in Proc. IEEE-IAS Int. Semiconductor Power Conversion Conference., Orlaando, FL, (1975), pp.389-398.
4. "Modern Power Electronics and AC Drives, by Bimal K. Bose. Prentice Hall Publishers".
5. "Power Electronics by Dr. P.S. Bimbhra. Khanna Publishers, New Delhi".
6. "DSPACE CP1104 installer guide and model executions".
7. DS1104 R&D Controller Board, "Hardware Installation and Configuration", CRC Press, Release 4.1 - March (2004).

Gate Driver Circuit Design, PWM signal generation using FEZ Panda III and Arduino for Inverter

Srinivasa Rao D[1], *Sandhya Rani* MN[1], *Sarfaraz Nawaz* Syed[2] and *Suresh Kumar* Tummala[3,*]

[1]Assistant Professor, EEE Department, GRIET, Hyderabad
[2]Associate Professor, EEE Department, GRIET, Hyderabad
[3]Professor, EEE Department, GRIET, Hyderabad

Abstract. This paper presents the gate driver circuit design and PWM signal generation using FEZ panda III and Arduino boards for Inverter. FEZ panda III works on .net micro framework and all its programming, debugging features are available through Microsoft's Visual studio whereas Arduino features are available through Arduino software. Inverters are used in various applications like in motor drives and solar and wind power applications. This paper is commenced by basic understanding of need of gate driver circuit for inverter and different types of PWM techniques for generating gate signals for inverter. Simulation of SPWM inverter in both open loop and closed loop operation is done and its respective gate voltage waveforms are studied and later hardware implementation of gate driver circuit using TLP250H IC is carried out and gate voltages at different duty cycles is studied with the help of FEZ Panda III and Arduino boards.

1 Introduction

PWM Inverters are commonly used in variable speed ac drives. The PWM Inverter could be implemented for use in single phase and three phase types. This thesis describes the design of gate driver circuit for single phase PWM inverter using a 12-bit FEZ panda III board and 8-bit Arduino controller. The basic inverter design approach is to convert the line voltage to DC as input to the inverter, a gate driver circuit to switch on the IGBT's and controller to generate required gating signals. A potentiometer can be used to vary the duty cycle of the pulses generated

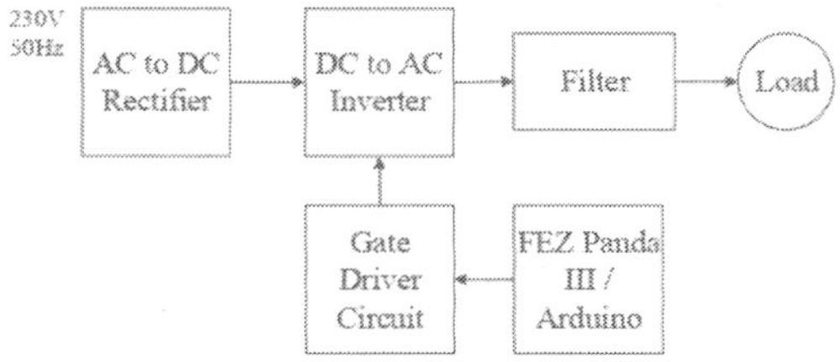

Fig. 1. Block diagram

The basic block diagram of this paper is shown in the figure. The overview of the paper is given through this block diagram. A rectifier, inverter, filter, gate driver circuit, microcontroller and load are the components used. Power supply for the inverter is given by the rectified dc supply from rectifier. The gate pulses required for the inverter to operate are generated from the microcontroller and then given to gate driver circuit. The gate voltage is checked at different duty cycles

2 PWM Inverter

Inverter is basically a circuit which converts electrical energy in DC form to AC form. Inverters usually work on Pulse Width Modulation (PWM) technique. PWM is a technique in which the width of the gate pulses are controlled by various mechanisms which is used to keep the output voltage of the inverter at the rated voltage irrespective of the load.

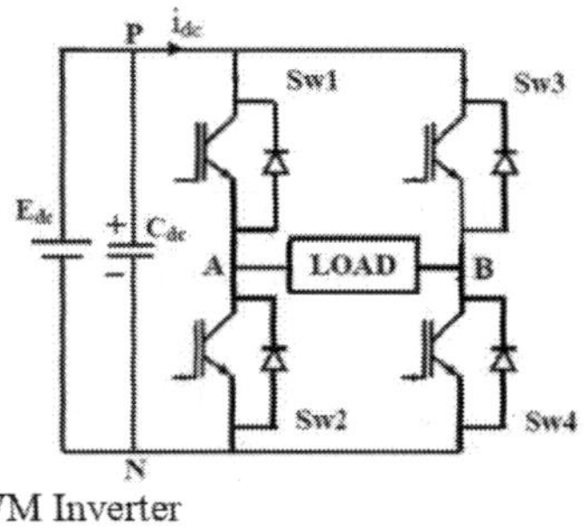

Fig. 2. PWM Inverter

The gating signals for Inverter are generated using microcontroller and a gate driver circuit is needed for switching purpose. First different PWM techniques were studied and later simulation of SPWM single phase inverter is done in which gating signals were observed and then hardware implementation is carried out.

* Corresponding author: sureshkumar.t@griet.ac.in

https://doi.org/10.1051/e3sconf/20198701003

3 SPWM Technique

Out of many PWM techniques SPWM is the most commonly used method. Gating pulses are generated by comparing the reference signal with the carrier wave form. In this method sinusoidal waveform is taken as the reference signal and triangular waveform is the carrier waveform. Gating pulses are generated whenever sine wave value is higher than the triangular wave value. The width of the pulses are not equal because the reference signal is sinusoidal waveform. The gate pulse generation is shown in the figure below.

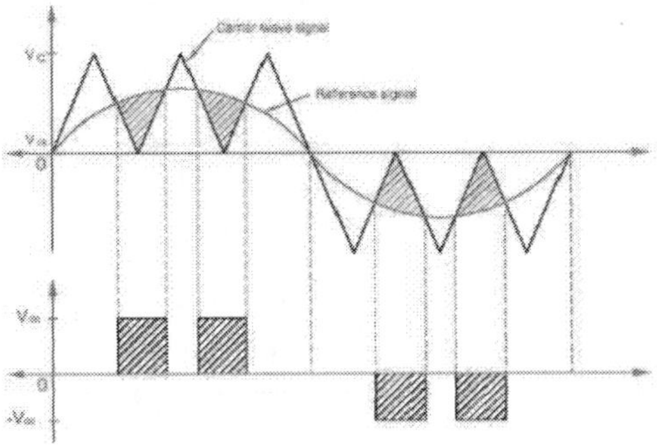

Fig. 3. Sinusoidal pulse width modulation pulses

The duty cycle of the gate pulses is varied which is given by the amplitude modulation ratio ma.

$$m_a = \frac{Vref}{Vcarrier}$$

If ma> 1 it is over modulation
If ma<1 it is under modulation

The output of the SPWM inverter is given by

$$Vo = Vs\sqrt{\frac{p\delta}{\pi}}$$

Where p= number of pulses

4 Simulation

In this section simulation of single phase inverter using SPWM technique is described and here induction motor is used as load on which open loop operation is performed

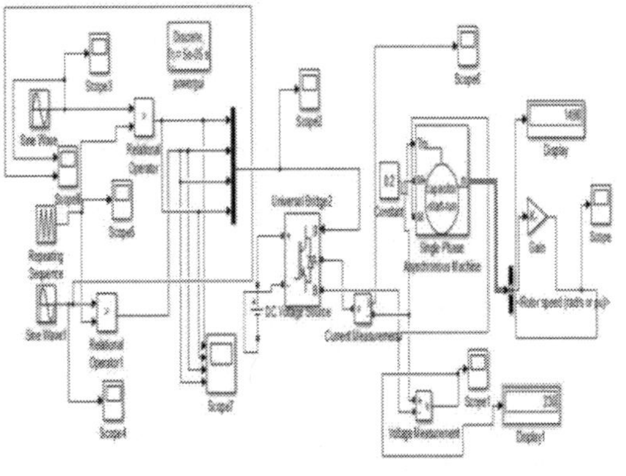

Fig. 4. Simulation diagram of single phase Inverter

In Figure 4 Simulink diagram of single phase inverter with induction motor as load is shown.

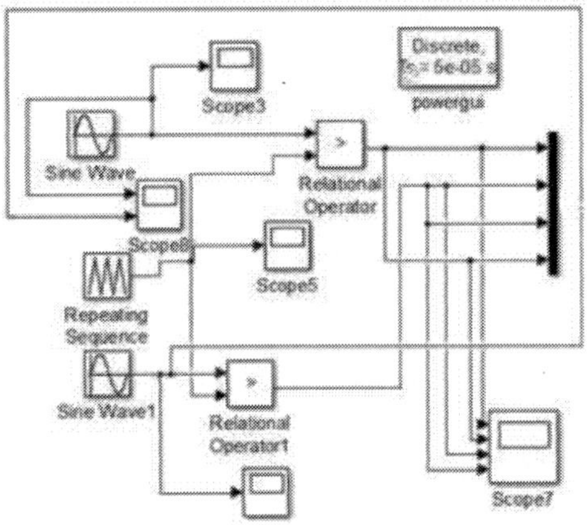

Fig. 5. Sinusoidal pulse width modulation

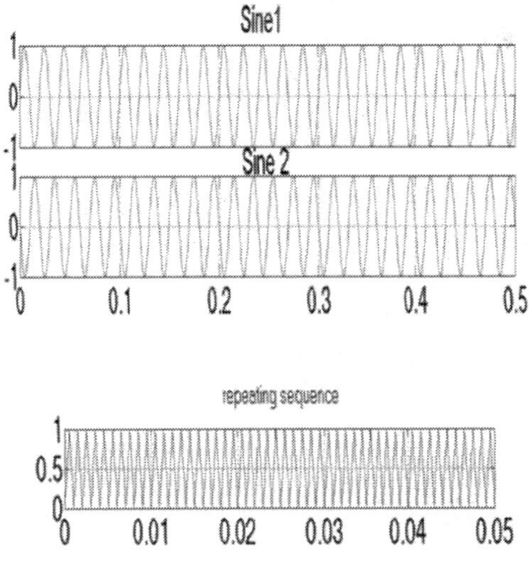

Fig. 6. SPWM Technique

The PWM technique used here sinusoidal pulse width modulation. In this technique two sine waves 180^0 out of

 https://doi.org/10.1051/e3sconf/20198701003

phase are taken as reference signal and a repeating sequence block is taken as carrier signal. The signals from these two blocks are compared using a relational operator block of greater than logic and the required gate signals are produced whenever the amplitude of sine wave is greater than triangular wave for the operation of single phase inverter. The obtained gate pulses are shown in the Fig. 6.

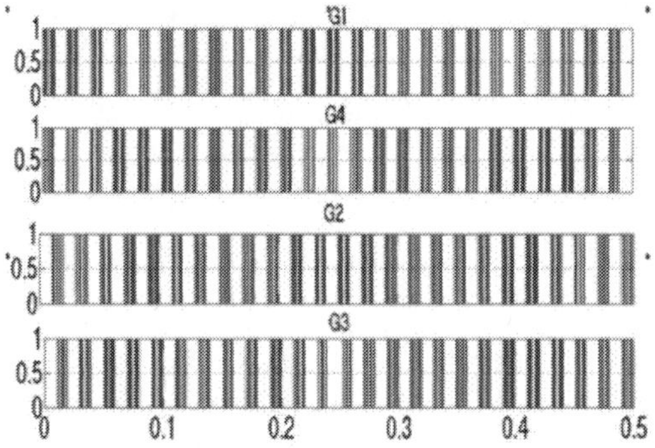

Fig. 7. Gate Pulses

5 Controller

In this chapter the controllers used in this project are discussed i.e. Arduino and FEZ Panda III controller.

A. Arduino

Arduino is an 8-bit microcontroller board based on ATmega348. It works on C programming language. It has 6 PWM pins.

B. FEZ Panda III

FEZ panda is a small low cost single board computer which runs on .net micro framework a tiny version of Microsoft .net framework. All the programming and debugging features are available through the visual studio allowing C# over a simple USB or serial communication. It has 51 IO and 21 PWM pins

6 Gate Driver Circuit

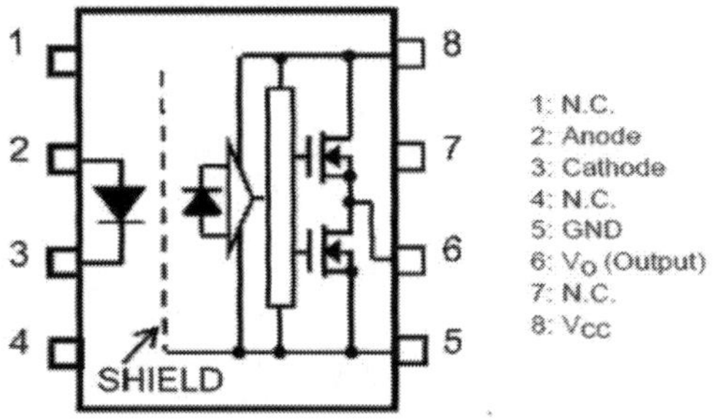

Fig. 8. Pin diagram of TLP250H

Power semiconductor devices have special requirements when it comes to switching which is fulfilled with the help of gate driver circuit. Generally micro controllers generate PWM signals of very low range which is not sufficient for the switches to turn on so a gate driver circuit is designed for that purpose. In this paper TLP250H IC is used for this purpose which is shown in the figure 8.

TLP250H photocoupler basically has a totem pole configuration. Gate driver IC is basically a half bridge with a common input and IGBT gate acts as a capacitor which needs to be charged up for switching. It has half bridge configuration at the top right, opto isolator in middle and LED at top left which activates the photo emitting device which in turn activates half bridge providing required gate voltage.

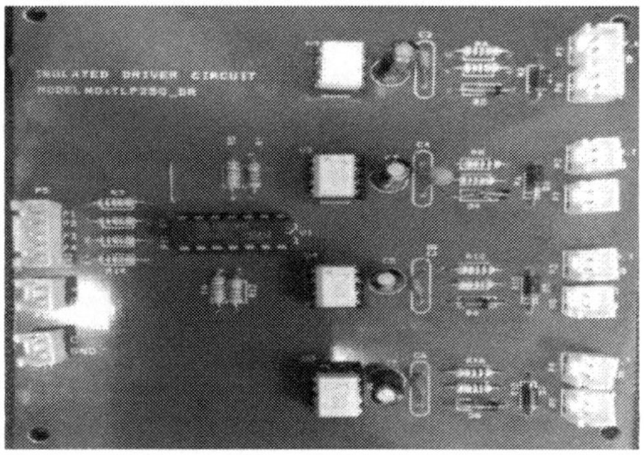

Fig. 9. Gate Driver Circuit

7 Results

The results obtained are shown in the below sections

A. Gate pulses

Gate PWM pulses are generated using both Arduino and FEZ Panda III controller.

Program flowchart

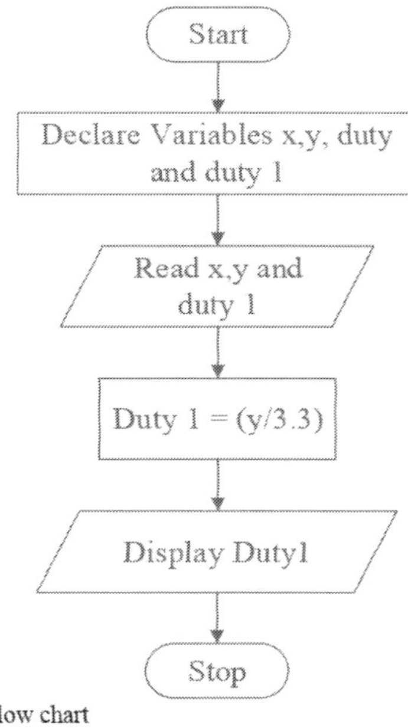

Fig. 10. Flow chart

 https://doi.org/10.1051/e3sconf/20198701003

Pulses generated using Arduino

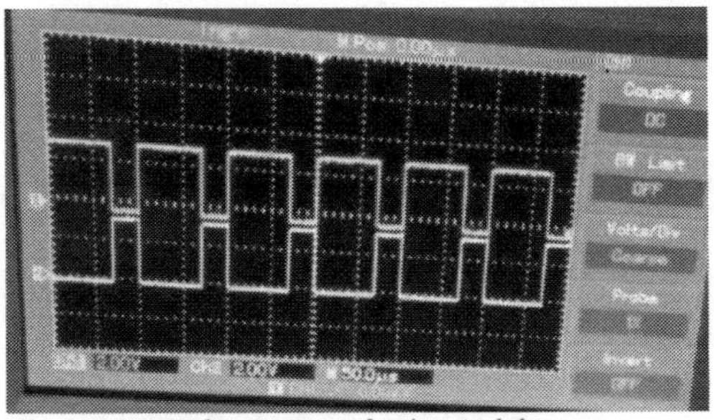
Fig. 11. Gate Pulses generated using Arduino

Pulses generated using FEZ Panda III

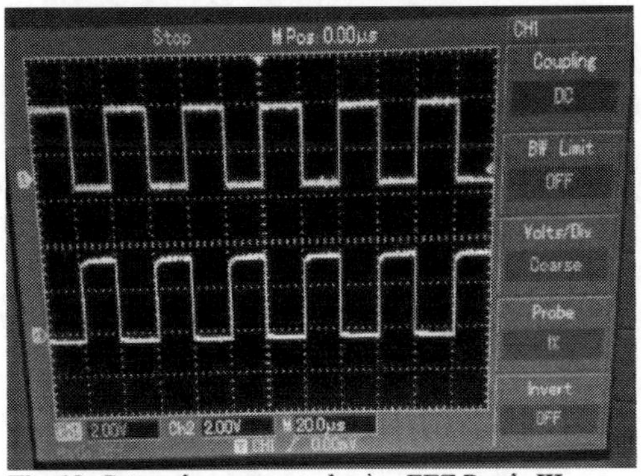

Fig. 12. Gate pulses generated using FEZ Panda III

Gate Voltage

PWM pulses are obtained at different duty cycles by varying potentiometer

Fig. 13. FEZ Panda III with potentiometer

Potentiometer is used for varying the duty cycle of the gate pulses. It consists of three terminals one for supply, other for ground and the third terminal to the analog input of the micro controller used. A program is written in such a way that as we vary the value of this potentiometer duty cycle of the PWM pulse generated changes. 3.3V or 5V of the microcontroller is used as supply for this.

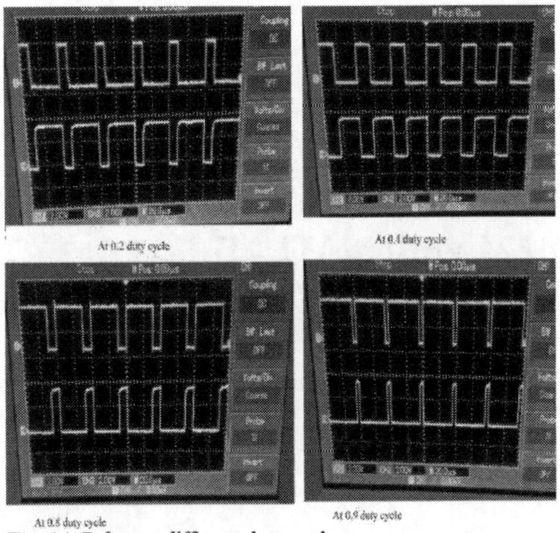

Fig. 14. Pulses at different duty cycles

8 Conclusions

Single phase PWM Inverter is designed using IGBT switches and the gate signals are generated using FEZ Panda- III and Arduino microcontroller boards. The required gate driver circuit for switching IGBT switches is designed using TLP250H IC. The simulation of sinusoidal pulse width modulation (SPWM) inverter is performed and the respective gate voltages is observed and hardware implementation is carried out using gate driver circuit and gate voltages at different duty cycle are obtained using FEZ Panda III and Arduino boards

References

1. Induction Motor speed control using IGBT's and constant V/F method, Department of Electrical & Electronics Engineering, P.D.A. College of Engineering, Gulbarga, Karnataka, India.
2. Microcontroller based PWM inverter for speed control of an Induction motor, Department of Electrical and Electronics Engineering, BUET, Dhaka, Bangladesh.
3. Speed control of Induction motor using PWM Technique, Department of Electronics and Telecommunication Engineering, Dr. Daulatrao Ahar College of Engineering, Maharashtra, India
4. L. Vinod Kumar and Syed Sarfaraz Nawaz, "Design and Implementation of Inverter System to Drive Induction Motor using DSPACE" International Journal for Research in Applied Science & Engineering Technology (IJRASET) Volume 4 Issue XI, November (2016), ISSN: 2321-9653
5. Power Electronics by Dr. P.S. Bimbhra. Khanna Publishers, New Delhi

Speed Control of DC Motor using Fuzzy Logic Controller by PCI 6221 with MATLAB

Anil Kumar Rajagiri[1], *Sandhya Rani* MN[2], *Syed Sarfaraz* Nawaz[3], and *Suresh Kumar* T[4]

[1,2]Assistant Professor, Department of EEE, GRIET, Hyderabad, India
[3]Associate Professor, Department of EEE, GRIET, Hyderabad, India
[4] Professor, Department of EEE, GRIET, Hyderabad, India

Abstract. This thesis demonstrates the importance of Fuzzy Logic Controller. The operation of a DC motor is performed using Fuzzy Logic Controller (FLC) in MATLAB environment. Fuzzy Logic is one of the most successful applications of fuzzy set in which the variables are linguistic rather than numeric. A Fuzzy Logic Controller (FLC) is based on a set of control rules (fuzzy rules) among linguistic variables. The proposed fuzzy controller results in a better response compared to the normal response of DC motor. This thesis consists of two parts; software and hardware implementation. The software part aims to design and develop a Fuzzy Logic Controller in MATLAB Simulink. The hardware Part Consist of DC motor Driver and PCI 6221. The DC drive is used to convert AC voltage into variable DC voltage PCI 6221 is used as the hardware interface between Hardware and Software.

Keywords: PCI 6221 68 pin Card, Fuzzy Logic Controller, DC motor, DC Drive, Proximity Sensor

1. INTRODUCTION

Recent developments in science and technology provide a wide range scope of applications of high performance DC motor drives in area such as rolling mills, chemical process, electric trains, robotic manipulates and the home electric appliances. They require controllers to perform tasks. Hence, a fuzzy based DC motor control system was designed using MATLAB.

Almost every industry and household has motors used in their equipment or appliances. Motors that are often controlled by computers have also become an essential part of many motion control systems[1].

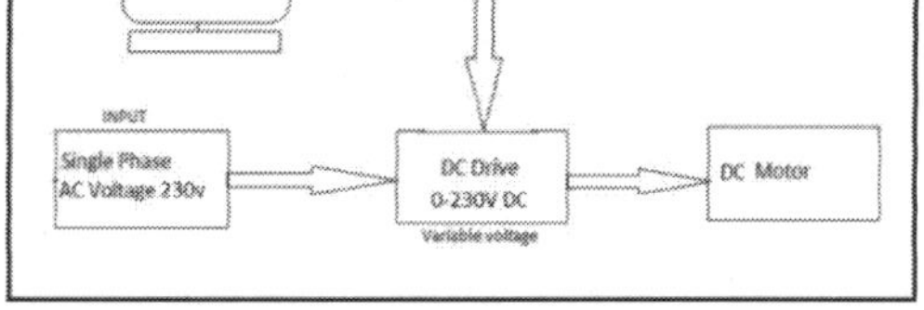

Fig 1. Block Diagram approach of the model developed

2 FUZZY CONTROL SYSTEM

Fuzzy Logic idea is similar to the human being's feeling and inference process. The output of a fuzzy controller is derived from fuzzifications of both inputs and output using the associated membership functions. A crisp input will be converted to the different members of the associated membership functions based on its value. From this point of view, the output of a fuzzy logic controller is based on its memberships of the different membership functions, which can be considered as a range of inputs. Fuzzy ideas and fuzzy logic are so often utilized in our routine life that nobody even pays attention to them. For instance, to answer some questions in certain surveys, most time one could answer with 'Not Very Satisfied' or 'Quite Satisfied', which are also fuzzy or ambiguous answers. Exactly to that degree is one Satisfied or Dissatisfied with some service or product for those surveys? These vague answers can only be created and implemented by human beings, but not machines. Is it possible for a computer to answer those survey questions directly as a human being did?

It is impossible. Computers can only understand either'0' or '1' and 'HIGH' or 'LOW'. Those data are called crisp or classic data and can be processed by all machines. Is it possible to allow computers to handle those ambiguous data with the help of a human being? If so, how can computers and machines handle those vague data? The answer to the question is yes. But to answer the second question, we need some Fuzzy Logic techniques and knowledge of fuzzy inference system. Applications in industrial manufacturing, automatic control, automobile production, Hospitals, elevator control system, automatic transmission control, washing machine etc[3]. Fuzzy logic techniques have been widely applied in all aspects in today's society[2].

2.1 Knowledge Base

The knowledge base can be divided into two sub-blocks namely the Data Base and Rule Base. The data base consists of the information required for Fuzzifying the crisp input and later Defuzzifying the fuzzy outputs to a crisp output. It consists of the membership functions for various fuzzy variables or sets used in the controller design. The rule base consists of a set of rules, which are usually formulated from the expert knowledge of the system. The rules are typically of the form "If...., then....." rules. The antecedent part of the rule may be a simple statement or a compound statement using connectives like "and", "or" etc. the consequent part may contain a fuzzy set (Mamdani type and Tsukamoto type) or a linear or a quadratic function of the crisp input variables (Sugeno type). The knowledge base is the heart of fuzzy logic based system it has to be designed without most care and require a lot of expertise in the knowledge of the system into which fuzzy logic controller is being incorporated[1].

2.2 Fuzzy Inferencing

The process of fuzzy reasoning is incorporated into what is called a Fuzzy Inferencing System. It is comprised of three steps that process the system inputs to the appropriate system outputs[4].

These steps are

1) Fuzzification
2) Rule Evaluation
3) Defuzzification

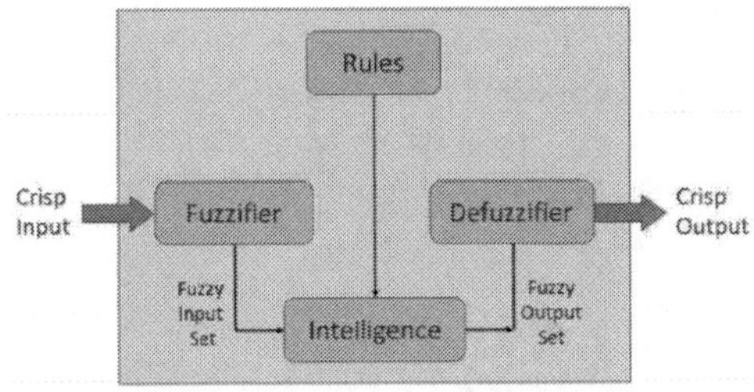

Fig 2. Structure of Fuzzy Logic Controller

2.2.1 *Fuzzification*

Fuzzification is the first step in the fuzzy inferencing process. This involves a domain transformation where crisp inputs are transformed into fuzzy inputs. Crisp inputs are exact inputs measured by sensors and passed into the control system for processing, such as temperature, pressure, rpm's, etc..

2.2.2 *Defuzzification*

Defuzzification involves the process of transposing the fuzzy outputs to crisp outputs.

There are a variety of methods to achieve this

- Centre of Area (CoA)
- Center of Sums (CoS)
- Center of Maximum (CoM)
- Mean of Maximum (MoM)

2.2.2.1 *Center of Area (CoA)*

Fuzzy logic controller first calculate the area under scale membership function and within the range of output variable

$$COA = \frac{\int_{x_{min}}^{x_{max}} f(x).x dx}{\int_{x_{min}}^{x_{max}} f(x) dx}$$

Where CoA is the center of area x is the value of the linguistic variable, and x_{min} and x_{max} represent the range of the linguistic variable.

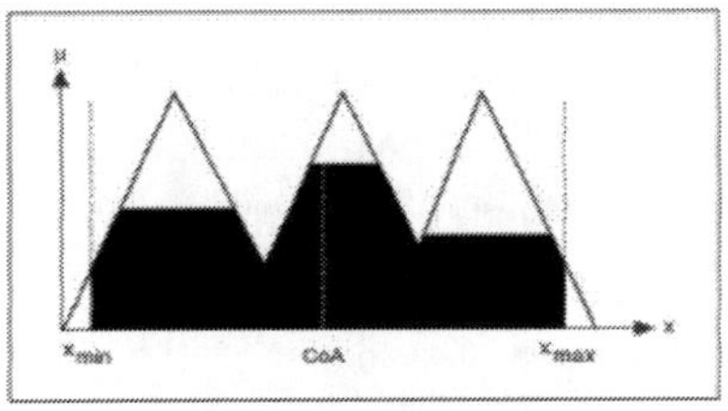

Fig 3. CoA Defuzzification method

2.2.2.2 Center of Sums

Fuzzy Logic Controller calculate the geometric center of area of each membership function and take see weighted average of center of area of all membership function.

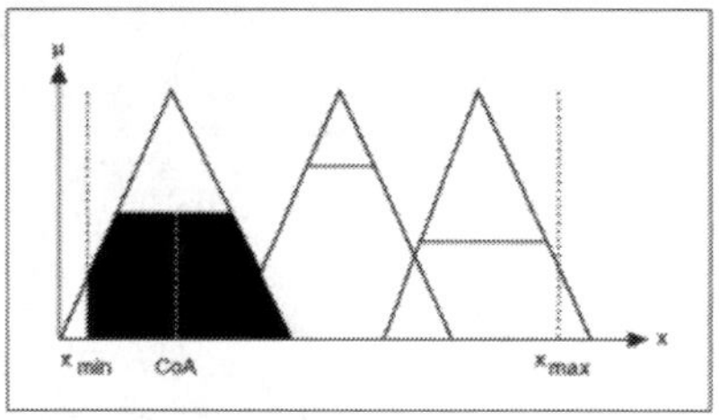

Fig 4. CoS Defuzzification method

$x_{\text{final}} = (CoA1 area1 + CoA2 area2 + \cdots + CoAn arean) / area1 + area2 + \cdots + arean$

Where CoA$_n$ is the geometric center of area of area of the scaled membership function n and area$_n$ is the area of the scaled membership function n.

2.2.2.3 Center of Maximum

In center of maximum method the fuzzy logic controller determines the typical numerical value for each scaled membership function and takes mean of numerical value corresponding to degree of membership function. when this is then placed in an external magnetic field, it will experience a force proportional to the current in the conductor, and to the strength of the external magnetic field.

2.3.1 Voltage and power equation of DC motor

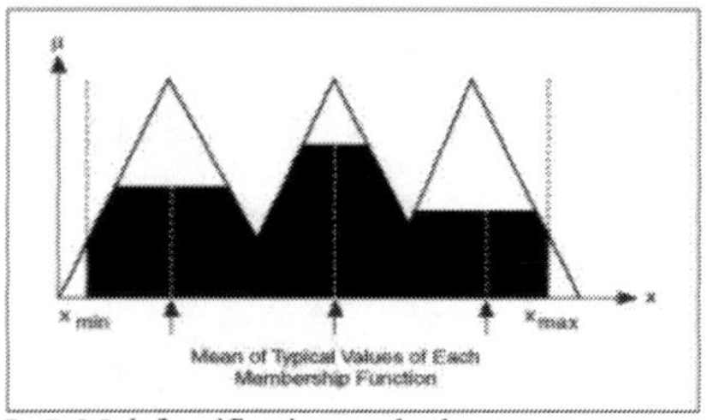

Fig 5. CoM defuzzification method

$$x_{\text{final}} = \frac{x_1\mu_1 + x_2\mu_2 + \cdots + x_n\mu_n}{\mu_1 + \mu_2 + \cdots + \mu_n}$$

Where x_n is the typical numerical value for the scaled membership function n and μ_n is the degree of membership at which membership function n was scaled.

2.2.2.4 Mean of Maximum

In mean of maximum method the scale fuzzy logic controller identify the membership function with the greater degree of membership function.

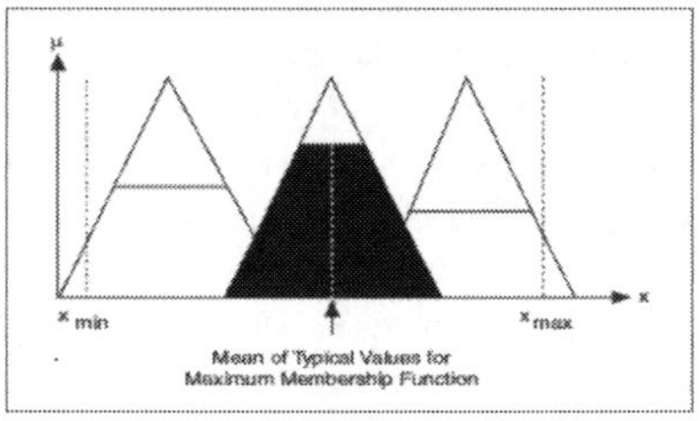

Fig 6. MoM Defuzzification method

2.3 DC Motor

Direct current (DC) motors are perhaps the most basic type of electrical motors. In any electric motor, operation is based on simple electromagnetism. A current-carrying conductor generates a magnetic field;

$$V = E_b + I_a R_a$$

If we multiply the above equation by I_a, we will get

$$VI_a = E_b I_a + I_a R_a$$

Vi_a = electrical power supplied to the motor

$E_b I_a$ = electrical equivalent of the mechanical power produced by the motor

$I_a^2 R_a$ = power loss taking place in armature winding

Thus,

$$E_b = VI_a - I_a^2 R_a$$

Input power – power loss

Thus, $E_b I_a$ = Gross mechanical power produce by the motor = rpm

2.3.2 Torque equation of DC motor

Mechanical power required to rotate the shaft

on mechanical side = Tω (1)

T = Torque in Newton – meter

ω = angular velocity in Radian/second

Gross mechanical power produce by the motor

on electrical side = $E_b I_a$ (2)

E_b = back emf in volts

I_a = armature current in ampere

Equating equation ① and ②, we get

$$E_b I_a = \mathrm{T}\omega \qquad (3)$$

$\omega = \frac{2\pi N}{60}$ .... $\{\frac{2\pi N}{60}$ = speed in rpm

And

$$E_b = \frac{N\phi ZP}{A60}$$

Thus, equation ③ become

$$\frac{N\phi ZP}{A60} I_a = \mathrm{T}\,\frac{2\pi N}{60}$$

$$\mathrm{T} = \frac{P\phi Z I_a}{2\pi A} = \frac{0.159 P\phi Z I_a}{A} = \left(\frac{0.159 PZ}{A}\right) \phi I_a$$

P,Z and A are constant, hence we can say

$$\mathrm{T} \propto \phi\, I_a$$

Thus torque produce by the DC motor is proportional to the main field flux ϕ and armature current I_a

Governing Equations

1. $E_b = \frac{P\phi NZ}{60A}$

   (Where, P = no. of poles, ϕ = flux /pole, N = speed in rpm, Z = no. of Armature conductors, A = parallel paths)

2. $E_b = V - I_a R_a$

   Thus,

3. $\mathrm{N} = \frac{E_b 60A}{P\phi Z}$ (A,P and Z are constants, hence)

4. $\frac{N \propto K\, E_b}{\phi}$ (where, K = constant)

2.4 Proximity Sensor

Proximity sensor is a sensor able to detect the presence of nearby objects without any physical contact. A proximity sensor often emits an electromagnetic field or a beam of electromagnetic radiation (infrared, for instance), and looks for changes in the field or return signal. The object being sensed is often referred to as the proximity sensor's target. Proximity sensors are also used in machine vibration monitoring to measure the variation in distance between a shaft and its support bearing. This is common in large steam turbines, compressors, and motors that use sleeve-type bearings.

Fig 7. Proximity Sensor

2.5 NI PCI-6221 (68 Pin)

NI PCI-6221 is used for acquiring the signals from the speed sensor and to generate the signal applied to the DC Drive. It is M Series Data Acquisition Device that provides the data Acquisition for some application of portable measurement. Data acquisition card is used as the hardware device. It has a total of 68 pins for various input/output purposes. Various characteristics of the card are[5]:

- 16-Bit, 250kS/s, 16 Analog Inputs
- Two 16-bit analog outputs (833kS/s)
- 24 digital I/O

Fig 8. NI PCI 6221 Card

The following are the specifications:

https://doi.org/10.1051/e3sconf/20198701004

Table 1. PCI 6221 specifications

Analog Inputs	:	16
Analog Outputs	:	02
Digital I/O pins	:	24
Resolution	:	16 bit
Input Range	:	±,0.2, ±1V, ±5V, ±10V
Output Range	:	±10V

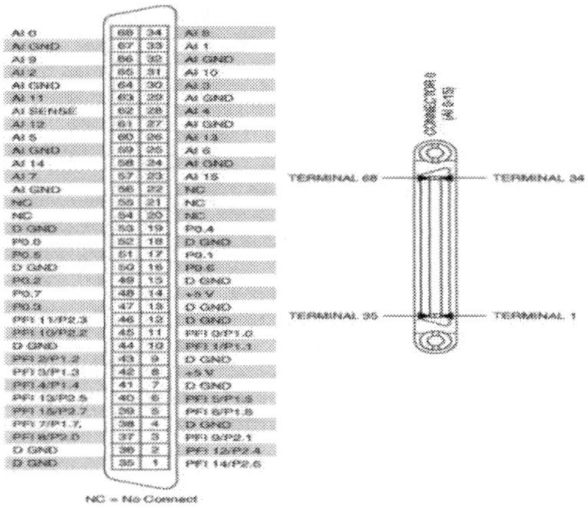

Fig 9. PCI 6221 Pin Diagram

NI CB-68LP is the daughter board used along with the PCI card. It is a low-cost termination accessory with 68 screw terminals for easy connection.

Fig 10. NI CB-68LP and its connector cable

2. Designing FLC in MATLAB

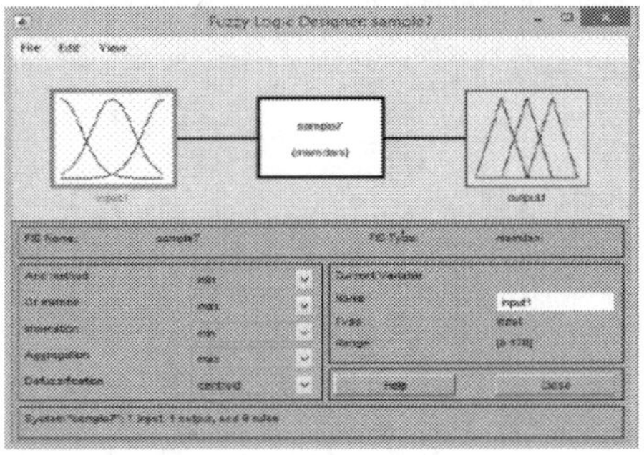

Fig 11. Fuzzy Block

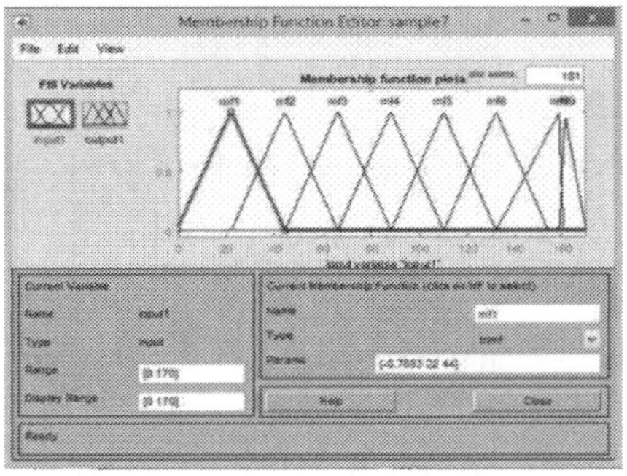

Fig 12. Fuzzy input membership function

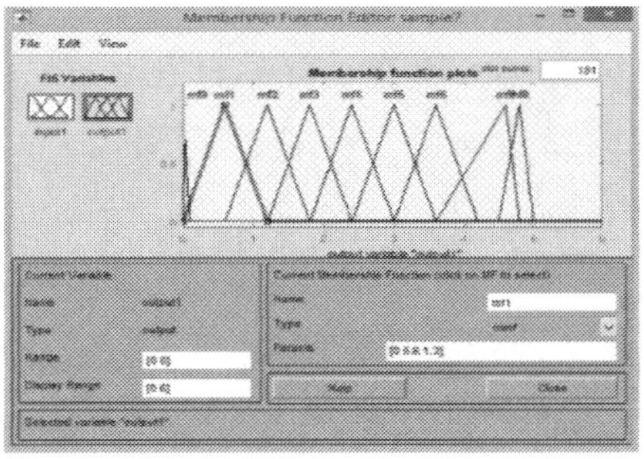

Fig 13. Fuzzy Output Membership Function

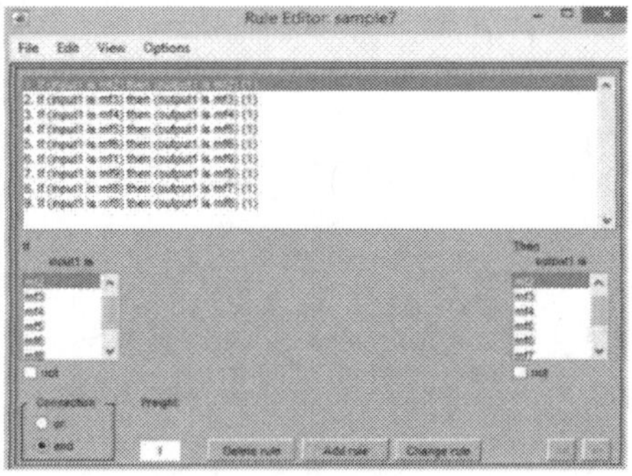

Fig 14. Fuzzy Rules set

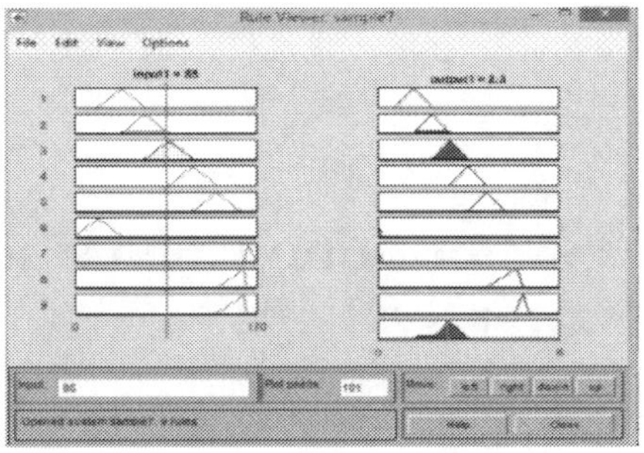

Fig 15. Fuzzy Rule Viewer

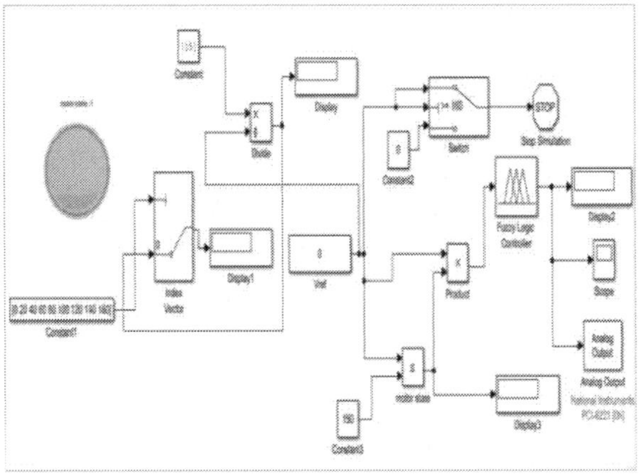

Fig 16. Circuit Diagram

Fig 17. Hardware

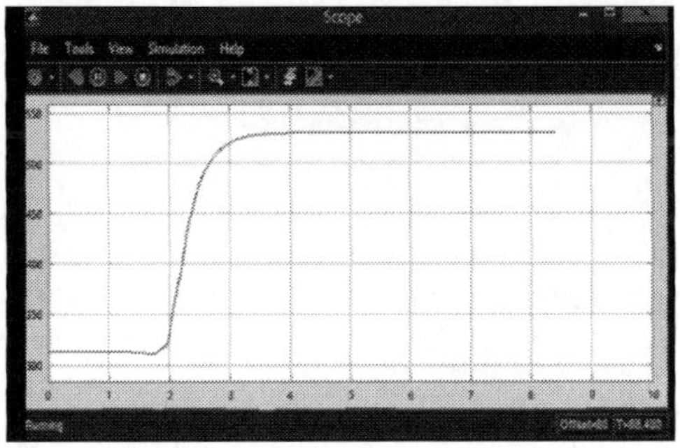

Fig 18. DC Machine speed

Table 2. DC Drive output with respective Fuzzy

Fuzzy Input (Voltage)	DC Drive Output (Voltage)	Speed (RPM)
0	0	0
20	31.53	265.7
40	47.37	414.9
60	65.05	579.1
80	86.95	782.9
100	105	952.3
120	125.9	1149
140	145.6	1346
160	159.3	1496

3. CONCLUSION

The fuzzy controller for the running of the DC motor is designed in MATLAB. The design of the FLC has been explained. The results of experiment on the real plant demonstrate that the proposed Fuzzy Logic Controller the response is very good and it is robust, faster and flexible. It shows control the speed. By using FLC application, shorter settling time can be achieved by tuning the control rules, membership function and universe of disclosure of the output variable. This presents an idea of speed control of DC motor with simple software application with good hardware results.

References

1. NPTEL Video lectures, Electrical Engineering-Intelligence System and control-introduction of Fuzzy Logic Control.
2. George J. Klir, Bo Yuan Fuzzy Sets and Fuzzy Logic.
3. J. Harries, An Introduction to Fuzzy Logic Applications.
4. Sivanandam, **SN.**, Sumathi, **S.**, Deepa, S. N, Introduction to Fuzzy Logic using MATLAB.
5. NI PCI 6221 (68 PIN) the M series user manual available at ni.com/manuals.

Application of Search Group Algorithm for Automatic Generation Control of Multi-area Multi-source Power Systems

Dillip Khamari, *Rabindra Kumar* Sahu*, and *Sidhartha* Panda

Department of Electrical Engineering, Veer Surendra Sai University of Technology (VSSUT), Burla-768018, Odisha, India

Abstract. This paper proposes a new Search Group Algorithm based PID controller, to deal with Automatic Generation Control of two-area with six unit power system. The supremacy of SGA tuned PID controller is being shown using the comparative study with Firefly Algorithm (FA) optimization method for the same test system using ITAE as an objective function. It has been demonstrated that SGA tuned PID controller improves the performance in a large compared with FA tuned PID controller. Furthermore variation in nominal values of operating load condition and system parameters with the position of step load perturbation is being carried out to achieve sensitivity analysis. From the result of sensitivity analysis it clearly depicts the robustness of the suggested method (SGA with PID controller) for two-area with six unit power system in AGC. Finally for better investigation, the proposed method is also examined by applying randomization in step load.

1 Introduction

Power systems arena is thriving day by day which leads to consider a major factor as Automatic Generation Control (AGC) for stable and secure power system operation. AGC plays a vital role to maintain the consistency in frequency with tie-line power. There is a great gap between electrical load demand and power generation which leads to deviation in outcomes. AGC nullifies the Area Control Error (ACE) using calculation with respect to corresponding load change in each area by adjusting automatically the generator set points. ACE is defined as the linear combination of frequency variation to the corresponding distortion in net tie-line power interchange [1-5]. Researchers are proposed number of control schemes in AGC of power systems to achieve optimized result.

The authors were investigated the different AGC based generic controllers in multi-area multi-source power system [6-9]. Sharma et al. have presented optimal design of AGC regulator controller in frequency regulation of multi-area power system with diverse power generating units [10]. Guha et al. have proposed Grey Wolf Optimization (GWO) based classical controller with PI/PID structure for AGC in multi-area power systems [11]. Saroj et al. have presented the supremacy of Firefly Algorithm tuned PID controller of two-area interconnected power system for AGC [12]. The effectiveness of AGC is not limited to artificial intelligence techniques simultaneously it depends on objective function as well the controller structure chosen which is clearly signifies from past literature.

Recently, metaheuristic based algorithm known as the Search Group Algorithm (SGA) is being proposed by Matheus Silva Goncalves et al. for the application to truss structure [13]. SGA is a population oriented search algorithm which maintains the balance in the design domain between the exploration & exploitation.

The novel contributions in this paper are:

(i) The demonstration of the superiority of new powerful computational intelligence technique like SGA over FA tuned PID controller for AGC

(ii) To show robustness of SGA based PID controller. Here, two area six unit like hydro, gas and thermal units are taken for investigation [14]. The empirical analysis result reflects the supremacy of the suggested method. Lastly, to show the effectiveness of the tuned controller parameters, variation in nominal values of system parameters, operating load condition with the position of SLP is carried out to achieve sensitivity analysis.

2 Proposed Method

2.1 Power System modelling:

Firstly,2-area with 6-unit power system is demonstrated in Fig.1. The nominal loading is being contributed based on the decision of the participation factor assigned by each unit. After summing each control participation factor the outcome should be equivalent to unity. Participation factor for hydro, gas and thermal units are evaluated as 32%, 13% and 55% respectively. The system parameters values are represented in Appendix. Regulation parameters R_1, R_2 and R_3 shown in Fig.1 denotes thermal, hydro and gas unit respectively.

* Corresponding author: rksahu_ee@vssut.ac.in

https://doi.org/10.1051/e3sconf/20198701005

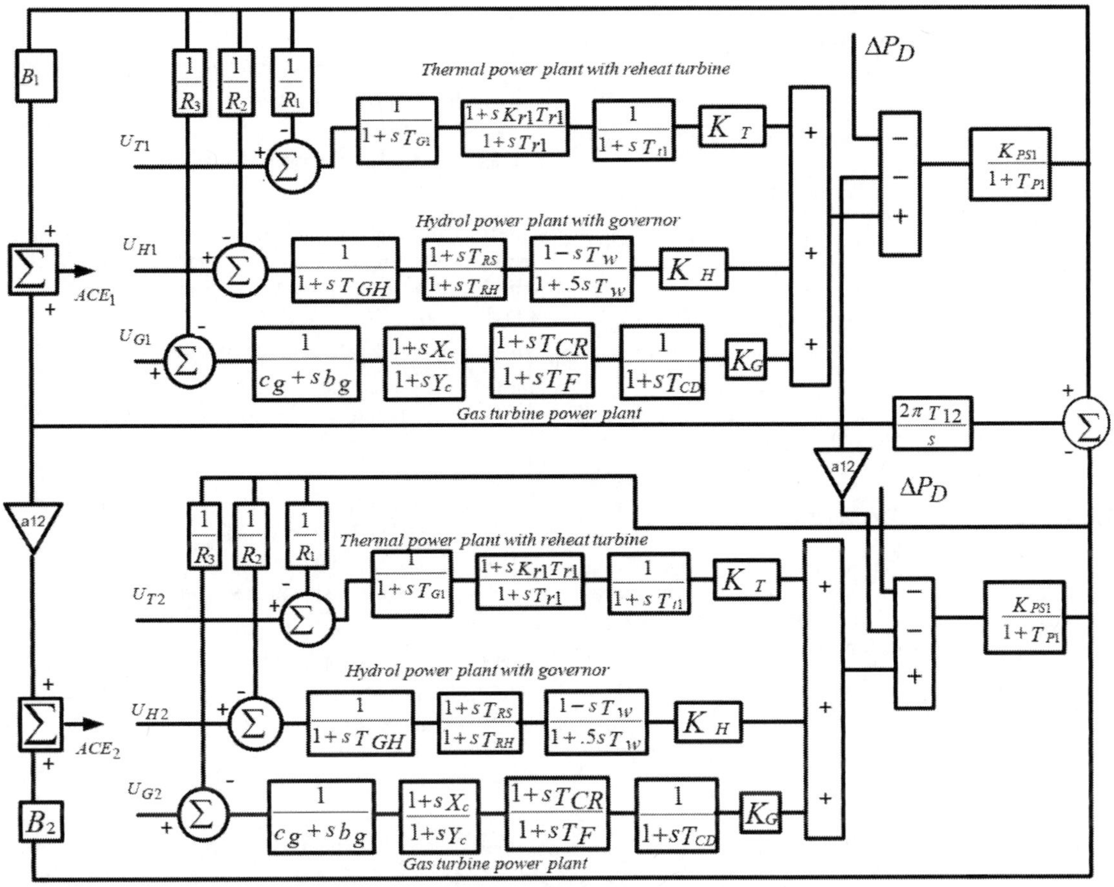

Fig. 1. Transfer function model of test system

The control outputs are represented as U_T for thermal, U_H for hydro and U_G for gas units. The participation factors are represented as K_T for thermal, K_H for hydro and K_G gas units. T_{SG}, T_T represents time constant of speed governor for thermal units and reheat steam turbine in second respectively. T_W represents penstock base starting time of water in sec. T_{RS} represents reset time for speed governor, T_{RH} represents time constant for governor droop, T_{GH} represents time constant for main servo of speed governor of hydro turbine in sec respectively. X_C represents lead-time constants and Y_C represents lag-time constants of speed governor for gas turbine in sec respectively. c_g represents gas turbine and b_g represents valve positioned based gas turbine constant. T_F represents time constant of fuel and T_{CR} represents time delay of combustion reaction in sec of gas turbine. T_{CD} represents time constant of discharge volume for compressor based gas turbine in sec. K_{PS} represents gain of power system in Hz/p.u.MW. T_{PS} represents time constant of power system in sec. ΔF, ΔP_D are the variation in frequency and load respectively.

2.2 Controller design with objective function

Circuit diagram of PID controller is shown in Fig. 2.

Where K_P: Proportional gain, K_I: Integral gain and K_D: Derivative gains. The general s-domain PID controller transfer function is given by.

$$TF_{PID} = K_P + \frac{K_I}{s} + K_D s \tag{1}$$

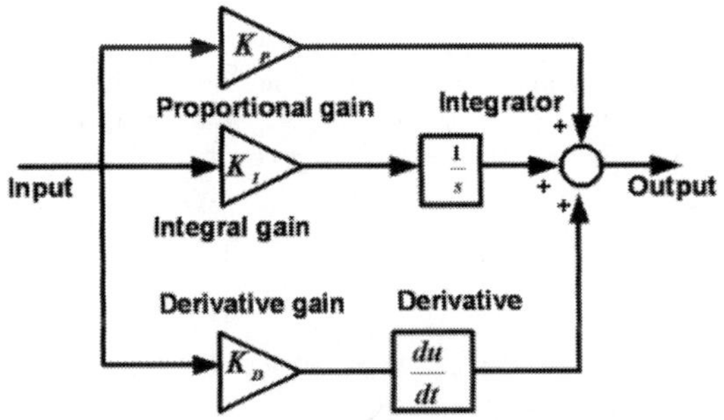

Fig. 2. PID controller structure

Errors inputs to the controllers of the corresponding ACE are mentioned in equations (2-3):

$$ACE_1 = e_1(t) = B_1 \Delta F_1 + \Delta P_{Tie} \tag{2}$$

$$ACE_2 = e_2(t) = B_2 \Delta F_2 - \Delta P_{Tie} \tag{3}$$

* Corresponding author: rksahu_ee@vssut.ac.in

The performance criteria suitable for AGC studies are Integral Time multiplied Absolute Error (ITAE) as reported in literature [15]. ITAE is used as objective function which is provided in equation (4):

$$J = ITAE = \int_0^{t_{sim}} \left(|\Delta F_1| + |\Delta F_2| + |\Delta P_{Tie}|\right) \cdot t \cdot dt \quad (4)$$

Equation (5-6) represents the optimization problem for the design issue

Minimize the value of J (5)

Subject to

$$K_{P\min} \le K_P \le K_{P\max}$$
$$K_{I\min} \le K_I \le K_{I\max}$$
$$K_{D\min} \le K_D \le K_{D\max} \quad (6)$$

The minimum parameters values are chosen as -2.0 and maximum value is 2.0 of PID controller.

3 Search Group Algorithm

A population-based optimization method as Search Group Algorithm was being proposed by M.S. Goncalves et al. [13].The important function of SGA is categorized into five steps is depicted in below.

3.1 Phase-1: Initial Population

Randomly initial population P is chosen based on equation (7)

$$P_{ij} = X_j^{\min} + (X_j^{\max} - X_j^{\min})U[0,1] \quad (7)$$

Where j = 1 to n, i = 1 to n_{pop}

P_{ij} represents as the jth design variable of population P for i^{th} individual. All design variables summation is n. n_{pop} signifies the total of population. The range of identical variable U [0, 1] is between 0 to 1 which is arbitrary in nature. The lower limit is $X_j^{\min}$ and the higher limit is $X_j^{\max}$ of jth design variable.

3.2 Phase-2: Initial search group selection

Initially the population has being formed after that objective function is evaluated, a benchmark tournament selection [13] is applied by selecting n_g individuals from the population P to build the search group R. In every cycle If R_i denotes the ith row of R, then R_1: denotes the finest design, R_{ng}: denotes the coarse design in R, n_g: denotes the members count in search group.

3.3 Phase-3: Selection of mutated search group

New offspring's (individuals) are generated by substituting n_{mut} individuals from R to increase the capabilities of global population search which is evolved from equation (8)

$$X_j^{mut} = E[R_j] + t\varepsilon\sigma[R_j],\ \text{for } j = 1,\ldots,n \quad (8)$$

Where $X_j^{mut} \rightarrow$ j[th] design variable of known mutated individual.

E $\rightarrow$ Mean, $\sigma \rightarrow$ Standard deviation, $\varepsilon \rightarrow$ random variable

Choosing of worst objective function is being replaced by "inverse tournament" selection.

3.4 Phase-4: Family generation of every search group member

Family defines as the generation of set of offspring's (individuals) from search group member using perturbation analysed by equation (9).

$$X_j^{mut} = R_{ij} + \alpha\varepsilon \quad \text{for } j = 1, \ldots n \quad (9)$$

Where $\alpha \xrightarrow[Controls]{}$ size of the perturbation

During the search process value of α reduces after subsequent iteration k. which is mentioned in equation (10):

$$\alpha^{k+1} = b\alpha^k \quad (10)$$

Where in every iteration α^k reduces in such a manner which is described by parameter b and controls the distance. Initially, value of α^k is chosen a maximum value to explore the design phase as a whole.

α^k ranges between $\alpha^0 \le \alpha^k \le \alpha_{\min}$

Where $\alpha_{\min} \rightarrow$ Minimum value of α^k to ascertain and nullify movement of new off springs (Individuals) also in last iteration of SGA.

$\alpha^0 \rightarrow$ Initial iteration value of α. Assume $F_i \rightarrow$ Family where i=1 to n_g.The objective function decides the creation of number of offspring by every search group members. Total count of offspring (Individuals) maximizes based on the superiority of objective function. Keeping constant by calculating total number of design in each iteration which is represents as (n_{pop} - n_g).

3.5 Phase-5: Selection of new search group

Global phase algorithm defines as the novel search group emerges by key (best) global member of individual family in the initial $it^{\max}$ iterations. Local phase of algorithm defines as the original search group is produced where $it_{global}^{\max}$ is less than iteration number by the better n_g individuals within the families. $it^{\max} \rightarrow$ Maximum iterations, $it_{global}^{\max} \rightarrow$ Maximum iterations in global phase.

4 Firefly Algorithm

Yang et al. [16] developed a meta-heuristic technique termed as Firefly Algorithm (FA) which is based on bioluminescence that is a biochemical process by the flashing characteristics (flashing light) of fireflies. For mating the flashing light may use as the main courtship signals [17, 18].It is based on the following three fire-flies characteristics:

Every firefly having unisex property gets attracted among them in spite of their sex.

Based on brightness the firefly gets attracted using distance parameters.

The optimized objective function decides the brightness of a firefly. The Firefly algorithm is elaborated detail in [12].

5 Analysis of results with discussion

In this paper MATLAB with SIMULINK is used to design and simulate the model for studying the system. SGA programs are written separately with MATLAB(.m) program file using control parameters and SGA parameter by taking step load perturbation of 2% at t=0.0 sec in area-1. SGA parameter plays a prominent role and it is chosen according to literature [13]. Table 1 includes the control parameter values of SGA which is applied in the algorithm.

Table 1. SGA Parameter.

Parameter	Value
α^0	2
n_{pop}	100
α_{min}	0.01
it^{max}	50
it^{max}_{global}	$0.5 \times it^{max}$
n_g	$0.2 \times n_{pop}$
n_{mut}	$0.03 \times n_{pop}$
t	t=1,t=2,t=3 are used for every mutated individual

SGA/FA is taken to tune parameters of the PID controller separately. The unknown parameters of PID controller are recorded at the end of it^{tmax}. A series of optimization process was conducted and the best final solution is chosen as final control parameters with minimum ITAE values. The optimized SGA based PID control parameter as well FA tuned PID control parameter are mentioned in Table 2.

Table 2. SGA and FA tuned controller Parameters.

FA:PID	SGA:PID
Unit1: Thermal KP1 = -1.9978 KI1 = 1.9978 KD1 = 1.0443	Unit1: Thermal KP1 = 1.9978 KI1 = 0.5126 KD1 = 1.991
Unit 2: Hydro KP2= 1.0443 KI2= -1.9959 KD2 = 1.6992	Unit 2: Hydro KP2= 1.9978 KI2= -0.1948 KD2 = 0.8454
Unit 3: Gas KP3 = 0.9696 KI3 = 1.2519 KD3= -0.4538	Unit 3: Gas KP3 = 1.6877 KI3 = 1.9978 KD3= -0.4538

At t = 0.0 sec a 2 % Step Load Perturbation (SLP) is applied in area-1, the SGA tuned PID controller and FA tuned PID controller system performance is demonstrated in Table 3. The superiority of SGA tuned PID controller with less ITAE value (ITAE=0.2947) than FA tuned PID controller (ITAE=1.6463) is depicted in Table 3 using the same controller with objective function. Again the settling times in frequency variation and deviations in tie power with SGA optimized PID controller are improved compare to FA tuned PID controller. Hence it is clear that SGA superior than FA. The system dynamic response deviations in frequency & deviations in tie-line power is depicted in Figs. 3-5.

Table 3. Comparative performance index values.

Techniques/ parameters	Settling times Ts			ITAE
	ΔF_1	ΔF_2	ΔP_{tie}	
FA:PID	2.34	2.33	1.91	1.6463
SGA:PID	1.13	0.94	1.07	0.2947

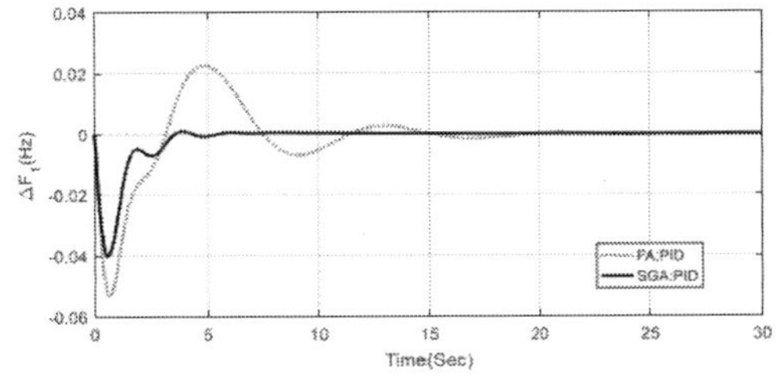

Fig. 3. ΔF_1 with 2% SLP in area-1.

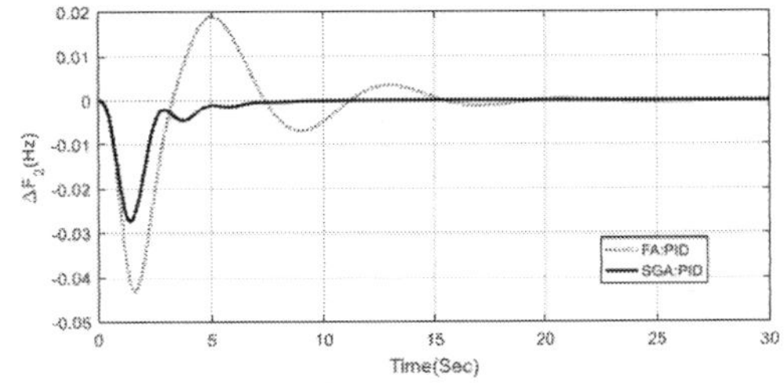

Fig. 3. ΔF_2 with 2% SLP in area-1.

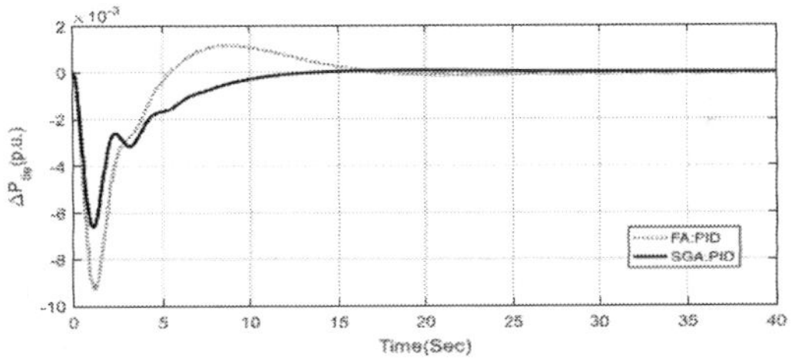

Fig .5. ΔP_{tie} with 2% SLP in area-1.

This clearly signifies from the Figs. 3-5 that best dynamic performance as well as better response is achieved by SGA compared to FA technique.

Further at t = 0.0 sec a simultaneous SLP of 2 %, 1 % in area-1 & area-2 respectively are applied. The dynamic performance with SGA tuned PID controller & FA tuned PID controller are demonstrated in Figs. 6-8. It is obvious from Figs. 6-8 that, proposed SGA optimized PID controllers based on the location of the disturbance changes also performs satisfactorily and gives robust result.

6 Sensitivity analysis

Sensitivity analysis is perform to test the robustness of the proposed approach with variation in the system parameters and operating conditions.

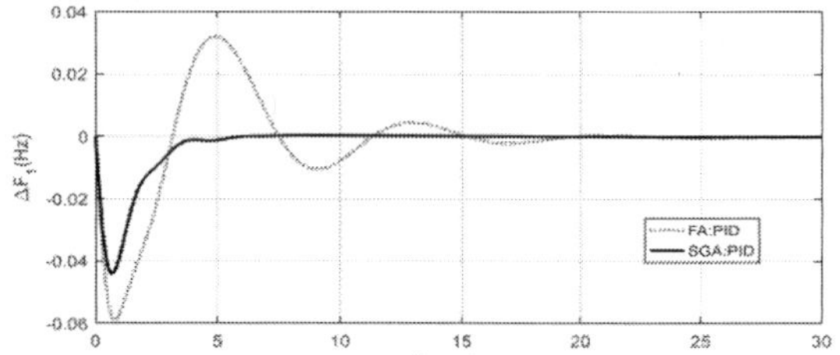

Fig. 6. ΔF_1 for 2 % & 1% SLP in area-1 and in area-2.

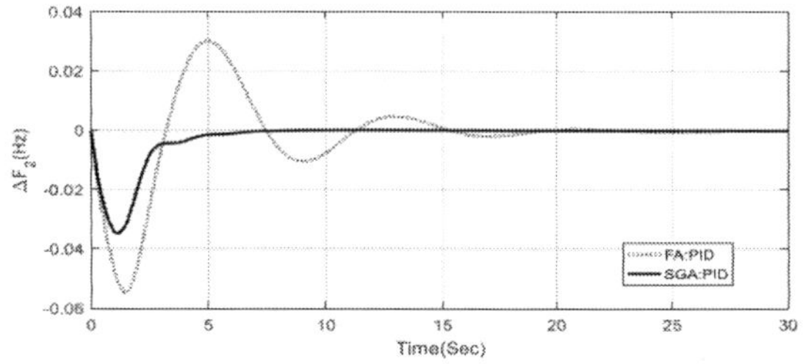

Fig. 7. ΔF_2 for 2 % & 1% SLP in area-1 and in area-2.

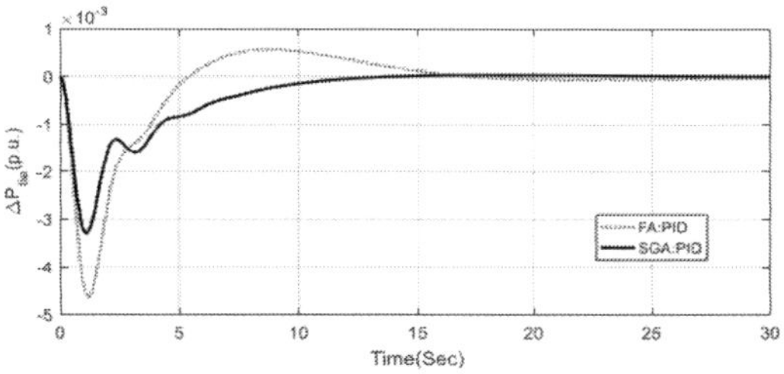

Fig.8. ΔP_{tie} for 2 % & 1% SLP in area-1 and in area-2.

The nominal values varying from +25% to -25% in the operating load condition and system parameters (given in appendix) mentioned in Table 4. It is clearly depicts the system time constants and variations on operating loading conditions over the system performance are negligible as well as reflect the similarity in performance indexes values.

Fig.9 shows ±25% variations in the loading conditions with nominal parameters for the deviation in frequency response of area-1. From Fig. 9 it clearly reveals that the deviation in loading condition over the system performance is evaluated as negligible. To evaluate the supremacy of the proposed approach a random step load change is applied in area-1. Fig.10 depicts the random step load based pattern (magnitude and duration) applied to the test system [9]. Fig.11 shows the transient responses for ΔF_1. From which it can be conclude that the proposed SGA tuned PID controller which provides superior damping compare to other.

7 Conclusions

In the present work, SGA/FA methods are used to tune PID controller parameters in a two-area six-unit power system. The superiority of the SGA technique is demonstrated by comparing the results with a FA optimization technique. The result reveals that SGA tuned

Table 4. Sensitivity analysis.

Parameter variation	%	Performance index: Settling time(2%band) T_s (Sec) ΔF_1	ΔF_2	ΔP_{tie}	ITAE
Nominal	0	1.13	0.94	1.07	0.2947
Loading condition	+25	1.12	0.96	1.06	0.2949
	-25	1.14	0.93	1.07	0.2947
T_G	+25	1.15	1.01	1.09	0.2948
	-25	1.21	1.02	1.16	0.2954
T_{GH}	+25	1.12	0.93	1.07	0.2956
	-25	1.11	0.95	1.04	0.2935
T_T	+25	1.37	1.19	1.31	0.3061
	-25	1.07	0.91	1.02	0.2949
T_{RH}	+25	1.12	0.86	0.99	0.3028
	-25	1.24	1.13	1.22	0.2965
T_{CD}	+25	1.13	0.93	1.06	0.2943
	-25	1.14	0.99	1.08	0.2950
B	+25	1.05	0.85	1.00	0.2601
	-25	1.33	1.16	1.15	0.3701
R	+25	1.18	0.90	1.06	0.3020
	-25	0.77	1.03	1.08	0.2923

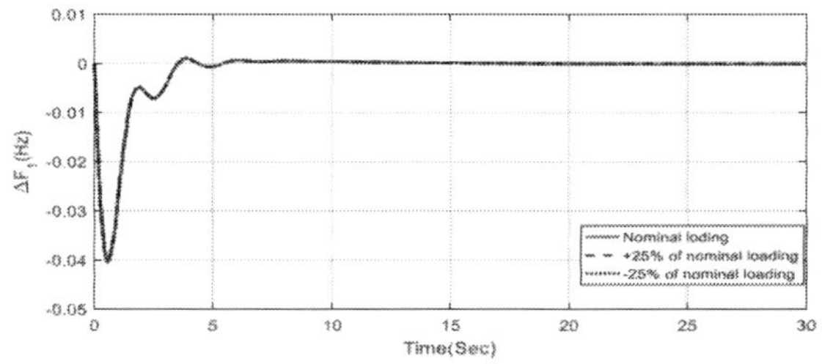

Fig. 9. ΔF_1 with change of loading.

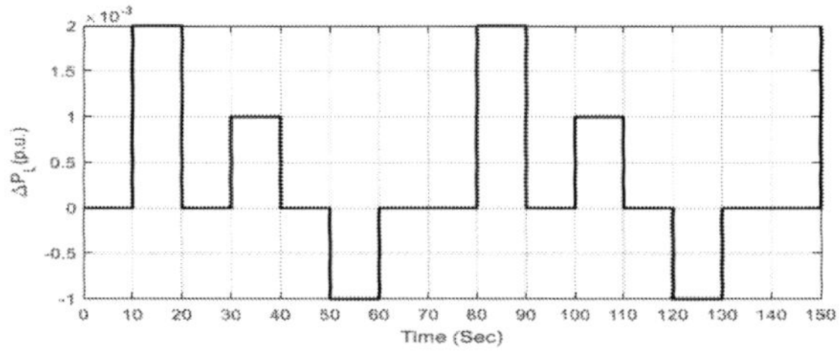

Fig. 10. Stochastic step load based pattern

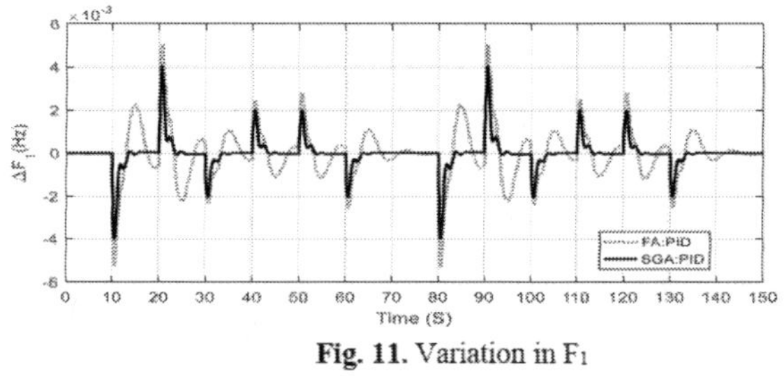

Fig. 11. Variation in F_1

PID controller offers significant improvement in the response than the FA tuned PID controller. After that sensitivity analysis is carried out to demonstrate the robustness of the proposed approach to wide variations of system parameter, operating loading conditions with respect to nominal values as well as random load disturbance. It is evident from simulation results that the proposed SGA optimized PID controller is much more effective, robust and furnish best system performance as comparison to FA tuned PID controller.

Appendix

The investigated system nominal parameters are: [14]
$R_1 = R_2 = R_3 = 2.4$ Hz/p.u; $B_1 = B_2$ 0.4312 p.u., $T_{12} = 0.0433$, $\alpha_{12} = -1$; $T_{SG} = 0.08$sec; $T_T = 0.3$s; $T_r = 10$ s; $K_r = 0.3$; $K_T = 0.543478$; $T_{PS} = 11.49$ s; $K_{PS} = 68.9566$ Hz/p.u. MW; $K_H = 0.326084$; $T_W = 1.0$s; $T_{RS} = 5$s; $T_{GH} = 0.2$s, $T_{RH} = 28.75$ s; $X_C = 0.6$s, $Y_C = 1$s, $T_F = 0.23$s; $T_{CD} = 0.2$s; $T_{CR} = 0.01$s, $K_G = 0.130438$; cg = 1, bg =0.05s.

References

1. H. Bevrani, *Robust Power System Frequency Control*, Springer, (2009)
2. H. Bervani, T. Hiyama, *Intelligent Automatic Generation control*, CRC Press, (2011)
3. O.I. Elgerd, *Electric Energy Systems Theory – An Introduction*, Tata McGraw Hill, New Delhi, (2000)
4. K.P. Ibraheem, D.P. Kothari, Recent philosophies of automatic generation control strategies in power systems, IEEE Trans. Power Syst., **20**, 346-357, (2005).
5. J. Nanda, A. Mangla, S. Suri, Some findings on automatic generation control of an interconnected hydrothermal system with conventional controllers, IEEE Trans. on Energy Conv.,**21**, 187-193, (2006)
6. L.C. Saikia, J. Nanda, S. Mishra, Performance comparison of several classical controllers in AGC for multi-area interconnected thermal system, Int. J. Elect. Power & Energy Systs., **33(3)**, 394-401, (2011)
7. R. K. Sahu, T.S. Gorripotu, S.Panda, Automatic generation control of multi-area power systems with diverse energy sources using teaching learning based optimization algorithm, Engg. Science & Technology, An International Journal, **19**, 113-134, (2016)
8. E. S. Ali, S. M. Abd-Elazim, Bacteria foraging optimization algorithm based load frequency controller for interconnected power system, Int. J. Elect. Power & Energy Systs., **33(3)**, 633-638, (2011).
9. R.K. Sahu, S. Panda, U.K.Rout, D.K. Sahoo, Teaching learning based optimization algorithm for automatic generation control of power system using 2-DOF PID controller, Int. J. Elect. Power & Energy Systs., **77(3)**, 287–301, (2016).
10. G. Sharma, I. Nasiruddin, K.R. Niazi, R.C. Bansal, Optimal AGC of a multi-area power system with parallel AC/DC tie lines using output vector feedback control strategy, Int. J. Elect. Power & Energy Systs., **81**, 22-31, (2016)
11. D. Guha, P.K. Roy, S. Banerjee, Load frequency control of interconnected power system using grey wolf optimization, Swarm and Evol. Compt., **27**, 97-115, (2016)
12. S. Padhan, R.K. Sahu, S. Panda, Application of firefly algorithm for load frequency control of multi-area interconnected power system', Elect. Power Compts and Syst., 42**(13)**, 1419-1430, 2014.
13. M.S. Gonçalves, R. H. Lopez, L.F. Fadel, Search group algorithm: A new metaheuristic method for the optimization of truss structures design, Computers and Structures, **153**, 165-184, (2015).
14. K.P.S. Parmar, S. Majhi, D.P. Kothari, Load frequency control of a realistic power system with multi-source power generation, Int. J. Elect. Power & Energy Systs., **42**, 426-433, (2012).
15. H. Shabani, B.Vahidi, M.A. Ebrahimpour, Robust PID controller based on imperialist competitive algorithm for load-frequency control of power systems, ISA Trans., **52**, 88-95, (2012)
16. X.S. Yang, Nature-Inspired Metaheuristic lgorithms, Luniver Press, UK, (2008)
17. X.S. Yang, Firefly algorithms for multimodal optimization, in: Stochastic Algorithms: Foundations and Applications, SAGA Lecture Notes in Computer Sciences, **5792**, 169–178, (2009)
18. X.S. Yang, Firefly algorithm, stochastic test functions and design optimization, International Journal of Bio-inspired Computation, **2**, 78–84, (2010).

Optimal Selection of Gear Ratio for Hybrid Electric Vehicles Using Modern Meta-Heuristics Search Algorithm

Vikram Kumar Kamboj[1], *Sobhit* Saxena[2], *Kamalpreet* Sandhu[3]

[1,2]School of Electronics and Electrical Engineering, Lovely Professional University, Punjab, INDIA
[3]Product and Industrial Design Department, School of Design, Lovely Professional University, Phagwara, INDIA

Abstract. Gear Train Design problem is most important design problem for machine tools manufacturers. *Recent* work on gear train improvement has been bound towards multi-shaft gear trains of the speed-change kind, where major focus is to maximize the range of operating speeds and to minimize the number of gears and spindles. In the proposed research, a hybrid meta-heuristic search algorithm is presented to design and optimize multi-spindle gear trains problem. The objective of the research is to optimize gear trains on the basis of minimum overall centre distance, minimum overall size, minimum gear volume, or other desirable criteria, such as maximum contact or overlap ratios. The proposed hybrid meta-heuristic search algorithm is inspired by canis lupus family of grey wolves and exploitation capability of existing grey wolf optimizer is further enhanced by pattern search algorithm, which is a derivative-free, direct search optimization algorithm suitable for non-differential, discontinuous search space and does not require gradient for numerical optimization problem and have good exploitation capability in local search space. The effectiveness of the proposed algorithm has been tested on various mechanical and civil design problem including gear train design problem, which includes four different gear and experimental results are compared with others recently reported heuristics and meta-heuristics search algorithm. It has been found that the proposed algorithm indorses its effectiveness in the field of nature inspired meta heuristics algorithms for engineering design problems for hybrid electric vehicles.

Keywords. Engineering Design Problems, Gear Train Design Problem, Hybrid Electric Vehicles, Meta-Heuristics.

1. INTRODUCTION

Multidisciplinary design optimization and multidisciplinary system design optimization are emerging area for the solution of design and optimization problems incorporating a number of disciplines. With the advancement in technology a new era of problem solving methods is emerging which make use of computers. Optimization techniques are considered to be one of the best tool for solving the engineering problems and to find the optimal results for the problem. These approaches consider the problem as black box and find the optimal solution. The optimization process initialize with random set for specified problem and then improving them over predefined steps. The engineering problems to be tackled consist of various difficulties such as constraints, uncertainties, local solution, multiple objective, etc. Optimization technique should be able to discourse these issues. While evaluating the optimization problem exploration and exploitation are the criteria to be taken into account based on these two features the algorithms are classified into two categories consisting of population based search algorithm which is exploration oriented and the other one is evolution based algorithms which are exploitation focused and there should be a good balance between them so as to enhance the working efficiency of the resultant algorithm. One of the method to achieve this balance is by using hybrid algorithm which enhance performance by combining two techniques the resulting technique is called memetic algorithm. In the recent years, various meta-heuristics search algorithms has been implemented such as Biogeography based Optimizer [1], Grey Wolf Optimizer [2], Ant Lion Optimizer [3], Moth Flame Optimizer [4], Multi Verse Optimizer [5], Dragon Fly Algorithm [6],Sine Cosine Algorithm [7],Lightning Search Algorithm [8], Seeker Optimization Algorithm [9],Virus Colony Search Algorithm [10], Whale Optimization Algorithm [11], Wind Driven Optimization [12]. However, some certifiable designing and logical improvement issues are exceptionally intricate and hard to settle, utilizing these techniques. On the off chance that there is more than one neighborhood minima in the problem, the outcome may rely upon the choice of an underlying point, and the acquired minima may not really be the worldwide minima. Moreover, the gradient search may end up plainly troublesome and unstable when the target work and, imperatives have different or sharp peaks. The computational disadvantages of existing numerical straight and nonlinear strategies have constrained analysts to depend on meta-heuristic calculations in light of reproductions to take care of building and logical streamlining issues. A few traditional techniques are accessible to take care of the unit commitment issue. Be that as it may, every one of these strategies require the correct numerical model of the framework and there is a shot of stalling out at the nearby optima. Also, The No-Free-Lunch theorem for optimization allow developers to develop new algorithm or to improve the existing algorithm because, it logically proves that there is no such optimization algorithm which can solve all the optimization problems with equal efficiency for all. Some algorithm work best for few problems and worst for the rest of the problems. So, there is always a scope or improvement to develop the algorithm which could work well for most of the problems.

2. PATTERN SERACH AGORITHM

Pattern search method, also known as black box method, is a derivative free method having local search capability and suitable for search problem, where the derivative of the objective function is inconvenient or unknown. The method involve two moves while performing its operation: one is exploratory search which is local search looking for improving the direction to be moved the other move is the pattern move which is a larger search for improving the direction in this move step size is increased unless the improvement is not altered.

Step-1: Initialize the input parameters for pattern search algorithm i.e. acceleration factor (υ), perturbation vector (P^0) and perturbation tolerance vector (τ).

Step-2: Initialize the current perturbation vector $P \leftarrow P^0$ and select the value for the starting point x^{int}.

Step-3: Update x^{int} to x^{iter} using exploratory search around x^{int} to find an improved point x^{iter} that has a better value of objective function.

if $x^{iter} > x^{int}$

DO $P \leftarrow P/2$.

if $P_i < \tau$

DO $x^{final} = x^{int}$

else

Go to step-3 and update the solution vector using exploratory move.

else

DO $x^{final} = x^{iter}$, $P \leftarrow P^0$ and go to step-4.

end

end

Step-4: Apply Pattern Move using following steps:

Step-4(a): Obtain tentative x^{iter+1} by a pattern move from x^{int} through x^{iter}.

Step-4(b): Obtain final x^{iter+1} by an exploratory search around tentative x^{iter+1}.

if $f(x^{iter+1}) > f(x^{iter})$

DO $x^{int} \leftarrow x^{iter}$ and go to step-3.

else

DO $x^{int} \leftarrow x^{iter}$, $x^{iter} \leftarrow x^{iter+1}$ and go to step-4

end

Fig.1. The PSEUDO code for the Pattern Search algorithm

The pattern move require two points one is the current point and the other one is some random point having better value of the objective function which guides the search direction the consideration of new point is guided by equation (1) $x^{iter+1} = x^{int} + \upsilon[x^{iter} - x^{int}]$

(1) Where, υ is positive acceleration factor, which is used to multiply the length of the direction improvement vector. The PSEUDO code for the Pattern Search algorithm is shown in Fig.1.

3. HYBRID GREY WOLF OPTIMIZER

Grey Wolf Optimizer [13] is swarm intelligence-based, recently developed, metaheuristics search algorithm inspired from the hunting mechanism and leadership hierarchy of grey wolves in nature and require few control parameters and was initially applied to solve 29 benchmark problems and three classical engineering design problems such as tension/compression spring, welded beam, pressure vessel designs problem and real world optical engineering. Further, it was successfully applied to solve various Engineering Optimization Problems such as Economic Load Dispatch Problem [14], Economic load dispatch problem with valve point effect [15], Unit Commitment Problem and training multi-layer perceptron [17]. Several algorithms have also been developed to improve the convergence performance of Grey Wolf Optimizer that includes parallelized GWO [18], hybrid GWO with Genetic Algorithm (GA) [19], Hybrid DE with GWO [20] , Hybrid Grey Wolf Optimizer using Elite Opposition Based Learning Strategy and Simplex Method [21], Modified Grey Wolf Optimizer (mGWO) [22], Mean Grey Wolf Optimizer (MGWO) [23], Hybrid Particle Swarm Optimization with Grey Wolf Optimizer (HPSOGWO) [24] and RW-GWO [25]. In the proposed research, the authors has further tried to improve the exploration phase of the existing grey wolf optimizer algorithm

3.1 PROBLEM FORMULATION

Primarily developed Grey Wolf Optimizer, is a transformative calculation algorithm, based on grey wolves, which recreate the social level and hunting component of grey wolves in view of three principle ventures of chasing: scanning for prey, encompassing prey and assaulting prey and its mathematically model was designed in view point of hierarchy level of different wolves. The fittest solution was designated as alpha (α). Accordingly, the second and third best solutions are named beta (β) and delta (δ) individually. Whatever is left of the hopeful solution are thought to be kappa (κ), lambda (λ) and omega (ω). For the fitness value calculation, the advancement (i.e. chasing) is guided by α, β and δ. The ω, κ and λ wolves trail these three wolves. In GWO, Encircling or Trapping of Prey was achieved by calculating $\vec{D}$ and $\vec{X}_{GWolf}$ vectors described by equations (2) and (3).

$$\vec{D} = \left|\vec{C}.\vec{X}_{Prey}(iter) - \vec{X}_{GWolf}(iter)\right| \quad (2)$$

$$\vec{X}_{GWolf}(iter+1) = \vec{X}_{Prey}(iter) - \vec{A}.\vec{D} \quad (3)$$

Where, iter demonstrates the present iteration, $\vec{A}$ and $\vec{C}$ are coefficient vectors, $\vec{X}_{Prey}$ is the position vector of the prey and $\vec{X}_{GWolf}$ shows the position vector of a grey wolf and the vectors $\vec{A}$ and $\vec{C}$ are calculated as follows:

$$\vec{A} = 2\vec{a}.\vec{\mu}_1 - \vec{a} \quad (4)$$

$$\vec{C} = 2.\vec{\mu}_2 \quad (5)$$

Where, $\vec{\mu}_1, \vec{\mu}_2 \in rand(0,1)$ and $\vec{a}$ decreases linearly from 2 to 0. The hunting of prey are achieved by calculating the corresponding fitness score and positions of alpha, beta and delta wolves using equations (6), (7) and (8) respectively and final position for attacking towards the prey was calculated by equation (9).

$$\vec{D}_{Alpha} = abs(\vec{C}_1.\vec{X}_{Alpha} - \vec{X}) \quad (6a)$$

$$\vec{X}_1 = \vec{X}_{Alpha} - \vec{A}_1.\vec{D}_{Alpha} \quad (6b)$$

$$\vec{D}_{Beta} = abs(\vec{C}_2.\vec{X}_{Beta} - \vec{X}) \quad (7a)$$

$$\vec{X}_2 = \vec{X}_{Beta} - \vec{A}_2.\vec{D}_{Beta} \quad (7b)$$

$$\vec{D}_{Delta} = abs(\vec{C}_3.\vec{X}_{Delta} - \vec{X}) \quad (8a)$$

$$\vec{X}_3 = \vec{X}_{Delta} - \vec{A}_3.\vec{D}_{Delta} \quad (8b)$$

$$\vec{X}(iter+1) = \frac{(\vec{X}_1 + \vec{X}_2 + \vec{X}_3)}{3} \quad (9)$$

In the proposed hybrid Grey-Wolf Optimizer-Pattern Search (hGWO-PS) algorithm, the randomly generated position vector $\vec{X}$ has been further modified using Pattern search method and the modified position vector $\vec{X}$, has been applied to grey wolves to evaluate alpha, beta and delta scores. In order to hybridize the GWO and PS, the heuristics procedure has been adopted. The exploration phase in hGWO-PS is similar to classical GWO. In order to explore the search space globally, vector $\vec{A}$ and $\vec{C}$ are used, which mathematically model divergence. The absolute value of $\vec{A}$ greater than 1 forces the grey wolves to diverge from the prey to optimistically find an adequate prey. The PSEUDO code of proposed hGWO-PS algorithm has been shown in **Fig.2.**

 https://doi.org/10.1051/e3sconf/20198701006

```
Initialize the grey wolf population X_i, where i ∈ 1,2,3,...,N
Initialize a, A and C
Calculate the fitness of each search agents using objective function
Assign X_α ← The best search agents
X_β ← The second best search agents
X_δ ← The third best search agents
Update the position of X_α using the Pattern Search method (refer to PSEUDO code of Fig.1)
while (iter < Iter_max)
    for each search agents
        Update the position of the current search agent by equation (9).
    end for
        update a, A and C using equations (4) and (5).
        update X_α, X_β and X_δ
    iter = iter + 1
end while
return X_α
```

Fig 2: PSEUDO code of proposed hGWO-PS algorithm

4. ENGINEERING OPTIMIZATION PROBLEMS

In order to validate the performance of the proposed hGWO-PS algorithm, six engineering design optimization problems are taken into consideration which includes Three-bar truss problem, Pressure vessel problem, Tension/compression spring design problem, welded beam problem, Cantilever Beam Design problem and Gear Train design problem (Fig.3 to Fig.8). In order to validate the stochastic nature of proposed algorithm, 30 trial runs has been performed and results are evaluated for mean, worst and best values of fitness including standard deviation. In the whole research study, 30 search agents are taken into considerations and algorithm is simulated for maximum iterations of 1500. For the effective analysis of results, the non-parametric statistical hypothesis test i.e. Wilcoxon rank sum test has been applied. The significance of the test is to examine the distribution of samples i.e. whether two dependent samples selected from populations having the same distribution or not. The results of the engineering optimization design problems are recorded with respect to best value, worst value, mean value, standard deviation and p-value and has been reported in Table-1. The comparison of results for hGWO-PS has been shown in Table-1 through Table-2. The results of aforementioned engineering problems strongly evidence the merits of the hGWO-PS algorithm in solving problems with unknown search spaces. The results also show that the proposed algorithm is suitable for constrained and discrete problems. The convergence curve and trial runs solutions for these engineering problems has been depicted in Fig.8 and Fig.9 respectively, which shows the superiority of hGWO-PS over GWO.

4.1 Three-Bar Truss Design Problem

In order to validate the results of proposed hGWO-PS algorithm for Engineering optimization problems, the first problem of a three-bar truss design has be undertaken, where the objective of the fitness function is to minimize its weight. The various constrains for three-bar truss design problem are stress constraint, deflection constraint, and buckling constraint. The mathematical model for the aforementioned problem has been shown below in equations (10) through (12).

Consider $\vec{x} = [x_1, x_2] = [A_1, A_2]$

Minimize $f(\vec{x}) = (2\sqrt{2}x_1 + x_2) * l$ (10)

Subject to $g_1(\vec{x}) = \frac{\sqrt{2}x_1 + x_2}{\sqrt{2}x_1^2 + 2x_1x_2} P - \sigma \le 0$ (11a)

$g_2(\vec{x}) = \frac{x_2}{\sqrt{2}x_1^2 + 2x_1x_2} P - \sigma \le 0$ (11b)

$g_3(\vec{x}) = \frac{1}{\sqrt{2}x_2 + x_1} P - \sigma \le 0$ (12)

Variable range $0 \le x_1$, $x_2 \le 1$

Where l = 100cm , P= 2KN/cm^2 , σ = 2 KN/cm^2

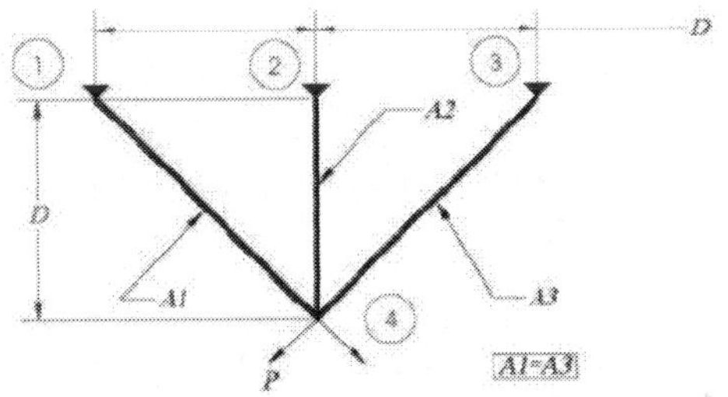

Fig 3: Three-Bar Truss Design Problem

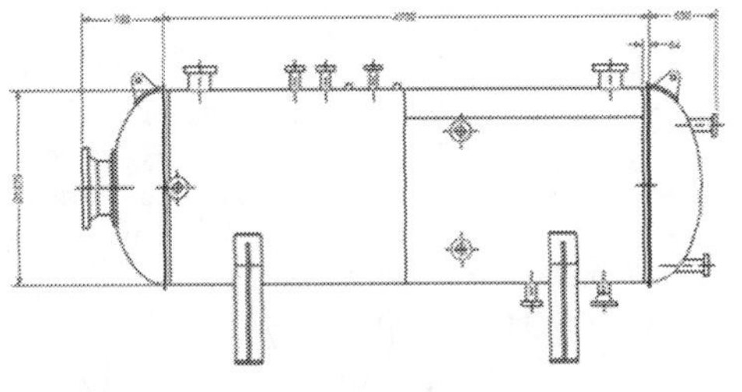

Fig 4: Pressure Vessel Design Problem

4.2 Pressure Vessel Design Problem

The second multidisciplinary design optimization problem, which has been taken into consideration is Pressure Vessel Design Problem. The objective function of this problem is to minimize the total cost, which is the cost of material, forming, and welding of a cylindrical vessel (Fig.4). For the designing of pressure vessel problem, four different variables are taken into consideration, which are Thickness of the shell (Ts), Thickness of the head (Th), Inner radius (R) and Length of the cylindrical section without considering the head (L). The head of the vessel has been taken in the hemi spherical shape and both ends of the vessel has been capped. The aforementioned design problem is subjected to four constraints. The mathematical formulation of the problem and its constrained is shown below in equations (13) through (14).

Consider $\vec{x} = [x_1 x_2 x_3 x_4] = [T_s T_h R L]$

Minimize

$$f(\vec{x}) = 0.6224 x_1 x_3 x_4 + 1.7781 x_2 x_3^2 + 3.1661 x_1^2 x_4 + 19.84 x_1^2 x_3 \quad (13)$$

Subjected to $g_1(\vec{x}) = -x_1 + 0.0193 x_3 \le 0$ (14a)

$$g_2(\vec{x}) = x_3 + 0.00954 x_3 \le 0 \quad (14b)$$

$$g_3(\vec{x}) = -\pi x_3^2 x_4 - \frac{4}{3}\pi x_3^3 + 1296000 \le 0 \quad (14c)$$

$$g_4(\vec{x}) = x_4 - 240 \le 0 \quad (14d)$$

Variable range $0 \le x_1 \le 99$

$0 \le x_2 \le 99$, $10 \le x_3 \le 200$, $10 \le x_4 \le 200$

4.3 The Tension/Compression Spring Design Problem

The third multidisciplinary design optimization, which is taken into consideration for engineering optimization is Tension/Compression Spring Design Problem (Fig.5), which is a kind of mechanical engineering problem. The objective of the spring design problem is to minimize the weight of the spring. To solve the problem, three design variables i.e. the number of active coils (N), mean coil diameter (D) and wire diameter (d) are taken into contemplation and the problem is subjected to surge frequency constraints, shear stress constraints and minimum deflection constraints. The mathematical model of the aforesaid problem is explained below in equations (15) through (16).

Consider $\vec{x} = [x_1 x_2 x_3] = [dDN]$,

Minimize $f(\vec{x}) = (x_3 + 2) x_2 x_1^2$, (15)

Subjected to

$$g_1(\vec{x}) = 1 - \frac{x_2^3 x_3}{71785 x_1^4} \le 0 \quad (16a)$$

$$g_2(\vec{x}) = \frac{4x_2^2 - x_1 x_2}{12566(x_2 x_1^3 - x_1^4)} + \frac{1}{5108 x_1^2} \le 0 \quad (16b)$$

$$g_2(\vec{x}) = \frac{4x_2^2 - x_1 x_2}{12566(x_2 x_1^3 - x_1^4)} + \frac{1}{5108 x_1^2} \le 0$$

$$g_3(\vec{x}) = 1 - \frac{140.45 x_1}{x_2^2 x_3} \le 0 \quad (16c)$$

$$g_4(\vec{x}) = \frac{x_1 + x_2}{1.5} - 1 \le 0 \quad (16d)$$

(16e)

Variable range $0.005 \le x_1 \le 2.00$

$0.25 \le x_2 \le 1.30$

$2.00 \le x_3 \le 15.0$

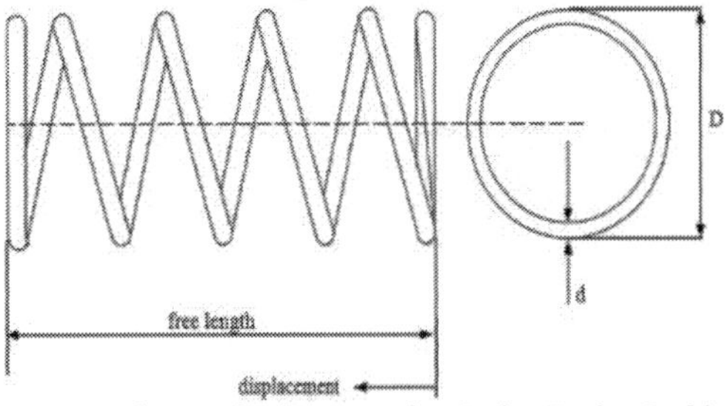

Fig 5: The Tension/Compression Spring Design Problem

4.4 Welded Beam Design Problem

The fourth engineering optimization problem, which is taken into consideration is welded beam design problem (Fig.6), where the objective of the problem is to minimize the fabrication cost of a welded beam. The problem consists of four variables, which are length of attached part of bar (l), thickness of the bar (b), the height of the bar (t) and thickness of weld (h). The problem is subjected to four constraints, which includes Side constraint, Buckling load on the bar (Pc), End deflection of the beam (d), Bending stress in the beam (h) and Shear stress (s). The mathematical formulation of the abovementioned problem is explained below in equations (17) through (19).

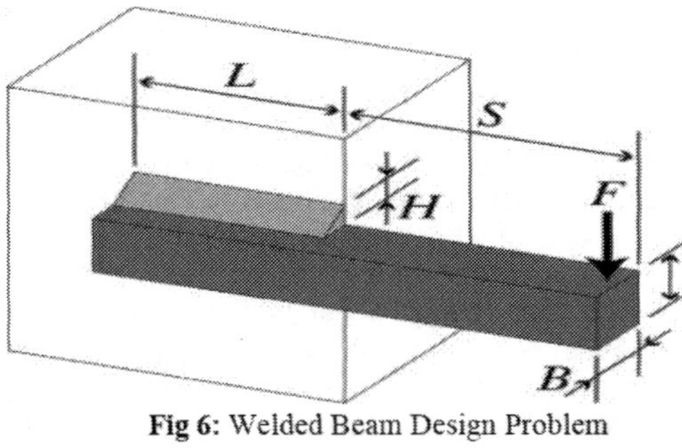

Fig 6: Welded Beam Design Problem

Consider $\vec{x} = [x_1 x_2 x_3 x_4] = [hltb]$, (17)

Minimize

$$f(\vec{x}) = 1.10471x_1^2x_2 + 0.04811x_3x_4(14.0 + x_2)$$

Subject to $g_1(\vec{x}) = \tau(\vec{x}) - \tau_{max} \le 0,$ (18a)

$$g_2(\vec{x}) = \sigma(\vec{x}) - \sigma_{max} \le 0, \quad (18b)$$

$$g_3(\vec{x}) = \delta(\vec{x}) - \delta_{max} \le 0, \quad (18c)$$

$$g_4(\vec{x}) = x_1 - x_4 \le 0, \quad (18d)$$

$$g_5(\vec{x}) = P - P_c(\vec{x}) \le 0, \quad (18e)$$

$$g_6(\vec{x}) = 0.125 - x_1 \le 0, \quad (18f)$$

$$g_7(\vec{x}) = 1.10471x_1^2 + 0.04811x_3x_4(14.0 + x_2) - 5.0 \le 0 \quad (18g)$$

Variable range $0.1 \le x_1 \le 2$, $0.1 \le x_2 \le 10$

$0.1 \le x_3 \le 10$, $0.1 \le x_4 \le 2$

Where $\tau(\vec{x}) = \sqrt{(\tau')^2 + 2\tau'\tau''\frac{x_2}{2R} + (\tau'')^2}$ (19a)

$$\tau' = \frac{P}{\sqrt{2}x_1x_2}, \tau'' = \frac{MR}{J}, M = P\left(L + \frac{x_2}{2}\right), \quad (19b)$$

$$R = \sqrt{\frac{x_2^2}{4} + \left(\frac{x_1 + x_3}{2}\right)^2}, \quad (19c)$$

$$J = 2\left\{\sqrt{2}x_1x_2\left[\frac{x_2^2}{4} + \left(\frac{x_1 + x_3}{2}\right)^2\right]\right\} \quad (19d)$$

$$\sigma(\vec{x}) = \frac{6PL}{x_4x_3^2}, \delta(\vec{x}) = \frac{6PL^3}{Ex_2^2x_4} \quad (19e)$$

$$P_c(\vec{x}) = \frac{4.013E\frac{\sqrt{x_3^2x_4^6}}{36}}{L^2}\left(1 - \frac{x_3}{2L}\sqrt{\frac{E}{4G}}\right) \quad (19f)$$

4.5 Cantilever Beam Design Problem

The fifth optimization problem of civil engineering has been taken into consideration, which is a cantilever beam consisting of five hollow elements with square-shaped cross-section (Fig.7). The objective of this optimization problem is to reduce the weight of the beam. In this design problem, each element is defined by one variable and overall structure consists of 5 structural parameters. The thickness of the beam is taken as constant. In the proposed design, the vertical load is applied to the free end of the beam (node 6) and the right side of the beam (node 1) is rigidly supported. While designing Cantilever Beam Design Problem, the vertical displacement constraint must be taken into consideration and it should not be violated in the final optimal design. The mathematical formulation of the aforementioned design problem has been illustrated below:

The problem formulation is as follows:

Consider

$$\vec{x} = [x_1x_2x_3x_4x_5]$$

Minimize

$$f(\vec{x}) = 0.6224(x_1 + x_2 + x_3 + x_4 + x_5) \quad (20)$$

Subject to

$$g(\vec{x}) = \frac{61}{x_1^3} + \frac{37}{x_2^3} + \frac{19}{x_3^3} + \frac{7}{x_4^3} + \frac{1}{x_5^3} \le 1 \quad (21)$$

Variable range $0.01 \le x_1, x_2, x_3, x_4, x_5 \le 100$

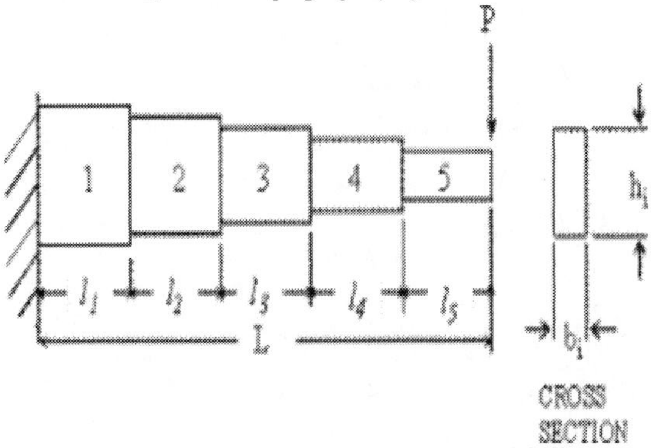

Fig 7: Cantilever Beam Design Problem

4.6 Gear Train design Problem

Gear train design problem is a discrete design problem with four parameters (Fig.8). The major objective this design problem is minimization of the gear ratio, which can be achieved by determining the optimal number of tooth for four gears of a train. In mathematics, the gear train design problem can be described as per the following mathematical equations:

Consider $\vec{x} = [x_1x_2x_3x_4] = [N_AN_BN_CN_D]$ (22)

Minimize $f(\vec{x}) = \left(\frac{1}{6.931} - \frac{x_3x_4}{x_1x_4}\right)^2$ (23a)

Subject to $12 \le x_1, x_2, x_3, x_4 \le 60$ (23b)

https://doi.org/10.1051/e3sconf/20198701006

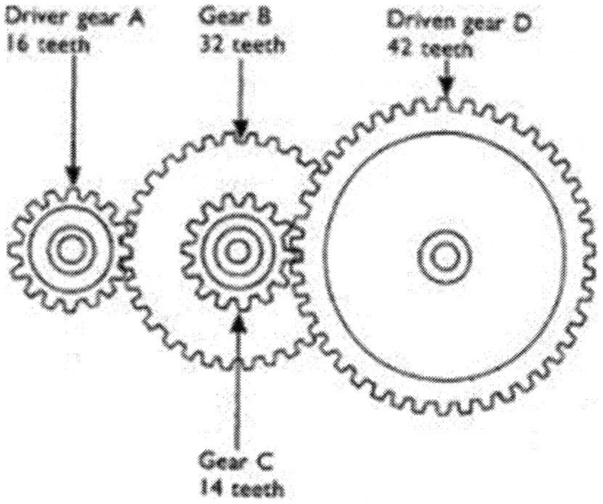

Fig.8: Gear Train design Problem

Table-1 : Solution of Engineering design problems using hGWO-PS

Engineering Design Problem	Best value	Mean value	Worst value	SD	median	Wilcoxon p-value
Three-bar truss problem	263.896	263.901	263.914	0.0044	263.900	1.73e-06
Pressure vessel problem	6160.106	7001.187	7320.343	391.305	7236.930	1.73e-06
Tension/compression spring design problem	0.01267	0.01385	0.01815	0.0010	0.01363	1.73e-06
Welded beam problem	1.72608	1.77333	2.44848	0.128052	1.7477	1.73e-06
Cantilever Beam Design	1.30363	1.30826	1.33052	0.0051	1.306	1.73e-06

Table-2 : Comparison analysis of Gear Train Design Problem

Algorithm	N_A	N_B	N_C	N_D	Optimal Fitness Value
MBA [27]	43	16	19	49	2.7009e-012
ISA [28]	---	---	---	---	2.7009e-012
GA [29]	---	---	---	---	2.33e-07
ABC [30]	19	16	44	49	2.78e-11
GA [31]	33	14	17	50	1.362e-09
ALM [32]	33	15	13	41	2.1469e-08
ALO	49	19	16	43	2.7009e-012
CS [26]	43	16	19	49	2.7009e-012
hGWO-PS	49	19	16	43	2.7009e-012

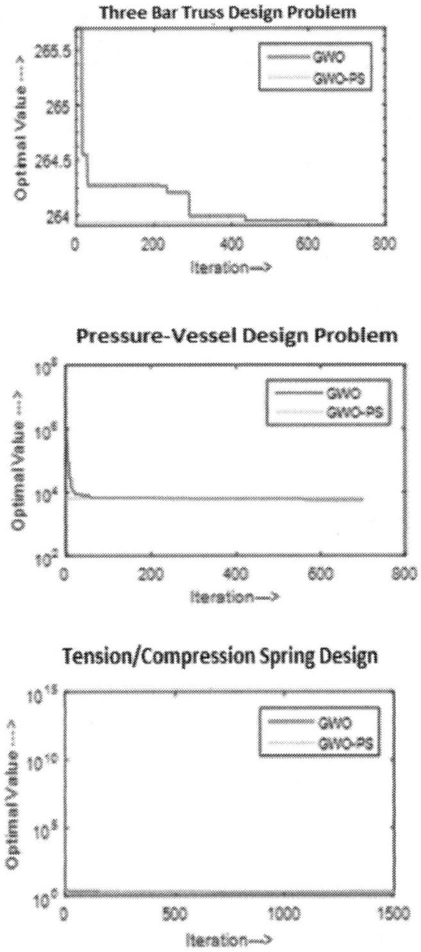

Fig.9: Convergence of GWO and hGWO-PS for various Engineering design problems

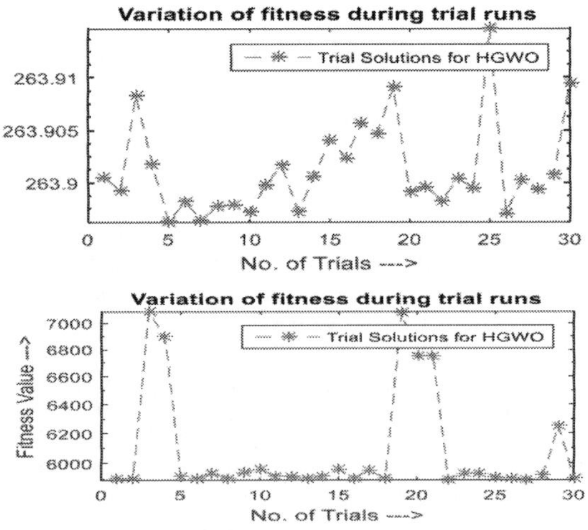

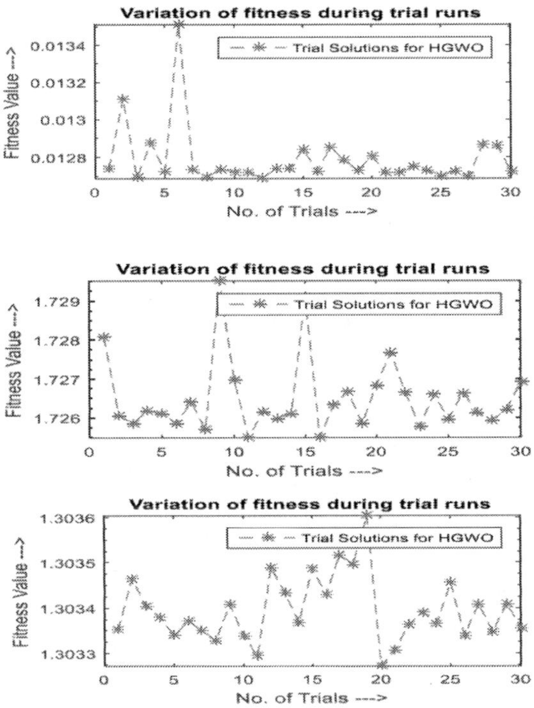

Fig.10: Trial runs solutions of GWO and hGWO-PS for various engineering optimization problems

5. CONCLUSION

In the proposed research, the exploitation phase of the existing grey wolf optimizer has been successfully improved using pattern search algorithm and the newly developed hybrid hGWO-PS has been tested for non-linear, highly constrained, non-convex engineering design and optimization problems. Experimentally, it has been found that the results of the proposed hybrid GWO-PS algorithm are better than existing Grey Wolf Optimizer algorithm, Ant Lion Optimizer algorithm, Moth Flame Optimization algorithm, sine-cosine optimization algorithm and others recently reported heuristics and meta-heuristics search algorithm and hence proposed algorithm indorses its effectiveness in the field of nature inspired meta heuristics algorithms for engineering optimization problems and for improved speed electric vehicle designing.

REFERENCES

1. D. Simon, "Biogeography-Based Optimization," *IEEE Trans. Evol. Comput.*, vol. 12, no. 6, pp. 702–713, Dec. 2008.
2. S. Mirjalili, S. M. Mirjalili, and A. Lewis, "Grey Wolf Optimizer," *Adv Eng Softw*, vol. 69, p. 46, 2014.
3. S. Mirjalili, "The ant lion optimizer," *Adv. Eng. Softw.*, vol. 83, pp. 80–98, 2015.
4. S. Mirjalili, "Moth-flame optimization algorithm: A novel nature-inspired heuristic paradigm," *Knowledge-Based Syst.*, vol. 89, pp. 228–249, 2015.
5. S. Mirjalili, S. M. Mirjalili, and A. Hatamlou, "Multi-Verse Optimizer: a nature-inspired algorithm for global optimization," *Neural Comput. Appl.*, vol. 27, no. 2, pp. 495–513, 2016.
6. S. Mirjalili, "Dragonfly algorithm: a new meta-heuristic optimization technique for solving single-objective, discrete, and multi-objective problems," *Neural Comput. Appl.*, vol. 27, no. 4, pp. 1053–1073, 2016.
7. S. Mirjalili, "SCA: A Sine Cosine Algorithm for solving optimization problems," *Knowledge-Based Syst.*, vol. 96, pp. 120–133, 2016.
8. H. Shareef, A. A. Ibrahim, and A. H. Mutlag, "Lightning search algorithm," *Appl. Soft Comput. J.*, vol. 36, pp. 315–333, 2015.
9. D. Chaohua, C. Weirong, and Z. Yunfang, "Seeker optimization algorithm," *2006 Int. Conf. Comput. Intell. Secur. ICCIAS 2006*, vol. 1, pp. 225–229, 2007.
10. M. D. Li, H. Zhao, X. W. Weng, and T. Han, "A novel nature-inspired algorithm for optimization: Virus colony search," *Adv. Eng. Softw.*, vol. 92, pp. 65–88, 2016.
11. S. Mirjalili and A. Lewis, "The Whale Optimization Algorithm," *Adv. Eng. Softw.*, vol. 95, pp. 51–67, 2016.
12. Z. Bayraktar, M. Komurcu, and D. H. Werner, "Wind Driven Optimization (WDO): A novel nature-inspired optimization algorithm and its application to electromagnetics," *2010 IEEE Int. Symp. Antennas Propag. CNC-USNC/URSI Radio Sci. Meet. - Lead. Wave, AP-S/URSI 2010*, no. 1, pp. 0–3, 2010.
13. S. Mirjalili, S. M. Mirjalili, and A. Lewis, "Grey Wolf Optimizer," *Adv. Eng. Softw.*, vol. 69, pp. 46–61, 2014.
14. J. G. Vlachogiannis and K. Y. Lee, "Economic load dispatch - A comparative study on heuristic optimization techniques with an improved coordinated aggregation-based PSO," *IEEE Trans. Power Syst.*, vol. 24, no. 2, pp. 991–1001, 2009.
15. M. Herwan Sulaiman, W. Lo Ing, Z. Mustaffa, and M. Rusllim Mohamed, "GREY WOLF OPTIMIZER FOR SOLVING ECONOMIC DISPATCH PROBLEM WITH VALVE-LOADING EFFECTS," vol. 10, no. 21, 2015.
16. A. Bhardwaj, V. K. Kamboj, V. K. Shukla, B. Singh, and P. Khurana, "Unit commitment in electrical power system - A literature review," in *2012 IEEE International Power Engineering and Optimization Conference, PEOCO 2012 - Conference Proceedings*, 2012.
17. S. Mirjalili, "How effective is the Grey Wolf optimizer in training multi-layer perceptrons," *Appl. Intell.*, vol. 43, no. 1, pp. 150–161, Jul. 2015.
18. T.-S. Pan, T.-K. Dao, T.-T. Nguyen, and S.-C. Chu,

"A Communication Strategy for Paralleling Grey Wolf Optimizer," Springer, Cham, 2016, pp. 253–262.
19. M. A. Tawhid and A. F. Ali, "A Hybrid grey wolf optimizer and genetic algorithm for minimizing potential energy function," *Memetic Comput.*, vol. 9, no. 4, pp. 347–359, Dec. 2017.
20. D. Jitkongchuen, "A hybrid differential evolution with grey wolf optimizer for continuous global optimization," in *2015 7th International Conference on Information Technology and Electrical Engineering (ICITEE)*, 2015, pp. 51–54.
21. S. Zhang, Q. Luo, and Y. Zhou, "Hybrid Grey Wolf Optimizer Using Elite Opposition-Based Learning Strategy and Simplex Method," *Int. J. Comput. Intell. Appl.*, vol. 16, no. 2, p. 1750012, Jun. 2017.
22. N. Mittal, U. Singh, and B. S. Sohi, "Modified Grey Wolf Optimizer for Global Engineering Optimization," *Appl. Comput. Intell. Soft Comput.*, vol. 2016, pp. 1–16, May 2016.
23. N. Singh and S. B. Singh, "A Modified Mean Gray Wolf Optimization Approach for Benchmark and Biomedical Problems.," *Evol. Bioinform. Online*, vol. 13, p. 1176934317729413, 2017.
24. N. Singh and S. B. Singh, "Hybrid Algorithm of Particle Swarm Optimization and Grey Wolf Optimizer for Improving Convergence Performance," *J. Appl. Math.*, vol. 2017, pp. 1–15, Nov. 2017.
25. S. Gupta and K. Deep, "A novel Random Walk Grey Wolf Optimizer," *Swarm Evol. Comput.*, Jan. 2018.
26. A. H. Gandomi, Y. X-s, and A. A. C. search algorithm:, "a metaheuristic approach to solve structural optimization problems," *Eng Comput*, vol. 29, no. 17, p. 35, 2013.
27. A. Sadollah, A. Bahreininejad, H. Eskandar, and H. M. M. blast algorithm:, "a new population based algorithm for solving constrained engineering optimization problems," *Appl Soft Comput*, vol. 13, no. 2592, p. 612, 2013.
28. G. Ah., *Interior search algorithm (ISA): a novel approach for global optimization*. ISA Trans, 2014.
29. W. S-j and C. P-t., "Genetic algorithms for nonlinear mixed discrete-integer optimization problems via meta-genetic parameter optimization," *Eng Optim+ A*, vol. 3524, no. 137, p. 59, 1995.
30. T. K. Sharma, M. Pant, and V. Singh, "Improved local search in artificial bee colony using golden section search. arXiv," 2012.
31. K. Deb and M. A. Goyal, "combined genetic adaptive search (GeneAS) for engineering design," *Comput Sci Inf.*, vol. 26, no. 30, p. 45, 1996.
32. B. Kannan and K. Sn., *No Title*. An augmented Lagrange multiplier based method for mixed integer discrete continuous optimization.
33. Seyedali Mirjalili,The Ant Lion Optimizer, Advances in Engineering Software,Volume 83,2015, Pages 80-98, ISSN 0965-9978, https://doi.org/10.1016/j.advengsoft.2015.01.010.

E3S Web of Conferences **87**, 01007 (2019) https://doi.org/10.1051/e3sconf/20198701007
SeFet 2019

Optimal Operation of Microgrid considering Renewable Energy Sources, Electric Vehicles and Demand Response

Salkuti Surender Reddy[1,*]

[1]Department of Railroad and Electrical Engineering, Woosong University, Daejeon, Republic of Korea.

Abstract. This paper proposes a new optimal operation of Microgrids (MGs) in a distribution system with wind energy generators (WEGs), solar photovoltaic (PV) energy systems, battery energy storage (BES) systems, electric vehicles (EVs) and demand response (DR). To reduce the fluctuations of wind, solar PV powers and load demands, the BES systems and DR are utilized in the proposed hybrid system. The detailed modeling of WEGs, solar PV units, load demands, BES systems and EVs has been presented in this paper. The objective considered here is the minimization of total operating cost of microgrid, and it is formulated by considering the cost of power exchange between the main power grid and microgrid, cost of wind and solar PV energy systems, cost of BES systems, EVs and the cost due to the DR in the system. Simulations are performed on a test microgrid, and they are implemented using GAMS software. Various case studies are performed with and without considering the proposed hybrid system.

1 Introduction

In recent years, the renewable energy/green energy technologies are growing rapidly and they are set to reach ambitious targets as they have the ability to reduce energy crisis with environmentally friendly manner. Denmark has set a target of 50% of its power demand from renewables by 2030, whereas Germany has set a target of 35% and UK has set a target of 30% of its power demand from renewables by 2020 [1]. Renewable energy sources (RESs) such as wind power and solar photovoltaic (PV) energy are often spatially distributed and inherently intermittent and unpredictable. The power output of RESs depends on the wind speed, solar irradiation, and other climatic factors, which are uncertain and intermittent in nature. Various countries are moving towards the microgrid with RERs and other locally available distributed generation sources to get the benefits of providing power to remote areas with high power quality and reliability. Microgrid (MG) is an active distribution network, which can operate in both island and connected modes. In island mode, MG is disconnected from the main grid, and it supplies their loads with existing distributed generations (DGs). The majority of DGs used in MGs are renewable types, which are connected to the grid by inverters as switch mode power supplies. Microgrids with RESs and electric vehicles (EVs) can reduce the emissions, loss reduction, costs of energy supply, peak shaving, enhances the reliability, voltage profile and improves power quality [2].
The integration of wind, solar PV energy systems, EVs, BES systems and demand response (DR) provides additional energy and storage options to the microgrid. The survey on existing research works on the optimal strategies for the operation of MGs is proposed in [3]. Reference [4] reviews and compares different optimization techniques applied for achieving optimal operation of MGs from diverse perceptions, in terms of objective functions, optimization techniques, constraints and computational performances. An operation optimization modeling for a MG considering the distributed generation, environmental factors and DR is proposed in [5]. Reference [6] presents an optimal power balancing model of distributed systems with distributed generation and EVs based on quadratic rotated conic programming. An optimal day-ahead scheduling and control models for MGs under uncertainty is proposed in reference [7]. A mathematical model to compute the optimal operation schedule that embodies DR in MG is presented in [8]. The impact of DR program intensity and BES system size on the operation of networked MGs is analyzed while considering the prevailing uncertainties is proposed in [9]. In [10], the DR program and battery are utilized to tackle the stochastic nature of wind, solar power generation and load demand to utilize an energy management system to manage hourly power of MG and optimally supply the demand. An approach to review the role of renewable energy as an energy producer in unit commitment scheduling problem is proposed in [11].
An optimal robust model for short-term operational scheduling of distribution network in the presence of uncertainties to minimize the cost of network operation is proposed in [12]. Utilization of PV systems for EVs charging for transportation requirements of smart cities is proposed in [13]. An economic dispatch problem of MG, i.e., to minimize the fuel cost of dispatchable DGs

* Corresponding author: salkuti.surenderreddy@gmail.com

by using four different optimization techniques - lambda iteration, lambda logic, direct search method, and particle swarm optimization is proposed in [14]. In [15], energy storage units are considered as an effective option for handling the imbalance of wind power generation. Reference [16] proposes an approach to integrate power to gas technology to the multi-source MG including wind power, electric boiler, micro turbine and gas boiler. In [17], the energy storage systems and DR are introduced to the traditional scheduling model of wind and thermal power with carbon emission trading for reducing the influence of random nature of wind power on system operation. From the above literature review, it can be observed that there is a pressing requirement for an optimal operation of MG considering the RESs, BES systems, EVs and DR.

The main contribution of this paper is to optimize the total operating cost of microgrid (MG) by considering the power exchange between main power grid and MG, renewable energy sources, battery energy storage system, electric vehicles and demand response. In this paper, an optimal scheduling methodology is presented for the MGs in large distribution systems. Here, the energy storage batteries and EVs are used to improve the operational flexibility of the system and to increase the penetration of RESs. The demand response that is considered in this paper will work as an operating tool for the improvement of reliability and efficiency of MG. The aim of demand response is to encourage the customers to become active partners of MG energy management system.

The remainder of this paper is organized as follows: Section 2 focuses on the modeling of MG components, i.e., modeling of wind energy generators, solar PV units, load demands, battery energy storage systems and EVs. The formulation of objective function and related constraints is introduced in Section 3. Numerical results and analysis has been carried out in Section 4. Finally, Section 5 presents the contributions with concluding remarks.

2 Modeling of Microgrid Components

Modeling of MG components such as WEGs, solar PV units, load demands, BES systems and EVs are presented in this paper. The detailed modeling of wind energy generators, solar PV units and load demands and their uncertainty handling approaches has been presented in references [18]-[20].

2.1. Modeling of Battery Energy Storage (BES) System

Generally, the battery is considered as constant voltage source. BES system is used to smooth out the energy supply by absorbing energy when the power generated by renewables is more than the scheduled power, and dispensing energy when the renewables do not fulfill the load demand. Charge/discharge power limits of battery are expressed as [21],

$$0 \le P_{B,k}^{ch,t} \le P_{B,k}^{ch,max} \tag{1}$$

$$0 \le P_{B,k}^{disch,t} \le P_{B,k}^{disch,max} \tag{2}$$

where $P_{B,k}^{ch,max}$ and $P_{B,k}^{disch,max}$ are the maximum charging and discharging rates of battery. The state of charge (SOC) of battery plays a major role in controlling the charging/discharging and in optimal scheduling of battery. The constraints on SOC of storage battery are expressed as,

$$SOC_k^{min} \le SOC_k^t \le SOC_k^{max} \tag{3}$$

$$SOC_k^t = \left(\frac{C_{BES,k}^t}{C_{max}}\right) \tag{4}$$

where SOC_k^{min} and SOC_k^{max} are minimum and maximum BES capacity limits, respectively. The $C_{BES,k}^t$ can be expressed as,

$$C_{BES,k}^t = C_{init} + \left(P_{B,k}^{ch,t}\eta_{ch}\right) - \left(P_{B,k}^{disch,t}\eta_{disch}\right) \tag{5}$$

where C_{init} is the initial capacity of battery storage system.

2.2. Modeling of Electric Vehicles (EVs)

EVs will play two roles when they plug into a MG. These EVs will appear as loads during the charging, and they also used as storage units to supply energy to the remaining load demands, if the batteries have the sufficient spare energy. The battery of EV acts as a source of stored energy to provide a number of ancillary services to supply primary and secondary control and voltage control. The battery systems of EV are the chemical storage devices, and they are exponential functions over time. The instantaneous charging status of the battery system of EV is expressed as [22],

$$P_{EV}^t = P_{EV}^{max}\left(1 - e^{\alpha t/t_{max}}\right) + P_{EV}^0 \tag{6}$$

where P_{EV}^0 is the current status of battery system, P_{EV}^{max} is the maximum power capacity of EV, t_{max} is the maximum charging time. Similarly, the discharging status of the battery system of EV is expressed as [22],

$$P_{EV}^t = P_{EV}^0 . e^{(\alpha t/t_{max})} \tag{7}$$

where α is a constant parameter and it is calculated by considering a fully empty battery system of EV that absorbs 97% of maximum power capacity, and its value is approximately equal to $(t_{max}/3)$.

3 Problem Formulation

The proposed structure of MG that has been considered in this paper is depicted in Figure 1. In this paper, the power exchange between main power grid and MG is considered. The components that are considered in the proposed MG structure are: distributed generators (i.e.,

WEGs and solar PV units), BES systems, EVs and demand response. The EV aggregator collects the information of charging/discharging strategies of EV batteries in electric vehicles [23]. The EV aggregator participates in the market clearing process to reduce the total operating cost (TOC) of the system by determining the mode of operation of EV and to provide an operational flexibility.

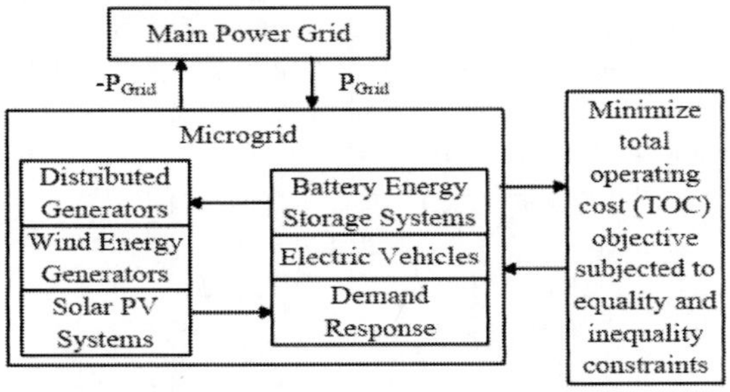

Fig. 1. Proposed Structure of Microgrid (MG).

The objective considered in this paper is to minimize the total operating cost (TOC) of the system, which is the sum of cost due to power exchange between main power grid and MG, cost due to power output from WEGs, solar PV units, BES systems, EVs and demand response. Mathematically, it can be expressed as,

Minimize TOC,

$$TOC = \sum_{t=1}^{T}\left[C_{Grid}^{t}(P_{Grid}^{t}) + \sum_{i=1}^{N_W} C_W^t(P_{W,i}^t) + \sum_{j=1}^{N_{PV}} C_S^t(P_{PV,j}^t) + \sum_{k=1}^{N_B} C_B^t(P_{B,k}^t) + \sum_{l=1}^{N_{EV}} C_{EV}^t(P_{EV,l}^t) + C_{DR}^t(P_{DR}^t)\right] \quad (8)$$

The first component in Eq. (8) represents the cost due to power exchanged between main grid and MG, and it can be expressed as [24],

$$C_{Grid}^t(P_{Grid}^t) = C_{Grid}^t \cdot P_{Grid}^t \quad (9)$$

where P_{Grid}^t is the power output from main power grid at hour t. P_{Grid}^t is positive if the main grid transmits power to the MG, and it is negative if the main grid absorbs power from the MG. Second term in Eq. (8) represents the direct cost associated with wind power and it is expressed as [25],

$$C_W^t(P_{W,i}^t) = C_W^t \cdot P_{W,i}^t \quad (10)$$

Third term in Eq. (8) represents the direct cost for the scheduled solar PV power, and it is expressed as,

$$C_S^t(P_{PV,j}^t) = C_S^t . P_{PV,j}^t \quad (11)$$

Fourth component in Eq. (8) represents the operating cost of BES system, which includes the installation and degradation costs, and it can be expressed as,

$$C_B^t(P_{B,k}^t) = C_{B,k}(P_{B,k}^{ch,t} + P_{B,k}^{disch,t}) \quad (12)$$

Fifth component in Eq. (8) represents the cost associated with the charge/discharge power of EV, and it can be expressed as,

$$C_{EV}^t(P_{EV,l}^t) = C_{EV}^t . P_{EV,l}^t \quad (13)$$

Last component in Eq. (8) represents the cost due to incentive based demand response (DR), and it can be expressed as,

$$C_{DR}^t(P_{DR}^t) = C_{DR}^t . P_{DR}^t \quad (14)$$

where P_{DR}^t is the amount of planned power/load reduction, C_{DR}^t is the incentive payment for load reduction.

The above objective function is solved subjected to the following equality and inequality constraints.

3.1. Power Balance Constraints

This equality constraint of MG system maintains the balance between power generation/supply and demand. This constraint is expressed as [26],

$$P_{Grid}^t + \sum_{i=1}^{N_W} P_{W,i}^t + \sum_{j=1}^{N_{PV}} P_{PV,j}^t + \sum_{k=1}^{N_B} P_{B,k}^t + \sum_{l=1}^{N_{EV}} P_{EV,l}^t + \sum_{m=1}^{N_L} P_{DR,m}^t = P_L^t + P_{loss}^t \quad (15)$$

where $t = 1,2, \dots, T$. P_L^t is the total system load at hour t.

3.2. Power Generation Capacity Constraints

For stable operation of MG, the active power output of each MG component is limited by lower and upper bounds, and they are expressed as [26],

$$P_{Grid,t}^{t,min} \le P_{Grid}^t \le P_{Grid,t}^{t,max} \quad (16)$$

$$P_{W,i}^t \le P_{W,i}^{max} \quad (17)$$

$$P_{PV,j}^t \le P_{PV,j}^{max} \quad (18)$$

$$P_{DR}^t \le P_{DR,max}^t \quad (19)$$

where $P_{W,i}^t$ is the power generated by i^th^ WEG at hour t, $P_{W,i}^{max}$ is its upper limit, $P_{PV,j}^t$ is the power generated by the jth solar PV unit at hour t, $P_{PV,j}^{max}$ is its upper limit, and $P_{DR,max}^t$ is the maximum allowed load reduction at hour t.

3.3.Constraints on BES System

where $P_{W,i}^t$ is the power generated by i^th^ WEG at hour t, $P_{W,i}^{max}$ is its upper limit...... The state of charge (SOC) balance equation of a BES system is expressed as,

$$SOC^t = SOC^{t-1} + I^t\Delta t \quad (20)$$

where I^t is the charge (+)/discharge (-) current (in kA). The charge/discharge current limits are expressed as [26],

$$I^{min} \le I^t \le I^{max} \quad (21)$$

3.4. Constraints on Transmission Line Capacity

As mentioned earlier, the capacity of transmission line that connects between the main power grid and MG can be expressed as,

$$-P_L^{max} \leq P_{L,t} \leq P_L^{max} \quad (22)$$

where P_L^{max} is the maximum capacity of transmission line.

4 Results and Discussion

The Microgrid (MG) considered in this paper has the WEGs, solar PV units, battery banks, electric vehicles (EVs), demand response and the power exchange between MG and main power grid. In this system, four wind mills with the capacity of 750kW each, four solar PV plants with the capacity of 250kW each, two battery units with the capacity of 250kW each and two electric vehicle aggregators with the capacity of 200kW each are considered. Hourly load demand, wind speed and solar irradiation data for 24 hour period has been presented in Table 1. The optimization problem formulated in this paper has been implemented by using Generalized Algebraic Modeling Systems (GAMS) software and solved by SNOPT solver [27]. All the developed programs were run on a personal computer with Intel Core i7, 3.0GHz and 8GB RAM. In this paper, v_{ci}, v_{co} and v_r are considered as 3m/s, 25m/s and 10m/s, respectively.

Table 1. Load demand, wind speed and solar irradiation data for 24 hour period.

Hour	Load (kW)	Wind speed (m/s)	Solar irradiation (W/m^2)	Hour	Load (kW)	Wind speed (m/s)	Solar irradiation (W/m^2)
1	2200	5.7	0	13	5300	5.9	833
2	2400	6.5	0	14	5200	4.9	850
3	2800	7.5	0	15	5000	3.5	680
4	3200	6.9	0	16	4600	3.4	595
5	3400	8.6	93.5	17	4000	2.8	255
6	3800	10.5	212.5	18	3700	3.1	212.5
7	4000	13.6	255	19	3400	2.3	153
8	4200	10.4	467.5	20	3200	2.9	68
9	4600	9.1	637.5	21	3000	3.5	42.5
10	5000	9.3	680	22	2800	3.8	0
11	5200	7.7	816	23	2600	3.8	0
12	5400	7.0	850	24	2400	4.8	0

In this paper, four different case studies are performed, and they are described next:

- Case 1: Considering the power output only from main power grid.
- Case 2: Considering the power output from main power grid and renewable energy sources (RESs).
- Case 3: Considering the power output from main power grid, RESs, BES systems and electric vehicles (EVs).
- Case 4: Considering the power output from main power grid, RESs, BES systems, EVs and demand response.

4.1. Case 1

In this case, the objective function consists only first term of Equation (8), (i.e., cost due to power output from main power grid. As mentioned earlier, this objective function is solved by using the SNOPT solver of GAMS software, and the obtained results are reported in Table 2. The total operating cost (TOC) obtained in this case is 590158.5$/day, and this cost is due to the power output from main power grid.

4.2. Case 2

In this case, the objective function consists only first three of Equation (8), (i.e., cost due to power exchange from main power grid and MG, and renewable energy sources (i.e., wind farms and solar PV plants)). The forecasted and actual power outputs of WEGs and solar PV units are depicted in Figure 2. Table 2 also presents the optimum objective function values for this case. The optimum cost incurred in this case for the 24 hours scheduling period is 515792.1$/day, which is the sum of cost of power exchange between main power grid and MG (i.e., 116604.3$/day), cost of wind power generation (301567.2$/day) and the cost of solar PV power generation (97620.6$/day). The total operating cost (TOC) obtained in this case is 12.6% less than the cost obtained in Case 1.

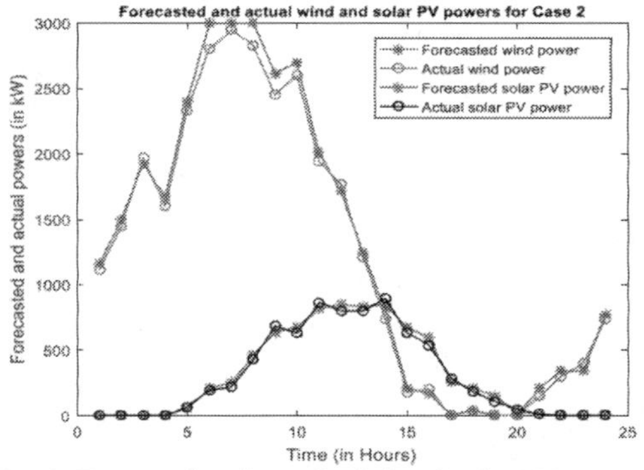

Fig. 2. Forecasted and actual wind and solar PV powers for Case 2.

Table 2. Optimum objective function values for cases 1, 2, 3 and 4.

Objective Function Values	Case 1	Case 2	Case 3	Case 4
Cost of power from main power grid (in \$/day)	590158.5	116604.3	71512.6	45762.5
Cost of wind power generation (in \$/day)	—	301567.2	295391.3	284691.8
Cost of solar PV power generation (in \$/day)	—	97620.6	91359.7	87368.4
Cost of battery storage and EVs (in \$/day)	—	—	42128.2	39850.5
Cost of demand response (in \$/day)	—	—	—	36484.2
Total operating cost (in \$/day)	**590158.5**	**515792.1**	**500391.8**	**494157.4**
Cost reduction (in %)	—	**12.6**	**15.2**	**16.3**

4.3. Case 3

In this case, the objective function consists all the terms of Equation (8) except last term, (i.e., cost due to power output from main power grid and MG, renewable energy sources (i.e., wind farms and solar PV plants) and battery energy storage (BES) system). In this case, to extend the lifetime of BES system, the SOC_k^{min} and SOC_k^{max} of the BES system are set to be 20% and 80%, respectively. This BES system stores the energy during the off-peak hours by charging the battery until it reaches SOC_k^{max}; whereas during the peak hours, it injects the stored energy to loads until it reaches SOC_k^{min}.
The TOC obtained in this case is 500391.8\$/day, which is the sum of cost due to power exchange between main power grid and MG (i.e., 71512.6\$/day), cost due to wind power generation (i.e., 295391.3\$/hr), cost due to solar PV power generation (i.e., 91359.7\$/day) and the cost due to BES system and EVs (i.e., 42128.2\$/day). From the simulation results, it can be observed that the TOC obtained in this case is 15.2% less than Case 1, and 3% less than Case 2.

4.4. Case 4

In this case, the objective function (i.e., TOC of the system) consists all the terms of Equation (8), (i.e., cost due to power output between main power grid and MG, renewable energy sources (i.e., wind farms and solar PV plants), BES system, EVs and demand response). The TOC obtained in this case is 494157.4\$/day, which is 16.3% less than Case 1, 4.2% less than Case 2, and 1.3% less than Case 3.

From the simulation results of four case studies, it can be observed that by operating the MG with RERs, BES system, EV and demand response simultaneously can reduce the total operating cost of the system, which leads to the optimal utilization of hybrid energy sources in the microgrid.

5 Conclusions

This paper proposes an optimal scheduling approach of Microgrid by optimally utilizing the renewable sources of energy such as wind and solar photovoltaic (PV) powers. The optimal operation of Microgrid is implemented by including the stochastic nature of wind energy generators, solar PV systems, battery storages, electric vehicles and optimal dispatch of schedulable loads, i.e., demand response. In order to reduce the intermittent nature of wind, solar PV powers, electric vehicles (EVs), battery energy storage (BES) systems and demand response are introduced in this paper. Minimization of total operating cost (TOC) (i.e., cost of power exchanged between main power grid and MG, wind, solar PV powers, BES systems, EVs and demand response) of the system is considered as an objective function. In this paper, the probability distribution functions are used to handle the uncertainties related to wind power, solar PV power and load demand. The proposed optimization problem has been solved using the GAMS software. The simulation results of this paper show that the total operating cost of the system has been reduced significantly by optimally utilizing various microgrid components simultaneously.

References

1. M. Meiqin, S. Shujuan, L. Chang, Economic analysis of the microgrid with multi-energy and electric vehicles, *8th International Conference on Power Electronics - ECCE Asia*, Jeju, 2011, pp. 2067-2072.
2. Renewable Energy Policy Network or the 21st Century, *Renewables 2017 Global Status Report*, Technical Report. Available. [Online]: http://www.ren21.net/wp-content/uploads/2017/06/17-8399_GSR_2017_Full_Report_0621_Opt.pdf
3. M. Modiri-Delshad and N. A. Rahim, Optimal operation of microgrid systems, *3rd IET International Conference on Clean Energy and Technology*, Kuching, 2014, pp. 1-6.
4. N. Karthik, A.K. Parvathy, R. Arul, Optimal Operation of Microgrids-A Survey, *International Journal of Applied Power Engineering*, vol. 7, no. 2, pp. 179-185, Aug. 2018.
5. Y. Wang, Y. Huang, Y. Wang, F. Li, Y. Zhang, C. Tian, Operation Optimization in a Smart Micro-Grid in the Presence of Distributed Generation and Demand Response, *Sustainability*, vol. 10, pp. 1-25, 2018.
6. J. Wu , Z. Wu, F. Wu, X. Mao, A power balancing method of distributed generation and electric vehicle charging for minimizing operation cost of distribution systems with uncertainties, *Energy Science and Engineering*, vol. 5, no. 3, pp. 167-179, 2017.
7. F. Farzan, M.A. Jafari, R. Masiello, Y. Lu, Toward Optimal Day-Ahead Scheduling and Operation Control of Microgrids Under Uncertainty, *IEEE Transactions on Smart Grid*, vol. 6, no. 2, pp. 499-507, Mar. 2015.

8. K. Ranjbar, M.M. Ghahderijani, S.M. Barakati, Effects of Time-Of-Use Demand Response Programs Based on Optimal Operation of a MicroGrid in Market Environment, *International Journal of Academic Research in Applied Science*, vol. 2, no. 5, pp. 1-12, 2013.
9. A. Hussain, V.H. Bui, H.M. Kim, Impact Analysis of Demand Response Intensity and Energy Storage Size on Operation of Networked Microgrids, *Energies*, vol. 10, no. 7, pp. 1-19, 2017.
10. H. Shayeghi, E. Shahryari, Optimal Operation Management of Grid-connected Microgrid Using Multi-Objective Group Search Optimization Algorithm, *Journal of Operation and Automation in Power Engineering*, vol. 5, no. 2, pp. 227-239, Dec. 2017.
11. S. Maghsudlu, S. Mohammadi, Optimal scheduled unit commitment considering suitable power of electric vehicle and photovoltaic uncertainty, *Journal of Renewable and Sustainable Energy*, vol. 10, no. 4, 2018.
12. M. Ghahremani, M.N. Heris, K. Zare, B. Mohammadiivatloo, Energy Management of Electric Vehicles Parking in a Power Distribution Network Using Robust Optimization Method, *Journal of Energy Management and Technology*, vol. 2, no. 3, pp. 22-30, 2018.
13. D. Thomas, O. Deblecker, C.S. Ioakimidis, Optimal operation of an energy management system for a grid-connected smart building considering photovoltaics' uncertainty and stochastic electric vehicles' driving schedule, *Applied Energy*, vol. 210, pp. 1188-1206, 2018.
14. A. Maulik, D. Das, Optimal operation of microgrid using four different optimization techniques, *Sustainable Energy Technologies and Assessments*, vol. 21, pp. 100-120, Jun. 2017.
15. A. Gheiratmand, E. Ayoubi, M. Sarlak, "Optimal operation of micro-grid in presence of renewable resources and compressed air energy storage," *Conference on Electrical Power Distribution Networks Conference*, Semnan, 2017, pp. 131-136.
16. W. Liu, D. Wang, X. Yu, H. Jia, W. Wang, X. Yang, Y. Zhi, Optimal Scheduling of Multi-Source Microgrid Considering Power to Gas Technology and Wind Power Uncertainty, *Energy Procedia*, vol. 143, pp. 668-673, 2017.
17. Y. Aiwei, X. Congwei, J. Liwei, A Joint Scheduling Optimization Model for Wind Power and Energy Storage Systems considering Carbon Emissions Trading and Demand Response, *Mathematical Problems in Engineering*, vol. 2016, pp. 1-10, 2016.
18. S.S. Reddy, J.Y. Park, C.M. Jung, "Optimal Operation of Microgrid Using Hybrid Differential Evolution and Harmony Search Algorithm", *Frontiers in Energy*, vol. 10, no. 3, pp. 355-362, Sept. 2016.
19. M. Meiqin, J. Meihong, D. Wei, L. Chang, Multi-objective economic dispatch model for a microgrid considering reliability, *The 2nd International Symposium on Power Electronics for Distributed Generation Systems*, Hefei, 2010, pp. 993-998.
20. C. Huang, D. Yue, S. Deng, J. Xie, Optimal Scheduling of Microgrid with Multiple Distributed Resources Using Interval Optimization, *Energies*, vol. 10, no. 3, pp. 1-23, 2017.
21. A. Arabali, M. Ghofrani, M.E. Amoli, Cost analysis of a power system using probabilistic optimal power flow with energy storage integration and wind generation, *Electrical Power and Energy Systems,* vol. 53, pp. 832-841, 2013.
22. J.G. Vlachogiannis, Probabilistic Constrained Load Flow Considering Integration of Wind Power Generation and Electric Vehicles, *IEEE Transactions on Power Systems*, vol. 24, no. 4, pp. 1808-1817, Nov. 2009.
23. R.A. Osama, A.Y. Abdelaziz, R.A. Swief, M. Ezzat, Microgrid self adequacy optimization using back tracking search algorithm, *IEEE Power & Energy Society General Meeting*, Denver, CO, 2015, pp. 1-5.
24. J. Wu, Z. Wu, F. Wu, X. Mao, A power balancing method of distributed generation and electric vehicle charging for minimizing operation cost of distribution systems with uncertainties, *Energy Science and Engineering*, vol. 5, no. 3, pp. 167-179, 2017.
25. A.A. Moghaddam, A. Seifi, T. Niknam, Multi-operation management of a typical micro-grids using Particle Swarm Optimization: A comparative study, *Renewable and Sustainable Energy Reviews*, vol. 16, pp. 1268-1281, 2012.
26. J. Chen, X. Yang, L. Zhu, M. Zhang, Study on Microgrid Multi-objective Economic Operation Optimization, *International Journal of Advancements in Computing Technology*, vol. 5, no. 8, pp. 1218-1229, Apr. 2013.
27. P.E. Gill, W. Murray, M.A. Saunders, Snopt: An sqp algorithm for large-scale constrained optimization, *SIAM review*, vol. 47, no. 1, pp. 99-131, 2005.

Implementation of OCPP Protocol for Electric Vehicle Applications

Thota Venkata Pruthvi[1], *Niladri* Dutta[2], *Phaneendra Babu* Bobba[3*], *B Sai* Vasudeva[4]

[1]VLSI Engineer, HCL Technologies, Chennai, India tv166@snu.edu.in
[2]Associate System Engineer, IBM India Pvt. Ltd., Bangalore, India nd933@snu.edu.in
[3]Dept. of Electrical& Engineering, GRIET, Hyderabad, Telangana., India bobbaphani@griet.ac.in
[4]Dept. of Electrical & Electronics Engineering, GRIET, Hyderabad, Telangana., India saivasudeva.b@gmail.com

Abstract. The ability of the software and hardware systems to interchange information is a key factor for the success of the electric vehicle industry. Standards have been developed and are in use to ensure base level interoperability of the front-end communication and signaling processes for smart charging between electric vehicles and charge stations. The Open Charge Alliance (OCA), a group of European industries, have developed an open source common back-end protocol, called Open Charge Point Protocol (OCPP), for charging stations to reduce and secure overall investment costs. OCPP intends to enable grid services based on smart charging. In this paper the authors provide a review of the functionalities OCPP offers and how it can be used in the electrical vehicle-charging infrastructure.

1 Introduction

For the budding EV charging market, assortment is something worth being thankful for. Rivalry drives down expenses and raise mechanical upgrades. Today, government or private charging suppliers look over a large number of charging station makers and system framework merchants. This wide decision brings up the issue of interoperability, or, as it were, the capacity of every EV accusing station to impart of a focal framework, paying little heed to the maker or IT back-end seller. That is the place the Open Charge Point Protocol (OCPP) comes in.

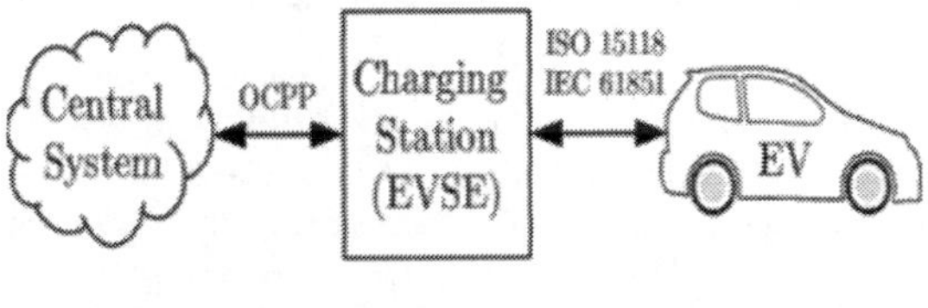

Fig. 1. Electric vehicle interface standards. [1]

Keep in mind when you initially purchased a mobile phone and it just worked with one transporter? You were secured to their administration bundle regardless. The Open Charge Point Protocol (OCPP) ensures that you can switch among equipment and programming suppliers without your speculation getting to be old. It's as basic as exchanging the SIM card in your telephone. The objective of OCPP is to empower a genuinely interoperable EV charging foundation - a framework that is versatile and simple to work both for EV drivers and system administrators. With the OCPP, clients can incorporate charging stations from a few merchants into a similar IT back-end framework. They can likewise choose the most fitting charging station vendor(s) and the most proper IT back-end provider freely.

OCPP was advanced by the Open Charge Alliance and has turned into an inexorably well-known convention and a true standard in excess of 50 nations and over 10,000+ charging stations. In contrast to restrictive correspondence conventions, OCPP is open and not associated with any permitting expense or prerequisite, making it simple to receive. Embraced by many charge point sellers and focal framework providers, OCPP-consistence is presently turning into an "absolute necessity have" prerequisite among speculators. They consider it to be an approach to keep their foundation merchant nonpartisan and to decrease costs.

Open Charge Alliance discharged another sort of convention called OSCP (Open Smart Charging Protocol) This convention can be utilized to impart a 24-hour forecast of the neighborhood accessible ability to the Charge Spot Operator. The Service Provider will fit the charging profiles of the electrical vehicles inside the

Table 1. Electric Vehicle Connector Standards

	USA	Europe	China/ India
Alternating current (AC)	SAE J1772/IEC 62196-2	IEC 62196-2	GB Part 2
Direct current (DC)	IEC 62196-3	IEC 62196-3	GB Part 2/ IEC 62196-3
Combined AC/DC	SAE J1772/IEC 62196-3	IEC 62196-3	

* Corresponding author

https://doi.org/10.1051/e3sconf/20198701008

limits of the accessible limit. It is a protocol between charge point management system and energy management system of the site owner.

2 Charging Infrastructure

Typically, the Charging Infrastructure consists of Electric Vehicle (EV), Charging Station, Charging Station Management system (CSMS) as Shown in Fig I. The Charging Station is the physical framework where an EV can be charged. A Charging Station has at least one Electric Vehicle Supply Equipment (EVSEs) [2]. An EVSE is considered as an autonomously worked and oversaw some portion of the Charging Station that can convey vitality to one EV at any given moment. EVSE needs to speak with Battery Management System (BMS) of battery pack in EV, to empower it to charge at right rate for keeping up SOH of batteries. Physical layer for this correspondence can be Power Line Communication (PLC) or Controller Area Network (CAN), CAN is ordinarily utilized by vehicle producers in India. There are different Electric Vehicle Connector Standards with respect to different countries, some of them are mentioned in Table II.

Table 2. Commonly used terms

Term	Meaning
Charging Station	The Charging Station is the physical system where an EV can be charged. A Charging Station has one or more EVSEs.
Charging Station Management System (CSMS)	Charging Station Management System: manages Charging Stations and has the information for authorizing Users for using its Charging Stations.
Electric Vehicle Supply Equipment (EVSE)	An EVSE is considered as an independently operated and managed part of the Charging Station that can deliver energy to one EV at a time.
CSO	Charging Station Operator
EV	Electric Vehicle
RFID	Radio-Frequency Identification

3 OCPP Version History

There are mainly three different versions of OCPP from the time of its first release. They are OCPP 1.5, OCPP 1.6 and OCPP 2.0 respectively.

3.1. OCPP 1.2

This version of OCPP is similar to OCPP 1.5, with fewer functionalities.

3.2. OCPP 1.5

There are 25 Operations described in OCPP 1.5[5], 10 are initiated by Charging Station and 15 are initiated by the Central management system. Authorize, Boot Notification, Data Transfer, Diagnostics Status Notification, Firmware Status Notification, Heartbeat, Meter Values, Start Transaction, Status Notification, Stop Transaction, are Initiated by charge point and Cancel Reservation, Change Availability, Change Configuration, Clear Cache, Data Transfer, Get Configuration, Get Diagnostics, Get Local List Version, Remote Start Transaction, Remote Stop Transaction, Reserve Now, Reset, Send Local List, Unlock Connector, Update Firmware are initiated by the central system.

3.3. OCPP 1.6

This adjustment of OCPP uses SOAP Framework to send messages between sections over the Internet. The upside of SOAP is that the workplaces for sending and tolerating messages are anchored by the standard [6]. This makes quick utilization possible. substance of a SOAP message are pulled in up according to the Extensible Markup Language (XML) standard. This dialect is identified with the HTML. Notwithstanding composed content, XML messages may likewise contain pictures and executable code. The enormous favorable position is that the messages are sent in decipherable content.

3.4. OCPP 2.0

OCPP 2.0 is the most recent adaptation discharged in April 2018, it contains a great deal of new highlights canvassed in 116 use cases. OCPP 2.0 [7] just backings JSON. There are many included and enhanced functionalities like Device Management, Improved Transaction taking care of, Added Security, Added Smart Charging functionalities, Support for 15118, Display and informing support, and numerous extra upgrades asked for by the EV charging network.
without title.

4 Implementation of OCPP 2.0

The OCPP protocol describes many use cases and messages. There are 16 Functional Blocks with one or more use cases in each of them. Out of these only some are needed to implement a basic Charging Station or CSMS. Table III lists messages that are typically implemented to deliver basic functionality for an OCPP managed Charging Station.

4.1. Booting, Configuration and Resetting a Charging Station

Booting, configuring and resetting of the Charging Station Functionalities goes under the Provisioning Functional block of OCPP 2.0. This Functional Block depicts every one of the functionalities that assistance a CSO arrangement their Charging Stations, enabling them to validate certifications and recover setup data from these Charging Stations over a system. Also, it comprises of the capacity to make changes to the Charging System configuration, as well. There are 12 use cases which gone under this functional block, among

 https://doi.org/10.1051/e3sconf/20198701008

them are– Cold Boot Charging Station, Cold Boot Charging Station - Pending, Cold Boot Charging Station-Rejected, Offline Behavior Idle Charging Station, Set Variables, Get Variables, Get Base Report, Reset-Without Ongoing Transaction and Reset-With Ongoing Transaction – are the mandatory use cases for the essential execution of OCPP 2.0.

4.2. Authorization options

This functionality comes under the Authorization Functional block of OCPP 2.0, This functional block describes all the authorization-related functionalities, it contains different ways of authorizing a user (online and/or offline), the AuthorizeRequest message handling, and the Authorization Cache functionality, etc. When a user wishes to unplug the electric vehicle from the Charging Station, the Charging Station needs to verify that the user is either the one that initiated the charging or that the user is in the same group and thus allowed to terminate the charging. Once authorized, the Charging Station informs the CSMS that the charging has been stopped. This functional block has 16 Use Cases in which any one of these – EV Driver Authorization using RFID, Authorization using a start button and Authorization using PIN-code are mandatory for the basic implementation.

4.3. Transaction mechanism

This functionality comes under Transactions Functional block of OCPP 2.0 and it describes the OCPP transaction related functionalities. Transactions can be started or stopped on the Charging Station with only one active transaction on an EVSE, at a time. This functional block has 15 use cases – Start Transaction Options, Start Transaction-Cable Plugin First, Start Transaction-IdToken First, Start Transaction-Id not Accepted, Stop Transaction Options, Transaction locally stopped by IdToken, Transaction stopped while Charging Station is offline one of when cable disconnected on EV-side: Stop Transaction and When cable disconnected on EV-side: Suspend Transaction, Connection Loss During Transaction, Inform CSMS of an Offline Occurred Transaction and Transaction related message not accepted by CSMS are mandatory for basic implementation.

4.4. Availability

This functionality goes under Availability Functional block of OCPP 2.0. It indicates how the Charging Station can educate the CSMS of its present accessibility for beginning new exchanges. It is vital for the CSO to know whether a Charging Station is accessible for EVs to be charged, with the goal that they can advise EV drivers whether the Charging Station is accessible for use. Therefore, the Charging Station ought to consistently send any status changes of itself or any of its EVSEs to the CSMS.

This transaction status is exceptionally helpful for an EV Driver when he encounters issues amid charging. At the point when a blame is recognized by the Charging Station it can communicate something specific telling the CSMS about the fault. At the point when the CSO needs the Charging Station to never again begin new exchanges, it can change the accessibility. For instance: they have to do upkeep on the Charging Station, and hence they don't need the Charging Station to be being used. The CSO can likewise change the accessibility for at least one EVSEs. For instance: A client calls, whining about a broken EVSE on the Charging Station. The CSO would then be able to set the Connector to inaccessible, making it impossible for an EV Driver to utilize that Connector. Clearly, with a direction from the CSMS it is likewise conceivable to make the Charging Station or a Connector accessible once more.

This functional block has 5 use cases, out of which 4 of the use cases are a mandatory, they are - Status Notification, Change Availability EVSE, Change Availability Charging Station and Lock Failure.

4.5. Sending transaction related Meter values

This functionality comes under Metering Functional block of OCPP 2.0. It describes the functionality that

Table 3. OCPP 2.0 Functionalities

Functionality	**Messages**
Booting a Charging Station	BootNotification
Configuring a Charging Station	Set Variables, Get Variables and GetReportBase (respond correctly to requests with reportBase = ConfigurationInventory, FullInventory, Summary Inventory).
Resetting a Charging Station	Reset
Authorization options	Authorize
Transaction mechanism	Transaction Event
Availability	Only Change Availability and Status Notification.
Monitoring Events	A basic implementation of the Notify Event message to be used to report operational state changes and problem/error conditions of the Charging Station, e.g. for Lock Failure. Also used for reporting built-in monitoring events.
Sending transaction related Meter values	Transaction Event
DataTransfer	Any OCPP implementations should at least be able to reject any request for Data Transfer if no (special) functionality is implemented.
Resetting a Charging Station	Reset
Authorization options	Authorize
Transaction mechanism	Transaction Event

enables a Charging Station to send periodic, possibly clock-aligned Meter Values. Extensive metering data relating to transactions can be recorded and transmitted in different ways depending on its intended purpose. This Functional Block has 3 use cases in which Sending transaction related Meter Values is the MUST Use Case to Implement Basic OCPP 2.0.

5 Sample Interface

The OCPP protocol ensures the standardized communication between the Central Management System (CSMS) and the Charging Station. A user interface will make it convenient to for the users to work with this protocol running in the back-end. Figure II shows possible user interfaces, which can be used by the

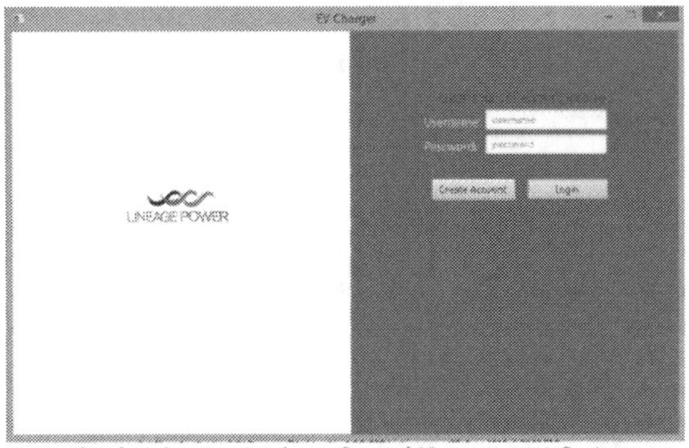

a) Administrator Login at Central Management System

b) Login interface at Charging Station for the customer

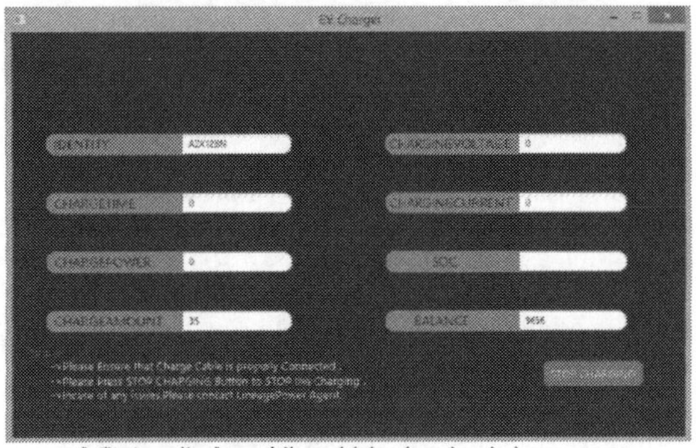

c) Status display while vehicle charging is in progress.

Fig. 2. Sample graphical user interface.

CSO in the CSMS and the consumer charge the electric vehicle at the charging station.

This sample implementation of the OCPP 2.0 protocol was developed using the Java programming language. Concepts like to object oriented programming and web sockets offered by the Java language makes it easy to implement the requirements of this protocol. The front-end design uses JavaFX to generate the graphical user interface (GUI).

6 Conclusion

This paper reviews the Open Charge Point Protocol (OCPP), which can be adopted as a standard for the back-end communication between electric vehicle charging stations and the central management system. The various versions of the protocol were discussed along with the improvements in its functionalities and features. Sample user interfaces of the possible implementation of the OCPP protocol were also shown for the central management system and the charging station, both.

References

1. D. Wellisch, J. Lenz, A. Faschingbauer, R. Pöschl, S. Kunze, *IFAC-PapersOnLine* (2015)
2. Open Charge Alliance - Global Platform For Open Protocols, *https://www.openchargealliance.org/* (2018)
3. Committee Report on Standardization of Public EV Chargers,*https://dhi.nic.in/writereaddata/UploadFile/Standardization%20of%20protocol.pdf* (2017)
4. Open Smart Charging Protocol 1.0, *https://www.openchargealliance.org/protocols/ocpp-10/* (2018)
5. Open Charge Point Protocol 1.5, *https://www.openchargealliance.org/protocols/ocpp-15/* (2018)
6. Open Charge Point Protocol 1.6, *https://www.openchargealliance.org/protocols/ocpp-16/* (2018)
7. Open Charge Point Protocol 2.0, *https://www.openchargealliance.org/protocols/ocpp-20/* (2018)

Extensive review on Laminated bus bar for low and high power applications

Prashant Singh B.T.[1,*], *Phaneendra* Babu Bobba[2], *Suresh* K.[3] and *Benny* J. Varghese[4]

[1,2] Dept of Electrical Engineering, GRIET, Hyderabad, Telangana, India
[3]CEO, MasterPCB PVT.LTD, KPHB, Hyderabad, Telangana, India
[4]PhD Student, UtahState, University, Logan, USA

Abstract. This paper explains about wide range of applications for Laminated Bus Bar used for high and low power applications. Authors also explains ways to effective utilize laminated bus bar when compared to conventional bus bar. Laminated bus bars are designed with low stray inductance and high capacitance with a uniform current distribution in conducting plate. Parameters like Length, Width, material thickness and other miscellaneous parameters effect Laminated bus bar performance. With a proper design of Laminated bus bar it can best utilized , laminated bus bar are suitable for many low and high power applications which are discussed in this paper. Replacing conventional bus bar with laminated bus bar additional advantages are achieved like lighter weight, less space and lower maintenance.

1 Introduction

Conventional bus bars used these days require lot of space and also create hazardous situations in many cases. Many hazardous situations are due to weak connections between points or insulation failure and so on. Apart from arrangement they posses more stray inductance and low capacitance which make the bus bar to create spikes in transient cases. Laminated bus bars overcome these problems by maintaing lower stray inductance and high capacitance on using different material and sandwiching them in a suitable and required manner. Laminated bus bar are lighter when compared to conventional bus bar. Laminated bus bar there is minimal chance of sparking, wiring errors and heating effects [1] which can obtained by using a proper material. Many applications today demands for low stray inductance, low weight, safe to operate, lower maintenance, low space requirement which can be fulfilled by using laminated bus bars.

This paper describes laminated bus bars uses in different range of power applications and also how to modify Laminated bus bar based upon the performance obtained after simulations. Material used while designing a Laminated bus bar makes the performance vary a lot so special care need to be taken so, as obtain better performance. By selecting a suitable material requitred for operating temperature, operating voltage and good reliable better performance can be obtained easily. Certain models are described in section III. Models are designed and are modified to obtain uniform current distribution.

2 Laminated Bus Bar for Different Applications

2.1 Inverters

Inverters are used in many applications these days like power flow control, speed control of motors, power factor correction and so on. But due to wiring and soldering of switches parasitic value arises distorting the switching waves. Due to these parasitic values stray inductance becomes higher and disturbs the switch operations an distort the waveform. When replaced conventional PCB with laminated bus bar the stray inductance gets reduced which make the inverter operate as better and additionally capacitance is also increased. Due to high capacitance and low stray inductance, a smaller snubber circuit can be preferred and heat sink requirement also declines [3]. As a result, inverter weight and space requirement goes down. A switched oriented inverter bus bar is shown in Fig.1. Internally in Laminated bus bar the positive link and Negative link are isolated to avoid short circuit.

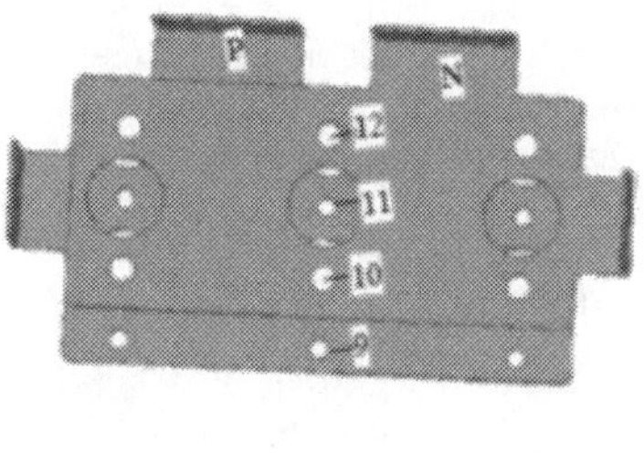

* Corresponding author:prash.5295@gmail.com

Fig.1. Inverter PCB with Laminated bus bar

2.2 Battery Bank

Many applications make use of energy storage device to fulfill the excess demand power requirement or as primary power sources like Electric vehicles. For a dynamic grid and a renewable plant with dynamic power provider gives a challenging issue. With the growth in battery technology [4], it is possible to store fluctuating energy in battery and utilize it later. But dealing with such high power requires huge batteries and also create the transient peaks in voltages. Battery banks serve as a primary source of energy for electric vehicles [5] which are in KWH. With growth in Electric Vehicle (EV) technology and awareness of global warming [6] automobile industries are shifting toward electric vehicle with batteries as a primary energy source. Use of laminated bus bars in EV make the power source free from wiring problems occurring due to vibrations in heavy vehicles[7] and also a good quality source free from large transient spikes. Using the Laminated bus bar for battery bank makes them safer and allows for longer operating time and posses a lighter weight.

2.3Super Capacitor Banks

Supercapacitor (SC) have high power density and low response time when compared to other power source like batteries, Alternators, fuel cells and so on. SCs are used as a power source to fulfill high power demand in a short time generally at transient state. SC can also be used to meet dynamic power demand of the grid to balance divergence of power between power plants and loads with laminated bus bars possessing lower stray inductance and high capacitance provide spike free voltage in a short span of time. By using Laminated bus bars spike free voltage is pumped to the grid in a short span of time. SC banks are utilized as hybrid energy sources where weight is reduced [8] by replacing certain range of battery with SC and life of primary power source is also increased.

2.4 Data Centers

Data centers where data is stored in a voltaic memory which demands for a pure and constant voltage [9]. So, laminated bus bar as a power source distributer can make the transient spike lower and in turn make data centers safe by protecting them from fluctuations and providing spike free voltage waves. Use of laminated bus bar in data centers lead to safe and risk free operation, free from wiring hazard situations which can also be observed in conventional bus bars.

2.5 DC Links

DC links used in industries or for inverters with a conventional bus bar contains ripples and also add a spike in supply by which the wave disturbance disturbs the load or motor performance. The spike in the DC links is due to the high stray inductance [10]. By using Laminated bus bars with low stray inductance and also high capacitance the spikes are reduced. These two parameters make the spike lower under disturbance or transient state of operation.

2.6 Power Distribution

Higher power systems which carry more power faces issues like spark at joints, transient spike from supply, more weight and more space. To avoid such problems laminated bus bars can be used. Due to low stray inductance and high capacitance in laminated bus bar spikes are observed lower in transients state [11]. In industries, high power requirement makes the conventional bus bar requirement huge and add addition problems. Laminated bus bars occupy less space for same power constraints.

2.7 Industrial Automation

Robots which runs with huge power in industries uses conventional bus bars [12]. But due to high power bus bar weights high and being dynamic in nature of work makes potential create a hazard after a certain operation period of time. On replacing conventional bus bars with laminated bus bars, weight reduction and hazardous situations reduce. Due to lower weight, industry energy demands lower power for the same task. This makes the industries demands for lower energy and also reduce for hazard situation due to dynamic body behaviour of robut. Save in energy due to lighter laminated bus bar of same constraints make the lower power consumption lower te energy bills and power demand.

3 Effects on Laminated Bus Bar due to Various Parameters

3.1 Effect of Physical dimensions

3.1.1 Model A

Generally models designed with lower dimension faces non-uniform crowing [13] of flux and current in conducting plates. Fig. 2a &b. shows a model and current plot for the model respectively. On observing the current distribution the current is non- uniformly distributed which can be seen in Fig. 2b.

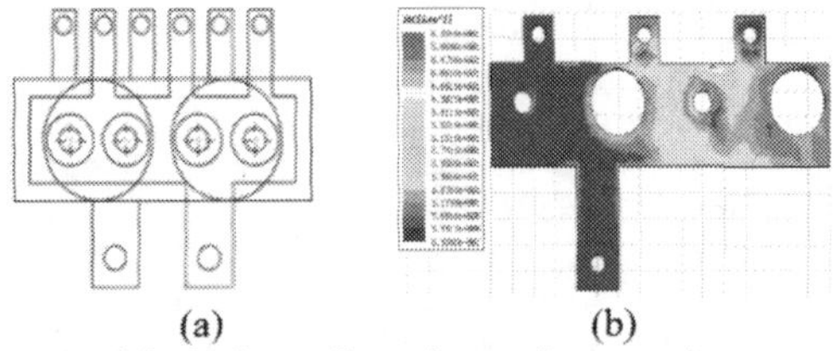
(a) (b)

Fig. 2. Model with lower dimensions and current plot.

A new model was designed to make the current uniformly distributed. Fig. 3a&b. shows the modified model and current plots respectively. The current

 https://doi.org/10.1051/e3sconf/20198701009

crowding is seen to be uniformly distributed and utilizing the whole conducting plate. Change in dimension make the plate to crowd uniformly and utilize effectively.

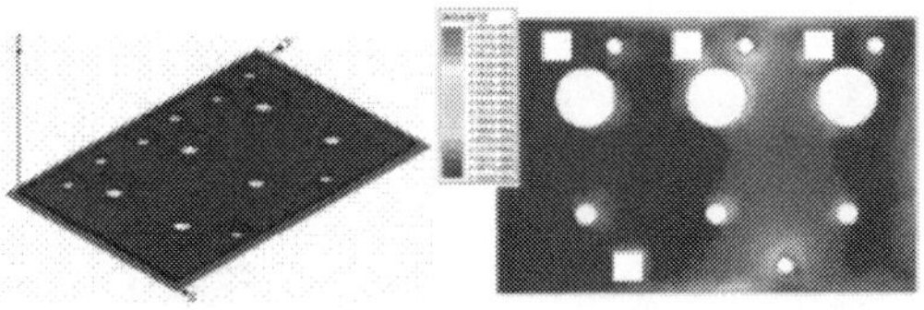

(a) (b)

Fig. 3. Modified model of lower dimension model and current plot

3.1.2 Model B

In another model based on current direction non uniform crowding can be observed [3]. In Fig.4. the current is seen to be distributed non-uniformly on positive and negative plates. The region nearer to the sink is crowded and remaining region is fewer crowded making the non-uniform crowding of the current.

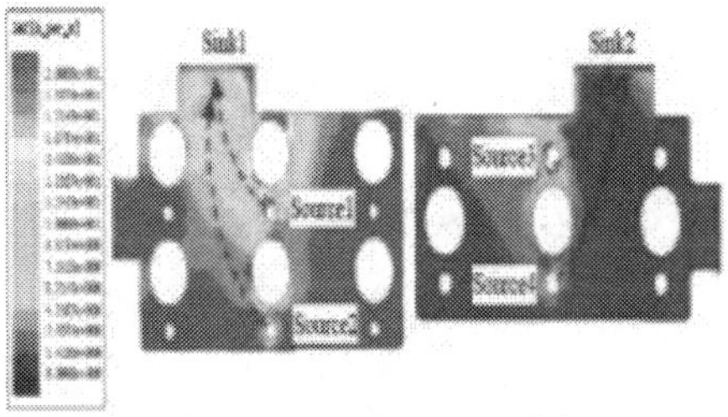

Fig. 4. Simulated current for plot model B

To make the density of current uniform, sink is made as source and vice-versa. The apertures seen in the model are bit farther from sink and creating a crowding of current. Either by designing the apertures as required or making the source as sink, the crowding of current problem can be reduced. Fig.5. describes the uniform current distribution as seen after operating source as sink and vice versa.

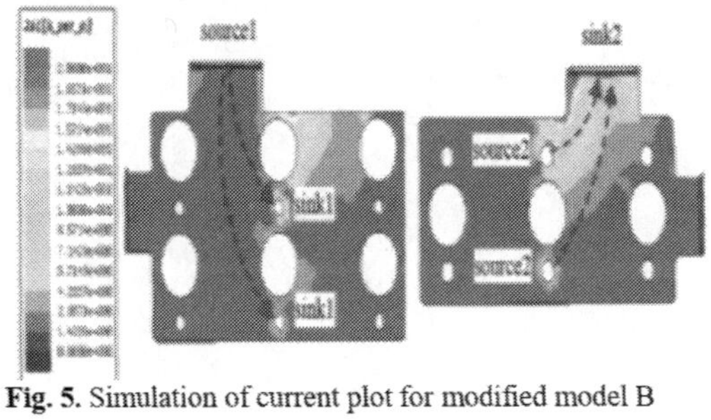

Fig. 5. Simulation of current plot for modified model B

3.2 Effects of Flexible Terminals

Apart from the physical structure and dimensions, some bus bar models have additional physical structures like flexible terminals acting as terminals of the bus bar, made of conducting material [14]. These terminals avoid current crowding and uniformly distribute the current. Flexible terminals are perpendicular to the plate so as make the terminal joints easy to connect. Fig.6. shows the flexible terminals perpendicular to the plate.

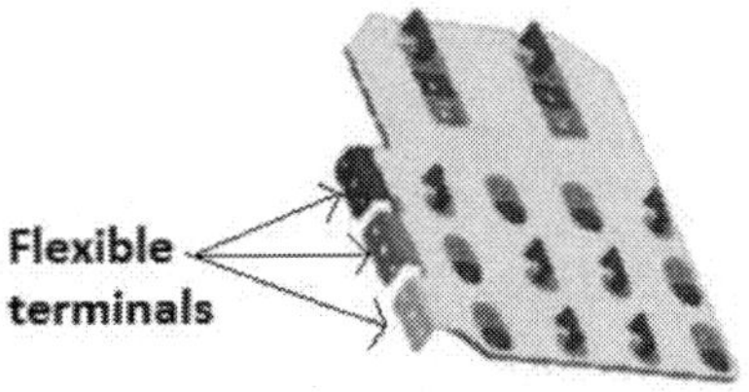

Fig. 6. Laminated bus bar with flexible terminals.

3.3 Effect of Apertures

3.3.1 Model A

Models are designed with apertures to maintain the uniform current flow, lower stray inductance and operate plate in heat limits. The apertures added in a model reduces material used, reduce operating temperatures, increase the durability and effectively use the plates [15].

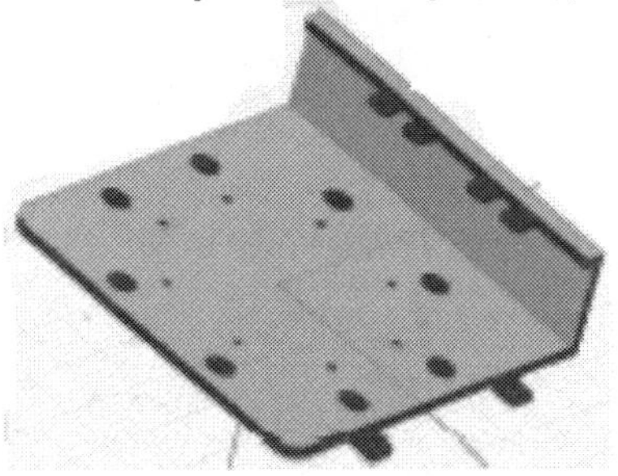

Fig. 7. Model A with no apertures

Fig.7. is a model with no aperture and a flexible terminal is included so as to make the model make uniform current distribution. Fig. 8. is a modified on Fig. 7. Which includes a big aperture in middle of plates. By adding a required aperture we could easily make the bas bar to uniformly current crowding.

Fig. 8. Model A with apertures

3.3.2 *Model B*

Generally including apertures in the model doesn't have a significant effect on its current and inductance parameters. Apertures included in the model affect the stray inductance directly but material reduction takes

https://doi.org/10.1051/e3sconf/20198701009

place. The stray inductance variations can be seen but the change is not expected to be much [14]. These can depend on position of apertures from edges and diameter of apertures. In Fig.9. current crowding near apertures is high and when the aperture is divided into many smaller apertures, a decrease in the crowding of current can be seen in Fig.9.

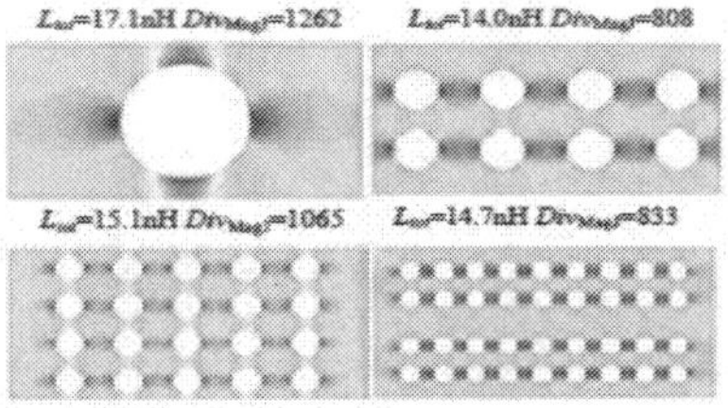

Fig. 9. Effect of stray inductance duet to apertures

To make the current uniform apertures were divided. On adding more apertures stray inductance is also affected, initially it was 17.1nH and it is reduced to 14nH. But a proper care need to be taken while adding a aperture so as to obtain a lower values of inductance and also uniform current distribution. Improper arrangement of aperture design make the inductance to increase rather than decreasing which can be seen in fig.8.

3.4 Effects of Stray Inductance

Laminated bus bars need to possess low stray inductance when compared to conventional bus bar. Laminated bus bar current and flux are not uniform. The stray inductance analytical formula [16] is given by

$$L = (31.9 \times l \times \frac{a}{d})\text{nH} \quad (1)$$

L= stray inductance of Laminated bus bar
l= length of Laminated bus bar in inches
a= dielectric thickness in inches
d= width of laminated bus bar in inches

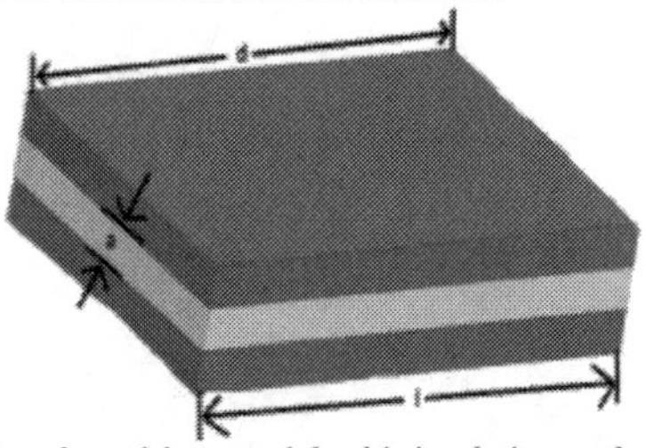

Fig. 10. Laminated bus model with insulation and conducting plate.

Fig.10. shows the Laminated bus bar arrangement red plates as conducting plate andd yelowas insulation material. Apart from stray inductance, mutual inductance also exists between the plates. Mutual inductance depends upon the distance between the plates. Fig.11 shows the variation of inductance due to change in distance between plates.

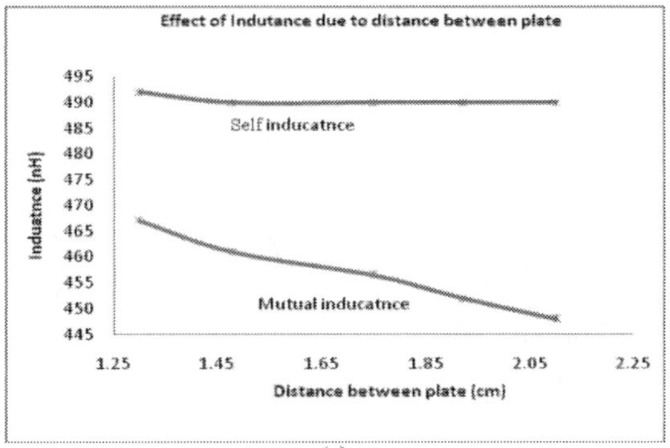

(a)

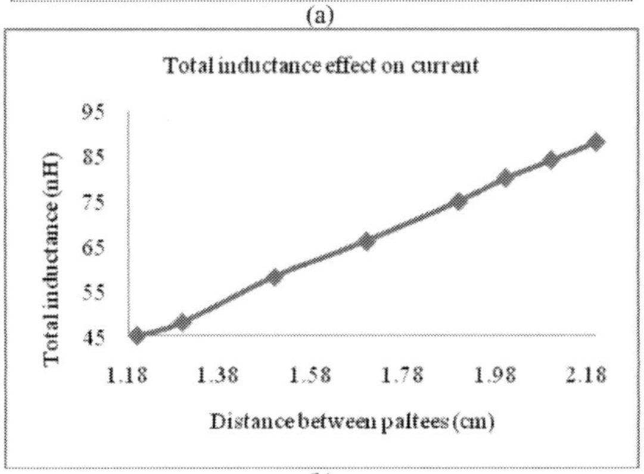

(b)

Fig. 11. Effect of distance between plate on stray inductance.

Apertures in model cause the inductance to reduce but only up to a certain limit. Fig. 9. shows a variation of stray inductance due to the apertures. Initially inductance had reduced to a 14.0nH from 17.1nH. In another model due to apertures, the effect of inductance variation can be seen due to change with frequency. Fig.12a &b. show the model and variation of inductance respectively between a apertures and a non-aperture model. Change in operating frequency make inductance to effect suitable material need to be designed to operated as required [15].

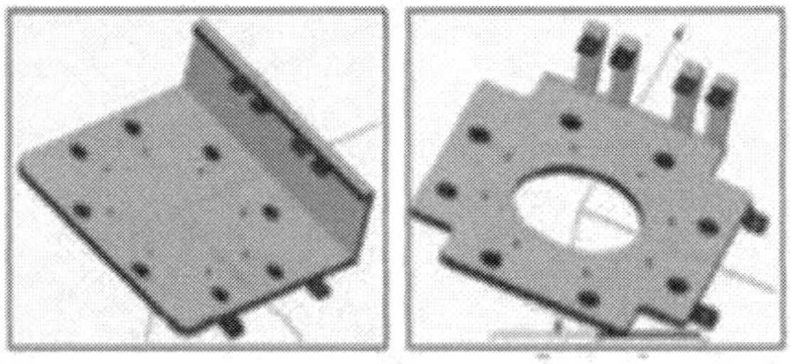

Design 1 Design 2

(a)

 https://doi.org/10.1051/e3sconf/20198701009

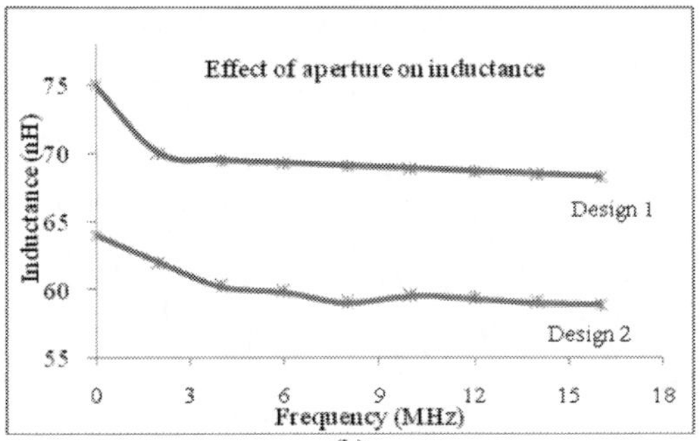

(b)

Fig. 12.Effect of aperture on inductance

3.5 Effects of Capacitance

The capacitance for laminated bus bar is given by an analytical [16] formula-

$$C = (.2225 \times l \times b \times \frac{Er}{a})\ \text{pF} \qquad (2)$$

C= stray inductance of laminated bus bar
l= length of laminated bus bar in inches
a= dielectric thickness in inches
b= Width of laminated bus bar in inches
E_r=Dielectric constant of dielectric material used in Laminated bus bar.

Change in area of plate, type of material used, thickness and distance between plate effects the capacitance [15]. Fig. 13a&b shows variation of capacitance due to width and distance between plates respectively. Apart from the plate area, the distance between the plates also affects the capacitance. Fig. 14. shows variation of capacitance due to change in material used in between conducting plates.

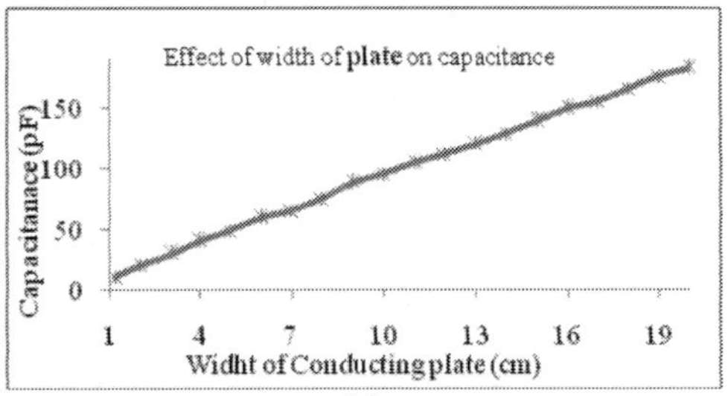

(a)

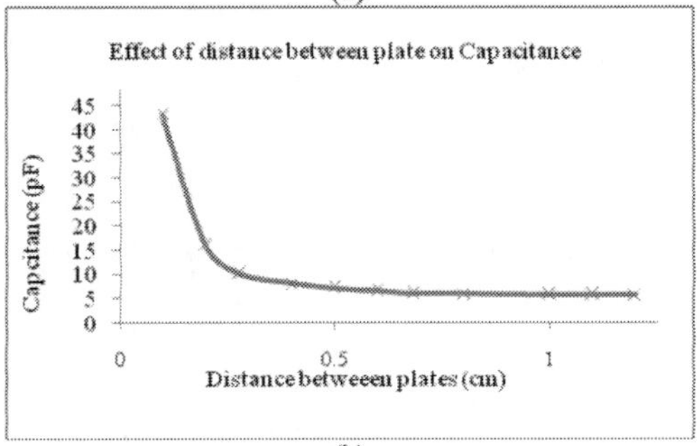

(b)

Fig. 13. Effect on capacitance due to width and distance between plate.

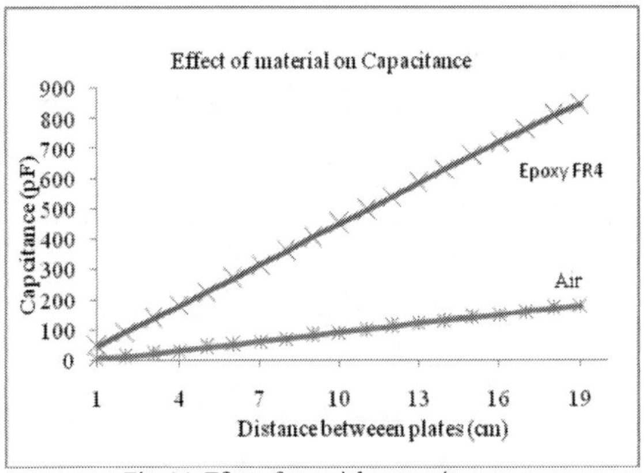

Fig. 14. Efect of material on capcitance

4. Effect of Other Parameters on Laminated Bus Bars

The stray inductance is mainly depended on the width to length (W:L) ratio. Current crowding is also indirectly related to W:Land also thermal effects [15]. Table 1. describes the effect of W:L ratio on stray inductance.

Table 1. Effect of W:L on Stray Inductance

Width (W) cm	Thickness (T) cm	W:T	Inductance (nH)
20	1	20	489.8
10	2	5	584.1
5	4	1.25	633.9
4	5	.08	633.7

Table 2. Effect of D:W on Stray Inductance and Capacitance

Diameter cm	D:W	Inductance (nH)	Capacitance (pF)
1	0.05	159.30	1.821
2	.1	159.27	1.751
4	.2	159.29	1.745
8	.4	160.18	1.618
10	.5	160.91	1.574
19.8	.99	182.81	1.101

5 Effect of Material on Laminated Bus Bar

Sometimes, laminated bus bar models designed perfectly even with optimal dimensions don't work as designed which could be due to material used. Based on working conditions like voltage, current, ambient temperature, working temperature and so on appropriate materials need to selected based on requirement [2]. Due to improper thickness used or excessive use of material may lead to improper respiration of Laminated bus bar. Many insulation materials are designed which have better properties than those currently used in wires or conventional bus bars. Table 4. lists different insulating materials used for laminated bus bars. Conducting materials used are generally copper and aluminum.

Table 3. Comparison of different materials used in laminated bus bar

Material	Dielectric constant	Dielectric strength V/mil	Minimum thickness	Continuous use temperature (° C)
Mylar	3.30	7500	.002	105
Epoxy glass	4	500	0.0025	~ 130
Kapton	3.70	4600	.001	400
Nomex	1.6	500	.003	220

CONCLUSION

Paper explains the applications of Laminated bus bar in modern world for low and high power applications. Design of Laminated bus bar plays a key role so as to effectively utilize its benefits. Aperture diameter and arrragement of the apperture in the plate make the non - current uniform and also disturb the stray inductance. Flexible terminal on adddition making the current uniform crowdinig. Change in frequency of the circuit effect the stray inductance. Suitable material used to make the model to operate in safer operating region. Change in material make the capcitance to increase. Proper design of model with apertures arrangement, material used, operating frequency, thickness and flexible terminals to operate effectively.

REFERENCES

[1] J. M. Allocco, "Laminated bus bars for power system interconnects," *Proceedings of APEC 97 - Power Electro. Conf.*, pp. 585-589 vol.2, 1997.
[2] M M. Stibgen, "Applying laminated busbars to enhance DC power distribution systems," *International Telecom. Energy Conf. (INTELEC)*, pp. 537-541, 2004.
[3] ChengfeiGengID ,Fengyou He, Jingwei Zhang and Hongsheng Hu- "Partial Stray Inductance Modeling and Measuring of Asymmetrical Parallel Branches on the Bus-Bar of Electric Vehicles". Energies **2017**, 10, 1519; doi:10.3390/en10101519
[4] M. H. Miles, "Recent advances in lithium battery technology," *GaAs IC Symposium. IEEE Gallium Arsenide Integrated Circuit Symposium*, pp. 219-222, 2001.
[5] K. W. E. Cheng, "Recent development on electric vehicles," *International Conf. Power Electron. Systems, Applications*, pp. 1-5, 2009.
[6] P. Gupta, R. Kumar, S. P. Singh and A. Jangid, "A study on monitoring of air quality and modeling of pollution control," *IEEE Conf. Human. Tech.*, pp. 1-4, 2016.
[7] Batterybank:Available:[Online]:https://www.mersen.com/products/solutions-power-management/bus-bars.
[8] Supercapacitor:Available:[Online]:https://www.mersen.com/sites/default/files/publications-media/8-markets-energy-electrical-energy-storage-mersen.pdf
[9] Dclinks:Available:[Online]:http://www.busbar.com/industries/computers/
[10] G. Zou, Z. Zhao and L. Yuan, "Study on DC bus bar structure considering stray inductance for the back-to-back IGBT-based converter," *IEEE conf. Applied Power Electron.(APEC)*, pp. 1213-1218, 2013.
[11] Powerdistribution:Available:[Online]:http://www.busbar.com/industries/industrial/
[12] Automobile:Available:[Online]:https://www.semic.cz/_obchody/semic.obchodak.net/prilohy/7384/busbar-1610110-287c0e.pdf
[13] H. Wen and W. Xiao, "Design and optimization of laminated bus bar to reduce transient voltage spike," *IEEE Inter. Symposium on Industrial Electron.* pp. 1478-1483, 2012.
[14] Jianing Wang, Shaolin Yu and Xing Zhang, "Effect of key physical structures on the laminated bus bar inductance," *IEEE Intern. Power Electron. and Motion Control Conf. (IPMEC)*, pp. 3689-3694, 2016.
[15] M. Khan, P. Magne, B. Bilgin, S. Wirasingha and A. Emadi, "Laminated busbar design criteria in power converters for electrified powertrain applications," *IEEE Transportation Electrification Conference and Expo (ITEC)*, pp. 1-6, 2014.
[16] Formulas:Available:[Online]:http://jp.methode.com/Documents/TechnicalLibrary/Bus_Bars_&_Capabilities_Brochure.pdf

Extensive review on Supercapacitor cell voltage balancing

Prashant Singh B.T.[1,*], *Phaneendra* Babu Bobba[2] and *Suresh* K.[3]

[1,2] Dept of Electrical Engineering, GRIET, Hyderabad,Telangana, India
[3]CEO, MasterPCB PVT.LTD, KPHB, Hyderabad, Telangana, India

Abstract. This paper explains about the supercapacitor cell voltage balancing circuits by comparing different topologies with regard to parameters like cost, balancing time, weight of the components used and control of switches. The advantage of supercapacitor over battery made to overcome weight and faster responding source problems. In supercapacitor bank cell voltages differ from each other which effects the performance of the device. Passive circuits consume power from cell for balancing but active circuits consume power from source. Many topologies are considered in this paper for different ratings and with different components. Balancing circuit is selected based upon total number of components in the balancing circuit, many components make circuit less reliable, complex and also increase the cost for balancing.

1 Introduction

Recent developments in graphene utilisation and manufacturing process made of the Supercapacitor (SC) demand rise [1]. SC a high power density source [2] when used in group like a bank for a high power requirement operation should be flawless. SC takes a span of seconds to pump the entire power to the load where time generally lies from 2 -15 seconds or so. Time of Power pumping demand is generally based on load requirement. Similar to alternators which pump power to power grid care is taken to make balance all alternators equally as required. Paralleling operation of alternator is dynamic in nature where voltage and frequency variations ±5% are observed. SC which deal with individual voltages of 2.6-2.85V [2] a small variation in voltage due to internal resistance or change in temperature of cell, change in a small part of composition of chemical and so on will lead to imbalance of power pumping. When used in a SC banks a small variations in any parameters mentioned above make them to unequal power pumping [4]. These parameters leads to discharge faster and SC cells to get overloaded due to imbalance and pump low power than their designed value. Adding a addition circuit which make all SC cells to balance and pump approximately equal power so that all cells are equally loaded and fulfill the load requirement for a given time. So, SC balancing circuit are designed to make effective and equally utilization of all cells in a SC bank and also improve life span of SC [5] by equally loading each cell.

2.Supercapacitor balancing using different topologies

SC balancing circuit are designed based on many parameters consideration taking in account like fast balancing, low source current to balance, easy to operate, effective voltage balance, bulk cells balancing, less components, and many more. But based on topologies selected the above parameters vary and decide the circuit functions and cost of module. Balancing circuit can roughly classified into two types namely

1) Passive circuits and

2) Active circuits

Passive circuits are those circuits which uses resistors and other elements to make balance of bank. But this method is a older and consume power from SC to balance. Active method includes many components on proper coordination between switches balancing takes place properly and effectively. But energy to balance cell voltage is taken from source.

2.2 Passive Balancing Circuits

2.2.*1* Topology 1

Topology 1 uses a resistors connected in a series with a switch connected in parallel with SC [6]. These circuit when used for a bigger banks becomes complex as each SC cell voltage differ from each other. So each switch need to operated independently Fig. 1. shows the circuit diagram used for topology 1 where resistor (R), Super Capacitor (C) and switch (S). These kind of topology discharges in resistors to balance by which SC is not effectively used and time to balance the SC is also more.

2.2.2 Topology 2

Topology 2 uses a Zener diode in parallel with SC. But use of Zener diode make use of power from SC to balance cells. Apart from that a strong temperature dependency of

* Corresponding author:prash.5295@gmail.com

Zener diode creates a drawback to the topology [7]. Time to balance the SC is also huge as no external source is used to balance it. Fig. 2. shows the balancing of SC using Zener diode.

2.2.3 Topology 3

Topology 3 is a advanced version of topology 1 where a proper coordination between cells are done by communicating between cell to cell. A controller senses cell voltage and controls according to the requirement [8]. A cyber layer is used in between to balance all cells. A closed loop operation is used in the topology making it complex and also costly. Fig.3. show the circuit diagram used for the topology 3. Cyber layer used to control the operaton make it to costly and also complex. Communication between cells plays a major role. Any mis-match of communication can damage the cell.

2.3 Active balancing circuits

2.3.1Topology 4

Topology 4 uses a DC/DC converter. Converters operates in buck/boost mode based on the cell voltage in the bank based on the average voltage of bank. But when used for bulk circuits it becomes more complex as individual circuits are required and each cell requires converter and it makes circuit complex, more weight and costly [9]. Fig.4. shows the balancing of topology 4. As the cells are balanced between each other it takes more time to balance. Topology used for many cells makes it costly and complex but high efficient.

2.3.2 Topology 5

Topology 5 is used for high power applications. Fig.5. shows the topology 5 balancing circuit. Toplogy 5 is high efficient reaching peak effficency of 91-95%. The charging current (I_{ch}) is used to charge the cells intially and balacing current (I_{bal}) is used to balance alternatively [10]. Time to balance all the cells is around the 160 secounds. But in fast charging mode where on pumping more current, makes time to balance lower. Balancing circuit designed uses a multi winding transformer in circuit. Windings required for the circuit is given by n+1. where n = number of SC cells in balancing circuit.

2.3.3 Topology 6

Topology 6 is a high efficient than any other topologies mentioned in this paper. Its peak efficiency lies near by 98.4%. Topology uses circuit in buck/boost modes to balance faster. Fig.6. shows the topology 6 balancing circuit [11]. Inductor is used as a part of converter. Circuit balances the voltage faster but it is costly when viewed from high end SC banks.

2.3.4 Topology 7

Topology is operated by a controller but with special source funtionality and with OCET switches. This make it different from other topologys but time to balance the cells voltage lies between 80 seconds. Fig.7. describes the balancing circuit of Topology 7. Controller operates according to cell voltage and other parameters [12]. But the balancing of cell is applied only between two cells. Fig. 7 shows circuit diagram for topology 7.

2.3.5 Topology 8

Topology 8 balancing is done on a hybrid power source. As the cells voltage of batteries also differs they need yet to be balanced [13]. These sources are placed in parallel to operate independently. Topology uses a same switch arrangement to balance batteries and also SC cells. Time to balance all those capacitors lies between 120 seconds which is more for when viewed from a automobile sector. As dynamic power pumping is needed to drive the electric vechile based on road profile for a urban traffic balancing need to be faster. Operating 120 secounds to balance cell voltage makes the vechile to decrease SC bank life span as balancing of cell voltage is not completed for a urban type road profile. Fig. 8 shows the circuit for topology 8.

2.3.6 Topology 9

Topology 9 is used for bulk operation, where the circuit designed is for 512 cells balancing. As many cells are used control becomes complex and balancing time is more. Fig. 9. shows circuit for Topology 9. Buck converter is used to provide a constant voltage of 24V. Active clamped circuit are used to make the balancing operation. Each cell voltage is balanced separately by a logic, by calculating a average voltage [14]. Later cells with higher voltages try to balance with lower voltages by sharing the voltage and reach the average voltage. For a bank of 512 cells 25A of balancing current is too small making it to balance slowly. Voltage variation of 1.2% was observed after balancing the cells.

2.3.7 Topology 10

Topology 10 is similar to the a passive controlled resistor circuit as mentioned in Topology 1. In these topology a capacitor is used instead of a resistor [15]. The advantage of the topology is that it tries to balance the circuit faster compared to topology 1. Fig. 10 shows circuit for Topology 10. Topology arrangement is complex where seen from controlling part. A small capacitor is placed parallel to the SC as "Equilibrium capacitor"(C'). Equilibrium capacitor make the SC to balance by transfer of power between them. Equilibrium capacitor make the control operation a complex to control.

2.3.8 Topology 11

Topology is a simple circuit and operates with lower currents. The circuit uses RLC tank elements to balance operating below resonant frequency [16]. All SC form a grid where balancing of all cells take place equally. Balancing of all cells with variation of 50-100mV. Fig. 11 shows the circuit diagram of topology 11. RLC tank circuit decides parameters like balancing time, balancing voltage, balancing current and switch frequency. So care

 https://doi.org/10.1051/e3sconf/20198701010

need to be taken while designing the RLC tank. As the voltage balancing is accurate it takes 350 seconds. The circuit developed can be applied to 1000 times higher rated than cells used in the circuit.

2.3.9 Topology 12

Topology 12 is a simple arrangement of cells in series and connected with the source and switches. Logics applied between the switches make circuit to balance in a different manner [17]. Some variation in control generates a high ripple which damages cells. A common logic between switches could be developed whatever the unbalancing of cell voltage might be. Perfect logic make the cells to balance safely and faster. Main advantage of the topology is that it can handle from low to high power cells. This topology uses many switches interconnection which makes the balancing operation faster and time to balance depends on the logic used and also on frequency of operation. Fig. 12 shows the circuit diagram used for topology 12.

All the 12 topologies are compared in Table. I & II on various parameters. Based on circuit arrangement and components used in topologies making the balancing operation different from each other which can be observed in Tables listed below. Based on the current flowing in the circuit between the elements decides the balancing. A higher voltage to the circuit make the SC cells to damage as SC are voltage sensitive elements. Lower the voltage makes the balancing slow. So a proper voltage need to be applied to make the balancing faster and operate in a safer region.

Table I. comparison of different topologies in various parameters

Topology	Switches used	Balancing time	External components	Control of balancing circuit	Weight of circuit	Cost of circuit
Topology 1	one / cell	High	one Power resistor/ cell	easy	light	Low
Topology 2	zero	High	Nil	No switch	light	Low
Topology 3	one / cell	Medium	Resistor	siimple	light	Moderate
Topology 4	one converter/ cell	Less than medium	dc/ dc converter /cell	easy	heavy	High
Topology 5	one/cell + four/ module	Medium	Multi winding transformer.	easy	heavy	High
Topology 6	two switches/ cell	Medium	one capacitor/ cell	easy	light	High
Topology 7	one switch/ cell	Low	2Resistor, 1 capacitors/ cell	complex	light	High
Topology 8	two switch/cell	Low	Nothing	complex	light	Moderate
Topology 9	Four switches/cell	Low	Forward converter/cell	complex	heavy	High
Topology 10	one switch/cell	Medium	Equilibrium capacitor/cell	complex	light	Moderate
Topology 11	Two swiitch/cell	Medium	RLC tank circuits	simple	light	Moderate

 https://doi.org/10.1051/e3sconf/20198701010

Table 2. Comparison of different topologies on Switch based parameters and circuit.

Topology	Application	Types of circuit	Type of Switch	Circuit arrangement
Topology 1	Low	Passive	Mosfet	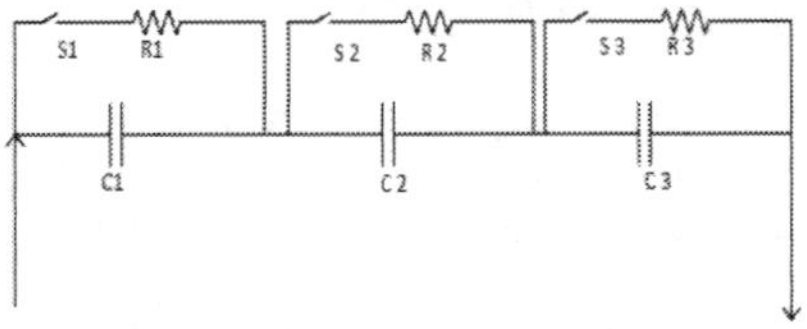 Fig. 1. Circuit diagram for Topology 1
Topology 2	Low	Passive	No switches	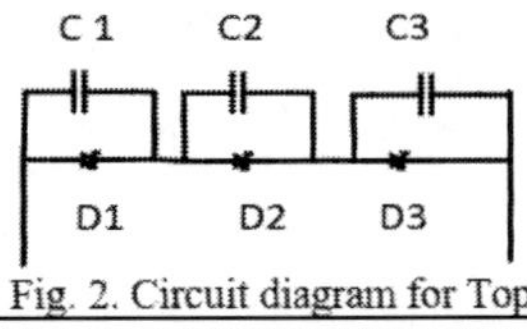 Fig. 2. Circuit diagram for Topology 2
Topology 3	Low	Passive	Mosfet	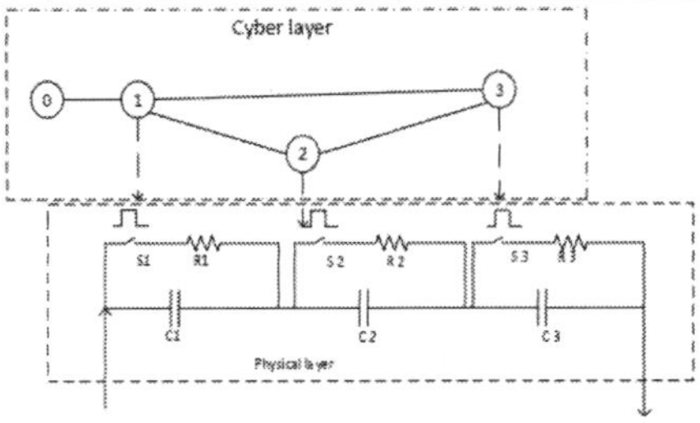 Fig. 3. Circuit diagram for Topology 3
Topology 4	Medium	Active	Mosfet	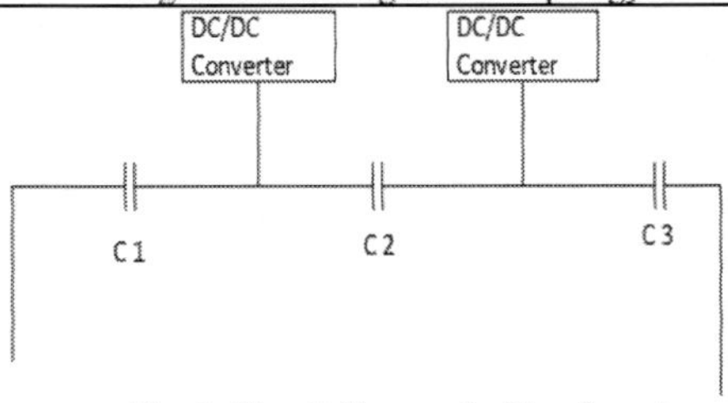 Fig. 4. Circuit diagram for Topology 4
Topology 5	Medium	Active	Mosfet	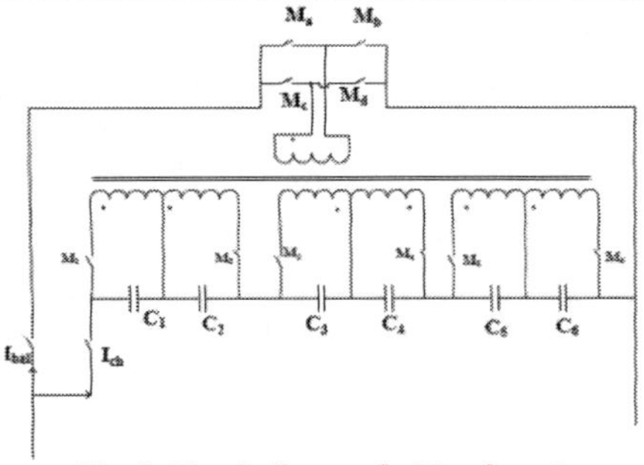 Fig. 5. Circuit diagram for Topology 5

Topology 6	Medium	Active	IGBT	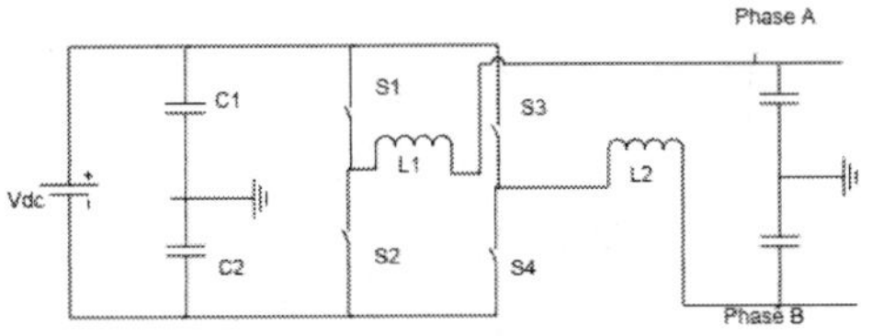 Fig. 6. Circuit diagram for Topology 6
Topology 7	Medium	Active	OECT	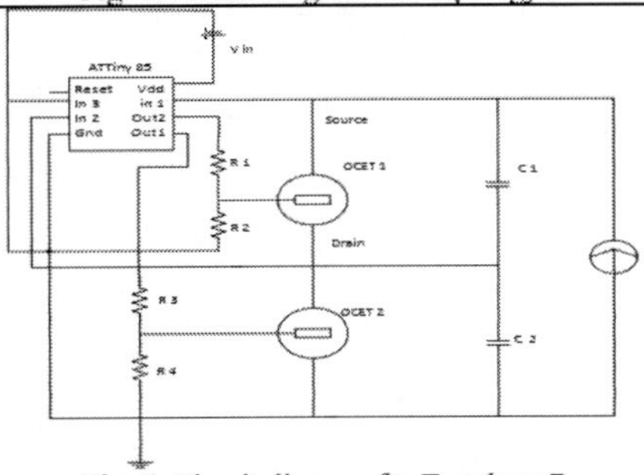 Fig. 7. Circuit diagram for Topology 7
Topology 8	High	Active	Mosfet	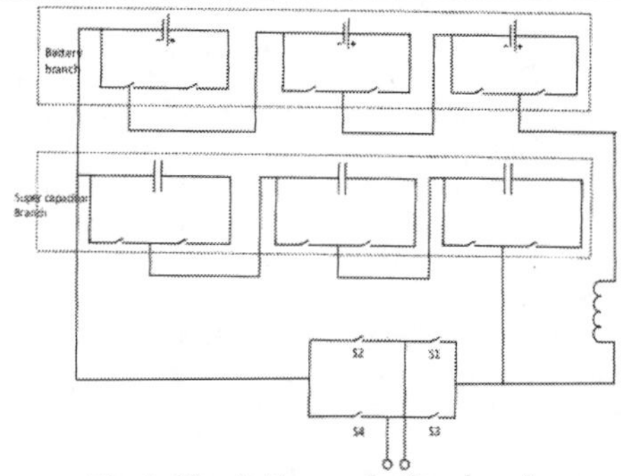 Fig. 8. Circuit diagram for Topology 8
Topology 9	High	Active	Mosfet, IGBT	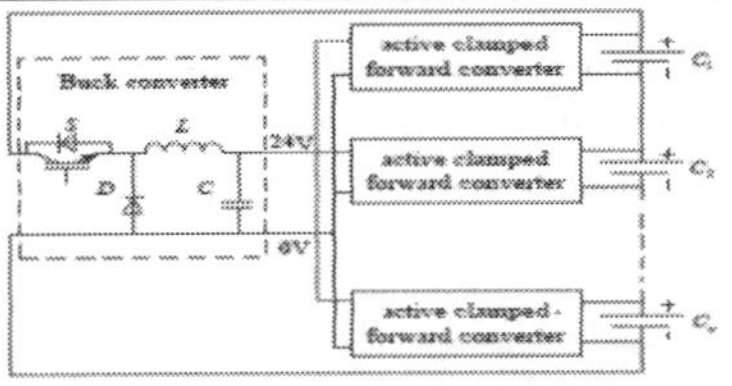 Fig. 9. Circuit diagram for Topology 9
Topology 10	High	Active	Mosfet	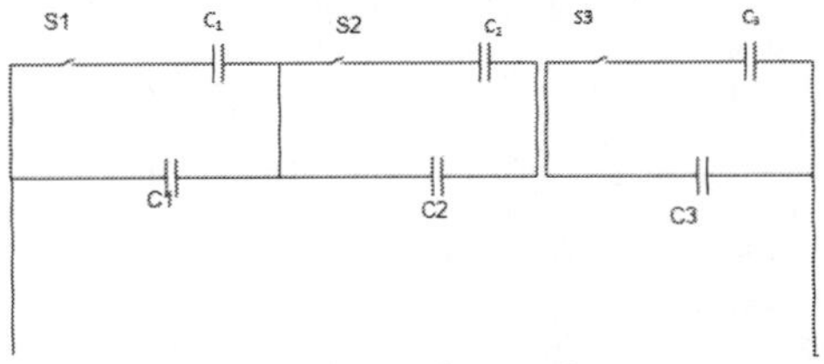 Fig. 10. Circuit diagram for Topology 10

Topology 11	Low	Active	Mosfet	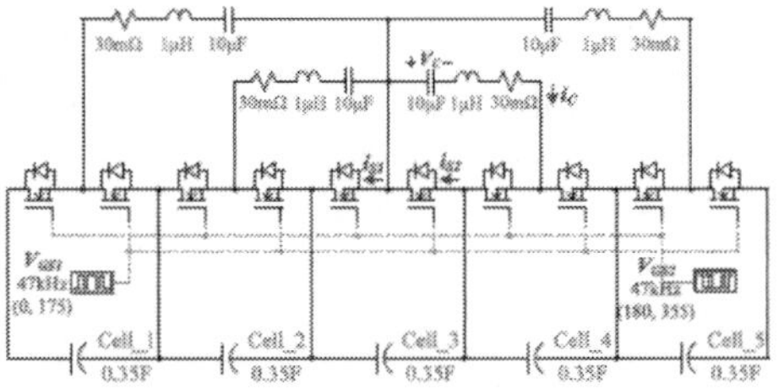 Fig. 11. Circuit diagram for Topology 11
Topology 12	Low - High	Active	Mosfet (or) IGBT	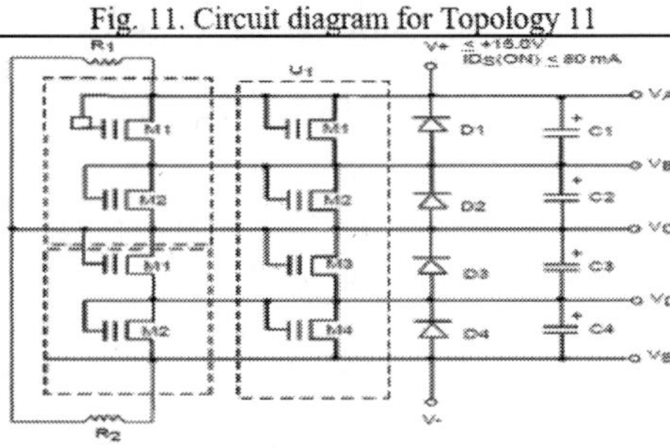 Fig. 12. Circuit diagram for Topology 12

3. Conclusion

This paper compares different topologies of Supercapacitor voltage balancing circuit considering balancing time, circuit complexity, elements used, weight of circuit, control strategy, cost and types of switches used. Based upon the arrangement and the logic used to balancing the circuit voltage balancing between cells behavior changes. Faster balancing, high efficient balancing, less complex and high power bank circuits are mention in the topologies. Major focus of the paper is on faster balancing of cells, less complex, lighter weight and cheap each topologies fulfill some parameters. But Topology 12 fulfill faster volatage balancing, lighter weight, cheaper and less complex parameters and can be used for low to high power applications.

References

[1] B. Xie, P. Zou and C. Yang, "Ultrahigh power graphene based supercapacitor," *Inter. Conf. on Electronic Packaging Tech.*, pp. 1211-1214, 2015.
[2] P. B. Bobba and K. R. Rajagopal, "Modeling and analysis of hybrid energy storage systems used in Electric vehicles," *IEEE International Conf. Power Electron., Drives and Energy Systems (PEDES)*, pp. 1-6, 2012.
[3] Dai, Keren et al. "Voltage Fluctuation in a Supercapacitor During a High-g Impact" *Scientific reports* vol. 6 38794. 13 Dec. 2016.
[4] Y. Qu, J. Zhu, J. Hu and B. Holliday, "Overview of supercapacitor cell voltage balancing methods for an electric vehicle," *IEEE ECCE*, pp. 810-814, 2013.
[5] S. Shili, A. Hijazi, A. Sari, X. Lin-Shi and P. Venet, "Balancing Circuit New Control for Supercapacitor Storage System Lifetime Maximization," *IEEE Trans. Power Electron.* vol. 32, no. 6, pp. 4939-4948, June 2017.
[6] F. M. Ibanez, "Analyzing the Need for a Balancing System in Supercapacitor Energy Storage Systems," *IEEE Trans Power Electron.* vol. 33, no. 3, pp. 2162-2171, March 2018.
[7] C. Ionescu, A. Vasile and R. Negroiu, "Investigations on balancing circuits for supercapacitor modules," *Inter. Spring Seminar on Electron. Techn.*, pp. 521-526, 2016.
[8] H. Li, J. Peng, J. He, Z. Huang, J. Pan and J. Wang, "Synchronized Cell-Balancing Charging of Supercapacitors: A Consensus-Based Approach," *IEEE Trans Industrial Electron.* vol. 65, no. 10, pp. 8030-8040, Oct. 2018.
[9] D. Linzen, S. Buller, E. Karden and R. W. De Doncker, "Analysis and evaluation of charge-balancing circuits on performance, reliability, and lifetime of supercapacitor systems," *IEEE Trans. Industry Applications*, vol. 41, no. 5, pp. 1135-1141, Sept.-Oct. 2005.
[10] Z. Zheng, K. Wang, Y. Li, H. Liu and F. Chang, "A Modular-Cascaded Active-Balanced Storage System for Electric Transportation," *IEEE Transportation Electrification Conf. Expo (ITEC)*, pp. 1-6 2018.
[11] L. Zhang, B. Gu, J. Dominic, B. Chen and J. Lai, "A capacitor voltage balancing method with zero-voltage switching for split phase inverter," *IEEE Applied Power Electron. Conf. and Exposition*, pp. 2394-2400 2014.
[12] V. Keshmiri, D. Westerberg, P. Andersson Ersman, M. Sandberg, R. Forchheimer and D. Tu, "A Silicon-Organic Hybrid Voltage Equalizer for Supercapacitor Balancing," *IEEE Journal on Emerging and Selected Topics in Circuits and Systems*, vol. 7, no. 1, pp. 114-122, March 2017.
[13] Z. Zheng, K. Wang, Y. Li, H. Liu and F. Chang, "A Modular-Cascaded Active-Balanced Storage System for Electric Transportation," *IEEE Transportation Electrification Conf.*, pp. 1-6, 2018.
[14] S. Fan, J. Duan, L. Sun and Y. Han, "The balancing system of super capacitor based on active clamped forward converter," *International Conf. on Circuits, Devices and Systems,* pp. 29-33, 2017.
[15] Jiangang Liu, Zhiwu Huang, Jun Peng and Jing Wang, "Distributed cooperative voltage equalization for series-connected super-capacitors," *American Control Conference (ACC)*, pp. 4523-4528, 2015.
[16] Y. Ye and K. W. E. Cheng, "Analysis and Design of Zero-Current Switching Switched-Capacitor Cell

https://doi.org/10.1051/e3sconf/20198701010

Balancing Circuit for Series-Connected Battery/Supercapacitor," *IEEE Trans. Vehicular Tech.*, vol. 67, no. 2, pp. 948-955, Feb. 2018
[17] Circuitdiagram:Available:[Online]:https://www.aldinc.com/pdf/SABMB6.pdf

Simulation of an aircraft radial engine misfire detection

Łukasz Grabowski[1], *Paweł* Karpiński[1,*], and *Konrad* Pietrykowski[1]

[1] Lublin University of Technology, Faculty of Mechanical Engineering, Department of Thermodynamics, Fluid Mechanics and Aviation Propulsion Systems, Nadbystrzycka 36 Str., 20-618 Lublin, Poland

Abstract. The misfire phenomenon is particularly unfavourable in aircraft engines because it affects the stability and reliability of work. This paper presents the algorithm for detecting ignition failure in a radial aircraft engine. The Crankshaft Velocity Fluctuation method was applied, which consists in analysing changes in the crankshaft speed signal as a function of time. A zero-dimensional model of the aircraft engine was developed in order to perform the research. The validation of the model was performed using the results from the test bench. The model was subjected to simulation tests in fixed operating conditions. Based on the engine speed signal obtained as a result of the simulation, the normalized second derivative of the signal was determined based on the adopted algorithm. On the basis of this derivative, a criterion was defined to assess the occurrence of the misfire phenomenon. The results of the calculations can be compared in future with the results of the real engine tests.

1 Introduction

The phenomenon of misfiring in piston engines has a negative effect on the emission of toxic substances. It also results in deterioration of the catalyst or may lead to its damage. In addition, it negatively affects the stability of the engine - there is a change in the distribution of forces acting on subsequent cranks. Misfire phenomenon causes variability and uniqueness of a combustion process. This can be seen by analysing measured cylinder pressure from radial engine with multifractal approach and illustrate this phenomenon by means of wavelet results like in the work [1]. Ignition losses in SI engines can be caused by various causes: failure of ignition coils, failure or wear of spark plugs, failure of the electronic ignition control module, malfunction of the fuel system. Also changing fuel from liquid state into gas state also can cause disturbances in air fuel mixture and affect in misfire phenomenon both in SI and CI engines. Gęca and Litak investigated fluctuations during the combustion of hydrogen-enriched gasoline an IC-SI engine in the paper [2]. The authors of the work [3] by applying the statistical and multifractal methods to the corresponding time pressure series showed the considerable changes in engine dynamics for a different equivalence ratio decreases from 0.781 to very lean conditions.

The other causes of misfiring in CI engines are different due to the self-ignition process. In the study [4] the misfire in the locomotive diesel engine was investigated using non-linear analysis based on the deterministic chaos theory. For the analysis, vibroacoustic signals obtained from an exhaust locomotive engine were used. A separate group of methods for analysing the phenomenon of misfire are methods based on neural networks. In work [5], based on the general work parameters of the supercharged diesel engine, the artificial neural network model (ANN) was designed to detect the misfire. A similar model of ANN was developed in paper [6], where investigated the phenomenon of misfire in an ethanol-powered HCCI engine. The experimental engine test results were used as input data to the model. Examples of other works in which neural networks are used to detect the misfire are papers [7-9].

One of the basic methods for detecting misfire is the Crankshaft Velocity Fluctuation (CVF) method. It involves the use of a crankshaft speed sensor to analyse engine speed fluctuations. The fluctuation is caused by the lack of combustion pressure in a given cylinder, and therefore the lack of torque generated in a given part of the crank-piston system. This type of method was used in work [10], which was verified by performing test runs on the track for various loads, gear ratios and accelerations of the vehicle.

In car engines, On-board Diagnostics System (OBD-II) controls the misfiring. In the work [11], various methods of misfire detection in accordance with the OBD-II system are presented. However, this standard does not apply to aircraft engines. The detection of the misfire phenomenon in the case of aircraft engines is particularly important from the point of view of obtaining stable and reliable operation of the engine.

Paper [12] presents a method meeting the OBD-II requirements for misfire detection based on a rotational acceleration analysis. In the algorithm developed by the authors, the "engine roughness" parameter was calculated based on the momentary engine speed and engine load. In addition, error correction functions were applied during fast-changing operating conditions.

* Corresponding author: pawel.karpinski@pollub.edu.pl

Paper [13] presents the concept of a system that allows misfire detection in SI engines based on the analysis of the crankshaft speed change. The tests were carried out on the basis of a 12-cylinder engine.

In addition, attempts have been made with other methods to detect misfiring. In the study [14] a method based on Ionization Current Monitoring was presented.

It is possible to use digital filtering to detect misfiring. An example of this is the research presented in work [15], where the misfire phenomenon in a high-speed four-cylinder engine was analysed using this method. In work [16], a model of ignition detection using Kalman filter was used.

The authors of the work [17] used in their work the Fourier transform to analyse the pressure pulsation of the exhaust gases in order to detect the misfire of the 12-cylinder V-engine. The method of misfire detection on the basis of analysis in the time-frequency domain was presented in the work [18]. The presented method has been experimentally verified using a spark-ignition engine.

The work [19] presents a combination of two methods: digital filtering with statistical recognition of patterns in real time. Periodic and non-periodic ignition misfires have been characterized by separate algorithms.

Most of the work devoted to investigating the phenomenon of ignition loss focuses on the analysis of in-line or V engines. Therefore, there is a need to investigate this process in the case of an aircraft radial engine. The authors of the paper [20] analysed IMEP in 9-cylinder radial engine. There is misfire especially at below 1200 rpm. At 500 rpm misfire occurs every second cycle. Results of CFD model of the aircraft radial engine presented in paper [21] show that combustion prolongs in time in the case of lack of ignition in one of two spark plugs.

One of the methods of testing internal combustion engines is to develop an engine model and perform simulation calculations. An example of such an approach is the research presented in paper [22], in which the model for determining the heat of the engine and heat flux penetrating through each of the engine's surfaces was used. The purpose of this work was to simulate the aircraft radial engine ASz-62IR model for misfire analysis. The analysis consisted in using the developed algorithm and defining the criterion for checking the occurrence of misfire based on the engine speed signal as a function of time. Section 2 presents the algorithm used to detect the misfire phenomenon. The next Section presents a zero-dimensional engine model used for research. In Section 4, the results obtained and their analysis are presented. The misfire condition for the proposed criterion was obtained.

2 Algorithm

The following algorithm is used to analyse the ignition failure in the cylinder. As a result of the simulation, the signal of engine speed was recorded as a function of time (1):

$$f = \omega(t) \tag{1}$$

Then the first derivative (2) was calculated, followed by the second derivative (3) of the engine speed as a function of time:

$$\varepsilon(t) = \omega'(t) \tag{2}$$

$$\gamma(t) = \varepsilon'(t) = \omega''(t) \tag{3}$$

The course of the second derivative was normalized according to equations (4) and (5) presented below:

$$\gamma_1(t) = \gamma(t) + |\gamma_{\min}| \tag{4}$$

$$\gamma_n(t) = \frac{\gamma_1(t)}{\gamma_{1,\max}} \tag{5}$$

In the next step, the signal was divided into k intervals with a length corresponding to 1/9 of the average duration of one cycle T_{mean}. The number of intervals k resulted from the duration of the measurement, which was 27 seconds. For the determined intervals, local maxima (6) for the normalized signal of engine speed were determined:

$$\gamma_{n,\max_i} \quad i = 1, \ldots, k \tag{6}$$

Finally, the following criterion was adopted to determine the occurrence of misfiring (7):

$$\gamma_{n,\max_i} \le X \text{ - misfire in cylinder for } i \text{ interval} \tag{7}$$

$$\gamma_{n,\max_i} > X \text{ - correct ignition in cylinder for } i \text{ interval}$$

The value of X was determined on the basis of the analysis of the values of local maxima for 1st cylinder (Section 4). Additional analysis of a signal comprising averaging, accumulation, or statistical filtering as potential different decision errors of process statistical error compensation should be included for further research.

3 Engine model

The object of this part of study was an ASz-62IR piston radial engine with spark ignition and supercharging performed by a single-speed centrifugal compressor.

Fig. 1. Aircraft radial engine ASz-62IR - research object.

The engine has 9 air-cooled cylinders. Thanks to the compressor, the charging pressure increases, which ensures that the nominal power is maintained up to a height of 1500 m.

The engine is equipped with a dual ignition system, an inertia starter and an electronically controlled injection system. The planetary gear allows to reduce the rotational speed of the propeller relative to the speed of the crankshaft. The construction of the engine enables driving various types of aggregates and tools, thanks to which it is suitable for installation in aircraft used in agriculture. ASz-62IR can be mounted on Antonov An-2, PZL M18 Dromader or DHC-3 Otter. The engine met the certification conditions in Poland, Canada, Brazil, China, Argentina and the USA. The basic engine parameters are presented in Tab. 1.

Table 1. Basic technical parameters of the tested engine (source: WSK PZL Kalisz, Technical description of aircraft engine ASz-62IR 16, Warsaw 1976).

Engine type	Piston engine, four-stroke, spark ignited, radial, supercharged
Type of supercharging	Centrifugal compressor
Capacity of cylinders	29.87 dm^3
Bore	155.5 mm
Stroke for 1st cylinder	174.5 mm
Compression ratio	6.4:1
Take-off power	745 kW (2200 rpm)
Firing order	1, 3, 5, 7, 9, 2, 4, 6, 8

Based on technical parameters of the engine, a zero-dimensional model was developed in the AVL BOOST program. This model was also used in the work [23], in which the emission of substances toxic to given working conditions was examined. This model is defined on the basis of the first law of thermodynamics. This means that the method of spreading the flame in the cylinder is assumed, and the time is an independent variable. The mass burning rate is based on literature or experimental data [24, 25].

The combustion model described in the Vibe function was adopted, which is determined by the beginning and duration of combustion, the shape parameter m and the parameter a. The Vibe function has the form (8):

$$\frac{dx}{d\alpha} = \frac{a}{\Delta\alpha_c} \cdot (m+1) \cdot y^m \cdot e^{-a \cdot y^{m+1}} \quad (8)$$

$$dx = \frac{dQ}{Q} \quad (9)$$

$$y = \frac{\alpha - \alpha_0}{\Delta\alpha_c} \quad (10)$$

Where:
Q – supplied heat,
α – crank angle,
α_0 – start of combustion expressed in crank angle,
$\Delta\alpha_c$ – duration of combustion expressed in crank angle,
m – shape parameter,
a – Vibe parameter, for perfect combustion is equal to 6.9. [26]

By integrating the Vibe function after the crank angle, a mass fraction burned (11) can be calculated:

$$x\int \frac{dx}{d\alpha} d\alpha = 1 - e^{-a \cdot y^{m+1}} \quad (10)$$

Where:
x – mass fraction burned, MFB. [26]

Parameters of the combustion model were determined based on experimental data loaded into the Burn module. This module allows to determine the parameters of the combustion model based on the actual operating characteristics of the tested engine. On the basis of the measured pressure signal read into the program, the parameters of the combustion model and the rate of heat release (ROHR) as a function of crank angle were determined. The indicated pressure was measured on the engine test bench belonging to the company WSK PZL Kalisz. OPTRAND optical pressure sensor installed in the central part of the cylinder head was used for the measurement. The signal from measuring sensor was recorded with a sampling frequency of 100 kHz using National Instruments measuring equipment and LabView software. The registered cylinder pressure signal (Fig. 2) has been loaded into the Pressure Trace tab. [27]

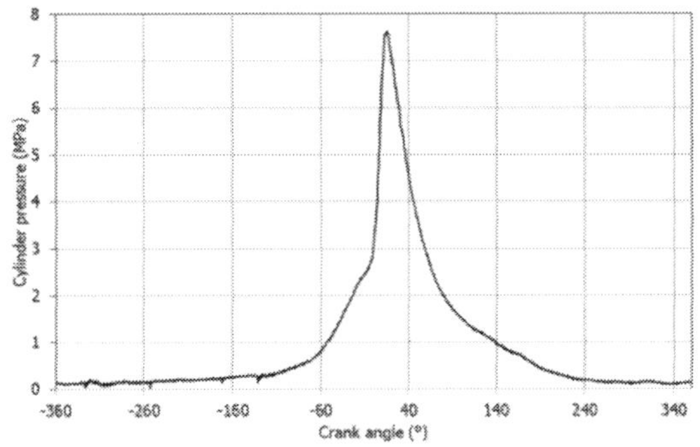

Fig. 2. Measured cylinder pressure [27].

In the settings of the Burn module, a two-zone combustion model and a classic model of species transport were chosen. The operation point has been defined for an engine speed of 2200 rpm. The model of Woschni 1978 was adopted for the heat exchange model. Detailed parameters adopted in the Burn module are presented in Fig. 3. [24, 27]

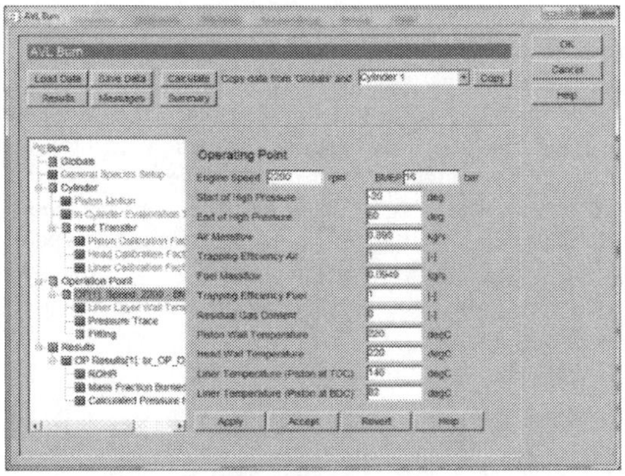

Fig. 3. Parameters set in the Burn module [27].

As a result of calculations, the following parameters of the combustion model were determined: start of combustion SoC = -4.74°, combustion duration CD = 28.4°, shape parameter m = 1.32. Based on these parameters and the Vibe function equation (8), the rate of heat release (ROHR) was determined as a function of the crank angle (Fig. 4). Although the curve determined on the basis of experimental tests does not ideally coincide with the curve determined on the basis of the Vibe function, the amount of heat released in one cycle is preserved. [27]

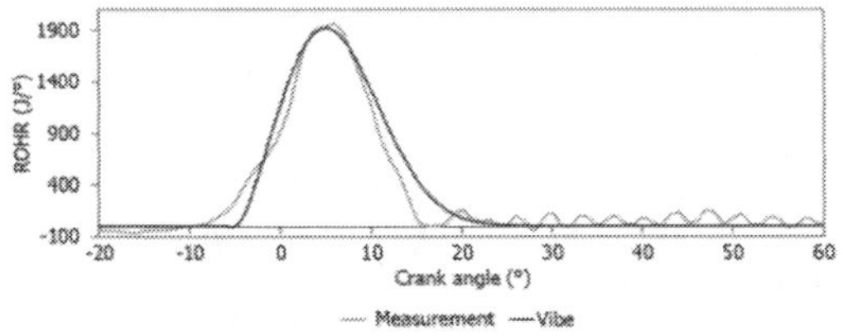

Fig. 4. Rate of heat release as a function of crank angle [27].

As a model of heat exchange, the Woschni model (1978) was adopted, which is a complex function of a number of parameters, including cylinder volume, cylinder temperature and pressure [28].

Heat exchange also takes place around the inlet and outlet valves. Therefore, a modified Zapf model was used, which takes into account heat exchange and temperature coefficients. The program defined the surface of the piston, head and temperature of these elements.

In modeling the air and fuel flow through pipes to the engine cylinders, only the diameters and lengths of the pipes connecting individual engine elements are taken into account. The timing phases have been determined based on the valve lift characteristics given by the engine manufacturer. The size of valve seats was also introduced and flow resistances were defined.

Fig. 5 shows the ASz-62IR engine model created in the AVL BOOST program.

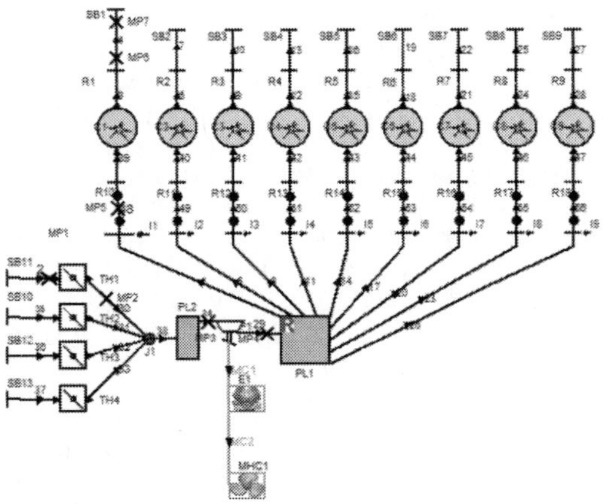

Fig. 5. The engine test model created in the AVL BOOST program.

The air flow in the developed model is realized through the intake pipes, the group of throttles, and then the centrifugal compressor. The compressor is driven by the transmission from the crankshaft. In the model this has been accomplished by using a mechanical connection MC1 with the engine block. Large volumes of small length in the intake duct have been replaced with Plenum type elements. The injection system is a set of 9 injectors located in the intake ducts. The mass flow of injectors varies dynamically depending on the mass flow of air and is adapted to the assumed composition of the fuel-air mixture. In selected places, measuring points and flow restriction elements were added on the connecting pipes. The elements used to create the model are presented in Table 2 [27].

Table 2. Components of the engine model.

Element	Name	Symbol
	System Boundary	SB
	Pipe	*number*
	Throttle	TH
	Junction	J
	Plenum	PL
	Turbo Compressor	TCP
	Engine	E
	Measuring Point	MP
	Injector	I
	Cylinder	C
	Restriction	R
	Mechanical Consumer	MHC

It was assumed that the engine is powered by gasoline with a calorific value of 43.5 MJ/kg and an air-

https://doi.org/10.1051/e3sconf/20198701011

fuel ratio of 14.5 kg of air/kg of fuel. It was assumed that the inlet air has a temperature of 24.85°C and a pressure of 1 bar. It was assumed that the engine works with an excess air ratio λ equal to 0.65. The chemical transport model in the working medium was defined as Classic Species Transport. The throttle operation has been defined based on data from the engine manufacturer. The firing order was specified in element E1. A compressor map from the manufacturer was loaded into the model. It was assumed that burned and unburned products have identical pressure and temperature, but different composition.

The model has been extended by the ability to perform calculations in dynamic states with a loaded engine. The power receiver was a propeller (MHC1 in Fig. 5) mechanically connected to the engine (reference MC2 in Fig. 5). Due to the fact that simulation studies relate to dynamic states, mass moments of inertia of elements have been defined. For the engine, the value of the polar moment of inertia being the equivalent torque of all rotating masses reduced to the axis of the crankshaft has been introduced. A mass moment of inertia of 69 kg·m2 was introduced in the Mechanical Consumer model element. It was assumed that the load on the propeller is 5400 Nm.

The aim of the research was to analyse the momentary engine speed change due to the lack of ignition in one cylinder. Burning in one cylinder has been intentionally switched off.

In order to perform the analysis of engine speed changes, calculations were made to determine the engine's operating conditions. The difference in the engine speed in subsequent cycles below 0.5% was assumed for determining the working conditions. These operating conditions were obtained after approximately 70 calculation cycles of the engine operating cycle.

The calculations were performed for the case of the take-off power at the engine speed of 2200 rpm and the pressure in the intake manifold equal to 140 kPa. Ignition in 1 cylinder was switched off.

4 Results and discussion

Fig. 6 shows the signal of the engine speed as a function of time obtained as a result of the simulation calculations.

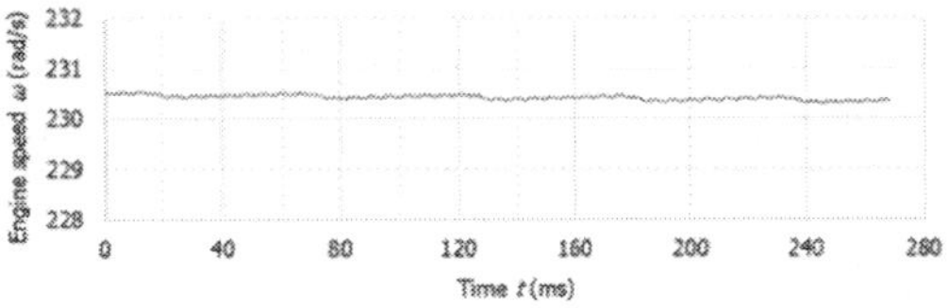

Fig. 6. Engine speed as a function of time.

The beginning of the signal coincides with the moment of ignition in the 4th cylinder. It is possible to observe temporary drops in rotational speed. This means that the ignition in one of the cylinders does not occur. After the temporary drop in speed, its value increases slightly. Local maxima correspond to the ignitions occurring in subsequent cylinders. The decreasing speed trend results from the fact that its value is still determined in subsequent calculation cycles.

Fig. 7 and Fig. 8 show respectively the first and second derivative of the engine speed as a function of time. On both signals you can observe a number of maxima corresponding to the work of the next cylinders. The maxima value remains constant, while in selected cases it drops significantly. The drop in value is cyclical, which means that the ignition in one of the cylinders does not occur.

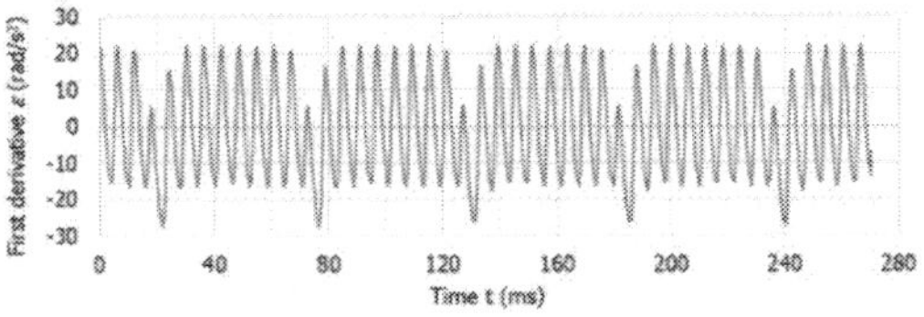

Fig. 7. The first derivative of engine speed as a function of time.

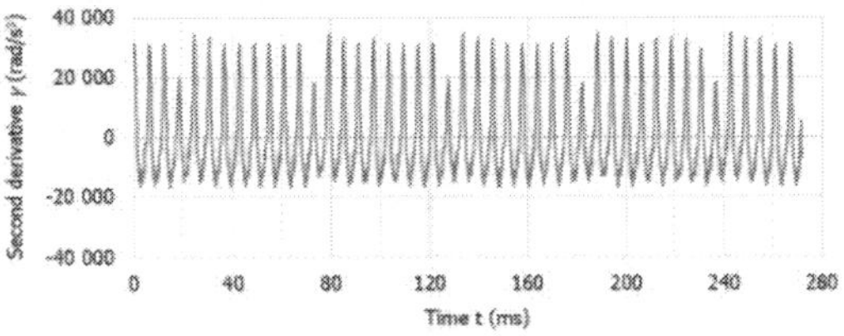

Fig. 8. The second derivative of engine speed as a function of time.

Based on the second derivative of the engine speed, the signal was normalized (Fig. 9), and then the misfire detection criterion was applied in accordance with the algorithm presented in the Section 2 of this paper.

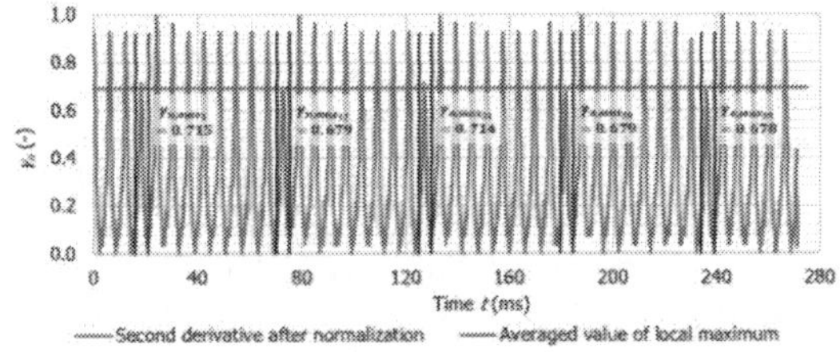

Fig. 9. The second derivative of engine speed as a function of time.

By comparing the time intervals with the firing sequence, it is possible to determine which cylinder corresponds to the specified maximum. In this way, the maxima and interval numbers to which 1st cylinder corresponds are determined. These data are presented in Tab. 3.

Table 3. Maxima corresponding to 1st cylinder.

Number of interval	Time interval	Value of local maximum
3	15.14 - 21.20 ms	0.715
12	75.72 - 81.78 ms	0.679
21	124.19 - 130.25 ms	0.714
30	178.73 - 184.79 ms	0.679
39	233.28 - 239.34 ms	0.678

On the basis of the maxima values, the threshold $X = 0.715$ for the criterion (7) presented in the algorithm in Section 2 is determined. This means that for maxima with the value equal to or less than 0.715 occur the misfire.

5 Conclusions

On the basis of the analysis of changes in rotational speed and its derivatives as a function of time, it is possible to evaluate the operation of individual cylinders in the absence of ignitions.

As a result of the simulation tests of the developed model of the radial engine, the signal of the rotational speed of the engine as a function of time was obtained. As a result of switching off the ignition, there was a cyclic rotational speed disturbance in 1 cylinder. It was caused by the lack of torque generated due to the rapid increase in pressure in a given cylinder. Based on the obtained signal, a developed algorithm was used to detect the misfire. The criterion introduced in the algorithm was determined after performing calculations and analysis (calculation of the second derivative of the signal and normalization of the speed signal).

During engine operation in real conditions external factors disturbing the engine speed in variable operating conditions may occur. Therefore, in further research, it will be necessary to verify the algorithm adopted based on bench tests and apply the necessary corrections to eliminate possible errors.

References

1. M. Gęca, M. Wendeker, J. Vibroeng. **14(2)**, 582- 591 (2012)
2. M. Gęca, G. Litak, Measurement **108**, 18-25 (2017)
3. G. Litak, M. Gęca, B. F. Yao, G. X. Li, Z. fuer Naturforsch. **64a**, 393-398 (2009)
4. P. Boguś, J. Merkisz, Mech. Syst. Signal Process. **19(4)**, 881-899 (2005)
5. B. Liu, C. Zhao, F. Zhang, T. Cui, J. Su, Appl. Therm. Eng. **55(1-2)**, 26-32 (2013)
6. B. Bahri, A. A. Aziz, M. Shahbakhti, M. F. M. Said, Appl. Energy **108**, 24-33 (2013)
7. Z. J. Wu, A. Lee, SAE Technical Paper No. 980515 (1998)
8. W. B. Ribbens, J. Park, D. Kim, ICASSP-94, 1994 IEEE International Conference on Vol. 2, pp. II-593 (1994)
9. H. Nareid, N. Lightowler, SAE Technical Paper No. 2004-01-1363 (2004)
10. D. Moro, P. Azzoni, G. Minelli, SAE Technical Paper No. 980521 (1998)
11. J. Merkisz, P. Boguś, R. Grzeszczyk, Journal of Kones **8(1-2)**, 326-341 (2001)
12. M. Klenk, W. Moser, W. Mueller, W. Wimmer, SAE Technical Paper No. 930399 (1993)
13. J. Förster, A. Lohmann, M. Mezger, K. Ries-Müller, SAE Technical Paper No. 970855 (1997)
14. A. Lee, J. S. Pyko, SAE Technical Paper No. 950003 (1995)
15. S. Kuroda, Y. Shimasaki, H. Igarashi, S. Maruyama, T. Asaka, JSAE Review **16(4)**, 387-390 (1995)
16. U. Kiencke, Control Eng. Pract. **7(2)**, 203-208 (1999)
17. M. Ceccarani, C. Rebottini, R. Bettini, SAE Technical Paper No. 980518 (1998)
18. N. Cavina, E. Corti, G. Minelli, G. Serra, SAE Technical Paper No. 2002-01-0480 (2002)
19. A. W. Osburn, T. M. Kostek, M. A. Franchek, Mech. Syst. Signal Process. **20(8)**, 2232-2258 (2006)
20. J. Czarnigowski, M. Wendeker, P. Jakliński, K. Pietrykowski, A. Nazarewicz, M. Gęca, T. Zyska, SAE Technical Paper No. 2007-01-2074 (2007)
21. K. Pietrykowski, SAE Technical Paper No. 2014-01-2578 (2014)
22. Ł. Grabowski, Z. Czyż, K. Kruszczyński, SAE International No. 2014-01-2883 (2014)
23. Z. Czyż, Ł. Grabowski, K. Pietrykowski, J. Czarnigowski, M. Porzak, Measurement **113**, 46-52 (2018)
24. T. Tulwin, Procedia Eng. **157**, 255-263 (2016)
25. T. Tulwin, R. Sochaczewski, ITM Web of Conferences **15**, 07016 (2017)
26. AVL BOOST Theory 11/2013.
27. P. Karpiński, Ł. Grabowski, Transactions of the Institute of Aviation **3(244)**, 305-317 (2016)
28. K. Pietrykowski, T. Tulwin, SAE International Journal of Engines **8(1)**, 82-88 (2015)

Complimentary Sensitivity Function based Novel Cascade Control Structure for Automatic Generation Control

B S Durga Kameswari [1], and *Dola Gobinda* Padhan[2*]

[1,2]Department of Electrical & Electronics Engineering,
Gokaraju Rangaraju Institute of Engineering & Technology, Hyderabad, India

Abstract. This paper introduces a series cascade control structure for automatic generation control. The control structure consists of two loops such as Primary loop and auxiliary loop (secondary loop). The secondary loop controller is designed using internal model control (IMC) approach. The primary loop controller is a PID controller which is tuned using desired complimentary sensitivity function. The beauty of the control structure is that it effectively nullifies the disturbances entering to the secondary loop as well as primary loop. The efficacy of the proposed controller is shown by comparing the simulation results with the existing methods in the literature.

1 Introduction

In electric power system design and operation Load Frequency Control (LFC) or Automatic Generation Control (AGC) is one of the most important issues for supplying sufficient and reliable electric power with good quality. The frequency of a power system is dependent on real power balance. A change in real power demand at one point of a network is reflected throughout the system by a change in frequency. Therefore, system frequency provides a useful index to indicate system generation and load imbalance. Any short-term energy imbalance will result in an instantaneous change in system frequency as the disturbance is initially offset by the kinetic energy of the rotating plant. Significant loss in the generation without an adequate system response can produce extreme frequency excursions outside the working range of the plant.

The successful operation of interconnected power systems requires the matching of total generation with total load demand and associated system losses. With time, the operating point of a power system changes, and hence, these systems may experience deviations in nominal system frequency and scheduled power exchanges to other areas, which may yield undesirable effects [1]
The primary notions of AGC for single area or multi areas power system are

- maintaining zero steady-state error for frequency deviations.
- limiting unscheduled tie line power flows between neighbouring control areas.
- better tracking for load demands and disturbances.
- maintaining less overshoot and settling time on the frequency and tie line power deviations.

There are two variables of interest, namely, frequency and tie-line power exchanges. Their variations are weighted together by a linear combination to a single variable called the ACE. The AGC problem has been augmented with the valuable research contributions from time to time, like AGC regulator designs incorporating parameter variations/uncertainties, load characteristics, excitation control, and parallel ac/dc transmission links. The microprocessor-based AGC regulator, self-tuning regulator, and adaptive AGC regulator designs have also been presented. The most recent advancement in this area is the application of concepts like neural networks, fuzzy logic, and genetic algorithms to tackle the difficulties associated with the design of AGC regulators for the power systems with nonlinear models and/or insufficient knowledge about the system required for its accurate modeling. Apart from advances in control concepts, there have been many changes during the last decade or more, such as deregulation of power industry and use of SMES, wind turbines, and PV cells as other sources of electrical energy to the system. Due to these, the control philosophies associated with AGC have changed to accommodate their dynamics and effects on overall system dynamic performance. The present study covers the critical review of a wide range of methodologies of AGC regulator designs of power systems with their salient features.
Ibraheem et al discussed a wide review of the recent philosophies in the area of AGC [1]. Also Several papers are avalable in the literature to address the concept of AGC [2–7].
In this paper, the Cascade control structure has been suggested for load frequency control for the first time. The approach is applied to single area power system non-reheated turbine only. The proposed control approach can be applied to other turbines as well as multi area power system.

The paper is organized as follows: Cascade control structure for AGC is discussed in section-2. Modeling of

* Corresponding author: dg.padhan@griet.ac.in

single area power system is discussed in section-3. Section-4 presents internal model control. Primary loop and secondary loop controllers design are discussed in section-5 followed by simulation results in section-6. Concluding remarks are given in section-7.

2 Cascade Control Structure for Automatic Generation Control

The basic block diagram of series cascade control structure is shown in Fig 1. $Gp1$ and $Gp2$ are the transfer functions of primary and secondary plant models whereas Gc_1 and Gc_2 represent the primary and secondary loop controllers, respectively. 'u_2' denotes the output of the secondary controller. Output of the secondary loop is denoted by y_2 whereas y denotes the output of the primary loop. The disturbances d_1 and d_2 are entering to the input and output of the secondary plant models. Generally, Set-point response is less priority response in the case of load frequency control. The output of the primary controller serves as the set-point for the secondary loop. It is discussed in the literature that Cascade control scheme is more effective when the secondary loop is faster than the primary loop and most of the disturbances enter the secondary loop. The governor, turbine, load and machine of the power system are collectively treated as primary process model. Secondary process is considered as first order dynamic process model.

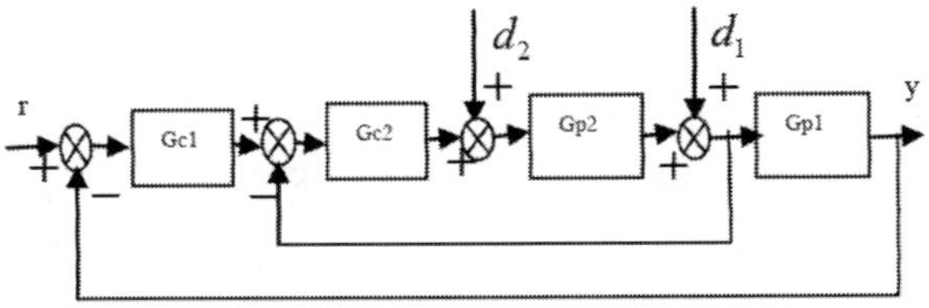

Fig. 1. Cascade control structure for AGC

3 Modeling of a Single area power System

Generally, power systems are highly non-linear and time-varying nature. However, for the purpose of frequency control analysis and synthesis in the face of load disturbances, a simple low-order linearized model is used. A linear model of a single-area power system is shown in Fig 2, in which a single generator is supplying power to a single-area.

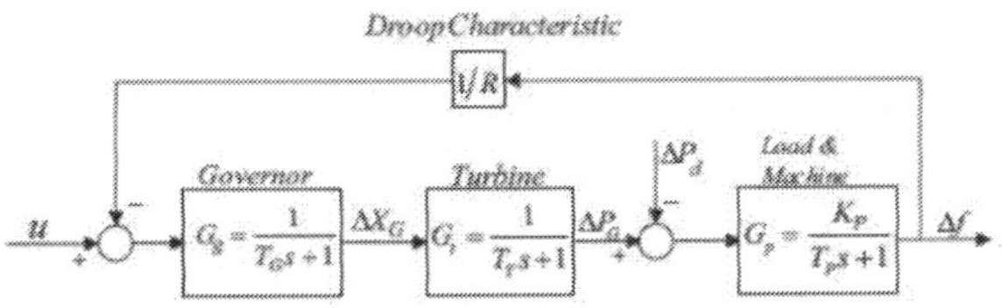

Fig. 2 Single area power system

In the present work, non-reheat turbine(NRT) is considered for LFC modeling. The plant model used for LFC without droop characteristicsis

$$G_{p1} = G_g G_t G_p \quad (1)$$

where Gg, Gt and Gp are the dynamics of the governor, turbine and load & machine, respectively.
The governor dynamics

$$G_g = \frac{1}{T_g s + 1} \quad (2)$$

and
theLoad and machine dynamics,

$$G_P = \frac{K_P}{T_S + 1} \quad (3)$$

Non-reheat turbines are first-order units.
The dynamics of the non-reheat turbine is represented as

$$G_t = \frac{1}{T_t s + 1} \quad (4)$$

4 Internal Model Control

4.1 Internal Model Control

The IMC design procedure is a two-step approach that, although sub-optimal in a general (norm) sense, provides a reasonable trade of between performance and robustness. The main benefit of the IMC approach is the ability to directly specify the complementary sensitivity and sensitivity functions and, which as noted previously, directly specify the nature of the closed-loop response. The IMC control structure is shown in Fig. 3.

The IMC design procedure consists of two main steps. The first step will insure that G_{IMC} is stable and causal; the second step will require G_{IMC} to be proper.

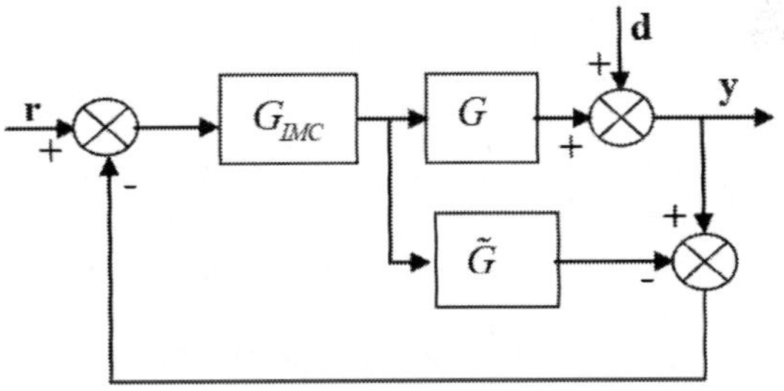

Fig. 3 IMC control structure

Decompose the plant model into two parts:

$$\tilde{G}(s) = \tilde{G}_+(s)\tilde{G}_-(s) \quad (5)$$

Where $\tilde{G}_+(s)$ contains all non-minimum phase terms and right half zeros.

Design a set point-tracking IMC controller as

$$G_{IMC}(s) = \tilde{G}_-^{-1}(s)F(s) \quad (6)$$

where $F(s)$ is a low pass filter having steady state gain of one. The filter is introduced for physical realisability of the IMC controller, $G_{IMC}(s)$. The simplest filter has the following form [9]

$$F(s)=\frac{1}{(\eta s+1)^n} \tag{7}$$

Where η is the tuning parameter of the filter.

5 Controller Design

5.1 Primary controller design

The primary controller is considered in the following form

$$G_{c1}=K_c\left(1+\frac{1}{T_i s}+\frac{T_d s}{\varepsilon T_d s+1}\right) \tag{8}$$

Where the derivative filter constant ε is chosen as 0.1 .
For load frequency deviation, the complimentary function is obtained as

$$T=\frac{GG_{c1}}{1+GG_{c1}} \tag{9}$$

In order to achieve the closed loop internal stability, the asymptotic stability

$$\lim_{s\to -\frac{1}{T_1},-\frac{1}{T_2}} (1-T)=0 \tag{10}$$

should be satisfied to reject a step change in load of power system.
In the present work, the desired closed loop complimentary function is suggested as

$$T=\frac{(\alpha_2 s^2+\alpha_1 s+1)}{(\lambda s+1)^4}e^{-\theta_m s} \tag{11}$$

Where λ is a tuning parameter to achieve satisfactory closed loop performance of power system.
Using the above constraints and complimentary sensitivity function, the primary controller is obtained as

$$G_{c1}=\frac{(T_1 s+1)(T_2 s+1)(\alpha_2 s^2+\alpha_1 s+1)}{k[(\lambda s+1)^4-e^{-\theta_m s}(\alpha_2 s^2+\alpha_1 s+1)]} \tag{12}$$

The resulting controller is not in the standard PID controller form.
So, using Laurent series G_{c1} can be expanded in the following form

$$G_{c1}=\frac{1}{s(\beta s+1)}\left(\cdots+\gamma(0)+\gamma'(0)s+\frac{\gamma''(0)s^2}{2!}+\cdots\right) \tag{13}$$

Now, the parameters of the primary controller can be obtained as

$$K_c=\gamma'(0)$$
$$\frac{K_c}{T_i}=\gamma(0) \tag{14}$$
$$K_c T_d=\frac{\gamma''(0)}{2!}$$

5.2 Secondary controller design

The inner loop controller (G_{c2}) is an IMC controller and is designed based on IMC approach. The detail of the approach has been discussed in the previous section. Using the IMC approach, the secondary loop controller is obtained as

$$G_{c2}=\frac{\tau_2 s+1}{k_2(\lambda_2 s+1)} \tag{15}$$

where λ_2 is the desired closed-loop time constant of the secondary controller.

6 Simulation Results

A non-reheated turbine studied by W Tan [8] has been considered for simulation. The model parameters of the reheat turbine are given by

Kp =120, Tp=20, Tt=0.3, Tg=0.08 and R=2.4

The second order identified model without droop characteristics is

$G_{p1}=\frac{120e^{-0.4626s}}{(28.4952s+1)(0.2202s+1)}$ and using the proposed tuning rules the controller settings are Kc=1.0326, Ti=1.2116 and Td=0.342.
The secondary loop controller is

$G_{c2}=(s+1)/(2s+4)$.

For nominal system, the frequency deviations and the corresponding control efforts are shown in Fig-4 and Fig-5.

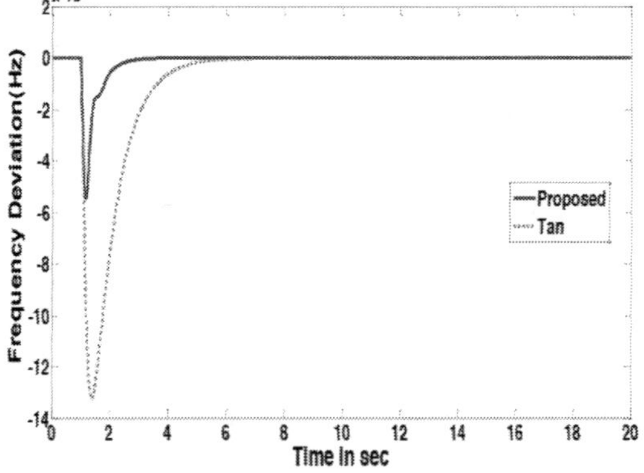

Fig. 4 Frequency deviation of the turbine without droop in nominal case

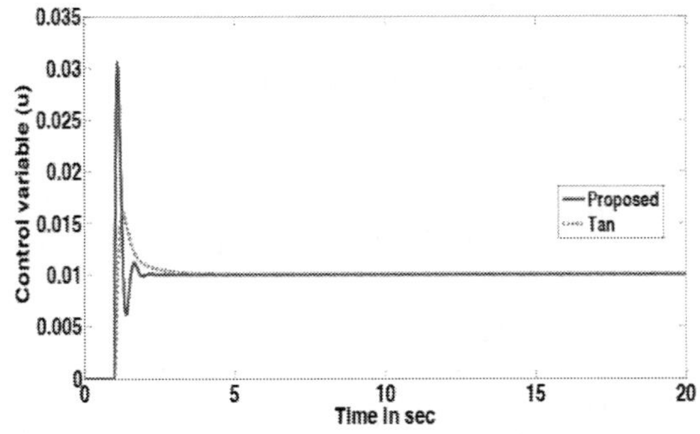

Fig. 5 Control Signal of the turbine without droop in nominal case

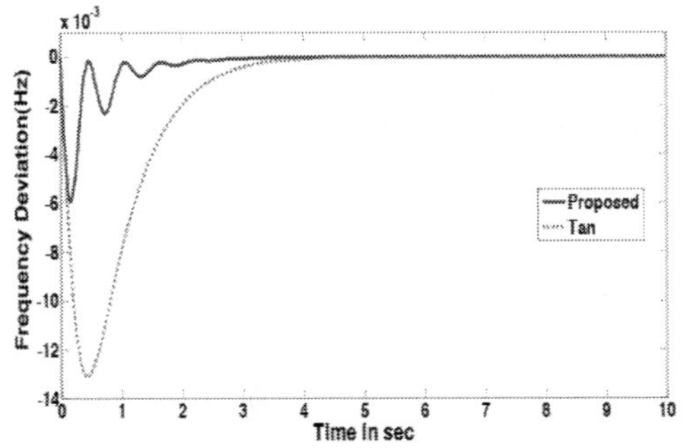

Fig. 6 Frequency deviation of the turbine without droop in perturbation case

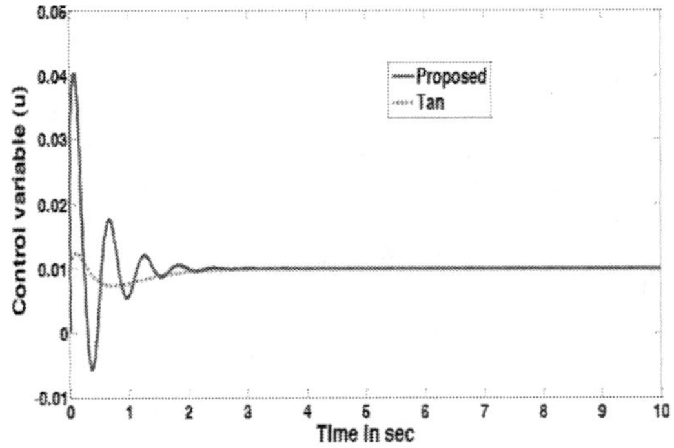

Fig. 7 Control Signal of the turbine without droop in perturbation case

The robustness of the proposed controller is tested by varying the plant parameters by 50%. For perturbed system, the frequency deviation and corresponding control effort are shown in Fig-6 and Fig-7 respectively. From the simulation results, it is seen that the proposed method gives closed loop performance than Tan's approach.

7 Conclusions

The issues of controlling the frequency and real power deviation have been tackled by proposing a cascade control scheme for single area power system. The main merit of the cascade control structure is that it improves regulatory response. The power system parameters have been varied to demonstrate the controllers are robust and stable. The complementary sensitivity method and IMC approach have been adopted to design the primary loop and secondary loop controller respectively. It is shown in the simulation results that the proposed control scheme gives improved results when compared with existing results in the literatures. The proposed approach can be readily extended to multi-input and multi-out power system.

References

1. Ibraheem, Prabhat Kumar, Dwarka P. Kothari., Recent Philosophies of Automatic Generation Control Strategies in Power Systems, IEEE Trans. On Power Systems, **20**, 1, (2005)
2. D.G. Padhan, S. Majhi, A new control scheme for PID load frequency controller of single-area and multi-area power systems, ISA Transactions, **52**, 242–251, (2013)
3. Shayeghi H, Shayanfar H, Jalili A., Load frequency control strategies: a state- of-the-art survey for the researcher. Energy Conversion and Management, **50**:344–53, (2009)
4. Agathoklis P, Hamza MH, Comparison of three algorithms for load frequency control, Elec. Power Sys. Research **7**,165–72, (1984)
5. Park YM, Lee KY. Optimal decentralized load frequency control, Elec. Power Sys. Research, **7**, 279–88, (1984).
6. Pan CT, Liaw CM. An adaptive controller for power system load frequency control. IEEE Tran. on Power Sys. **4**(1), 122–8, (1989)
7. Rubaai A, Udo V. An adaptive control scheme for load-frequency control of multiarea power systems Part I. Identification and functional design. Elec. Power Sys. Research, **24**,183–8 (1992).
8. Tan W., Unified tuning of PID load frequency controller for power systems via IMC. IEEE Trans. on Power systems **25**(1), 341–50, (2010)
9. Ibrahim Kaya, IMC based automatic tuning method for PID controllers in a Smith predictor configuration, Computers and Chemical Engineering **28**, 281–290, (2004)

DSTATCOM with Improved LMS based IRP theory

D. Suresh[1], *G.* Sravanthi[2] and *R.*Chander[3]

[12]Department of Electrical and Electronics Engineering, Vignan institute of Technology and Science, Deshmukhi
[3]Department of Electrical &Electronics Engineering, Osmania university college of Engineering, Hyderabad

Abstract. In this paper, grid connected DSTATCOM with three phase three wire has been proposed for harmonics elimination, reactive power compensation and active power injection into the grid. An improved instantaneous reactive power theory based on the least mean square is computed. The least mean square (LMS) algorithm is combined with instantaneous reactive power theory (IRPT) to improve its dynamics performance and eliminates the use of low pass filter requirement for computation of reference current. The harmonics component of the active power is estimated from LMS algorithm generated load harmonics currents. The performance characteristic of DSTATCOM is estimated with control scheme using computer simulation. The technique is based on the instantaneous transformation of instantaneous signals and taking advantage of the subsequent calculation of power from supply side to the loads. The neural network based method to estimate the harmonic online, DSTATCOM can compensate the harmonics. The computer simulation is carried out with MATLAB Simulink power system block sets.

1. Introduction

In recent time many researchers working on renewable energy source based power quality improvement in the electrical distribution system. There are different types of small power sources such as photovoltaic cell, fuel cell and wind energy [3–9].

The power electronics converter almost used in every consumer electronics and home appliances. The power electronics converters cause harmonics and flow of reactive current. Due to harmonics, AC supply tends to non-sinusoidal at point of common coupling. Also, current and voltage harmonics propagate into utility, which is not useful for developing active power. The generated harmonics causes overheating of conductor, interference with consumer, malfunction of sophisticated instrument and creates serious problems of electromagnetic interference and electromagnetic compatibility [1]. In literature, the performance investigation of active power filters based on different topologies and control schemes are widely studied. Mostly, the shunt active power filter is used for compensation of harmonics and reactive power demand of the load. The shunt active power filters also known as dynamic static compensator (DSTATCOM). The DSTATCOM is the family of custom power devices, which is used to mitigate the abnormalities due to the current harmonics and it is connected in parallel with load. Unified power quality controller (UPQC), dynamic voltage restorer (DVR), dynamic static compensator etc. are the family of custom power devices. The load contributes the multiple problems such as current harmonics, reactive power burden, voltage harmonics and these problems can be compensated with custom power devices. In addition, DSTATCOM can also be used for integration of the renewable energy sources with grid [2]-[5].

In literature, many numbers of controllers are devised for control of voltage and current of DSTATCOM inverter [3]. The DSTATCOM is connected in shunt with load to compensate the harmonics and reactive power demand of the nonlinear load. This task is accomplished by inject compensating current into the utility system via buffer inductors. In fact, several control methods have been reported in the literature, among them conventional controllers and artificial intelligent controllers are well established. Some of them are model reference adaptive controller, sliding mode control [4], fuzzy control [6, 7], adaptive neural network control [8] and neuro-fuzzy control [9].

However, these controllers can improve the dynamic performance as well as mitigate the harmonics and reactive power caused by the nonlinear load. Most of the controllers are used for regulating dc voltage of DSTATCOM and the training data is obtained from the conventional proportional plus integral controllers.

In literature many control schemes were proposed to compute the reference current of the DSTATCOM to track desired source current. Among them, the most widely used schemes are IRPT and synchronous reference frame theories. Conversely, these control strategies requires more number of transformation to estimate the compensating currents [6, 7, 10][11]. Consequently, the numbers of transformation also requires more number of multipliers and increase the level complexity for estimation of reference currents to the DSTATCOM [9].

In this article, control scheme is proposed to the integrate the least mean square with IRPT scheme. This control strategy is used for computations of desired current for tracking the source current of diode clamped multilevel inverter (DCMLI) based DSTATCOM [6]. However, conventional control scheme implantation on the digital platform is less complex and can achieve good results. Moreover, conventional IRPT require low pass filter for estimation of oscillating active power. The oscillating active power along with reactive power is used for estimation of compensating current of DSTATCOM. The low pass filter which is used for estimation of oscillating power is not adaptive and parameter adjustment of low pass filter is tedious task. In the proposed control scheme, the low pass filter is replaced with LMS algorithm to estimate the harmonics active power and subsequent calculation of the reference current for the DSTATCOM.

2. Topology of three level DSTATCOM

The DCMLI based DSTATCOM is shown in Fig.1. The DSTATCOM circuit consists of interfacing inductor, dc link capacitor and twelve insulated gate bipolar transistor. The DSTATCOM is connected at load side to compensate the abnormalities caused by the load such as harmonics and reactive power burden. The photo voltaic modules are integrated on the DC side of the DSTATCOM.

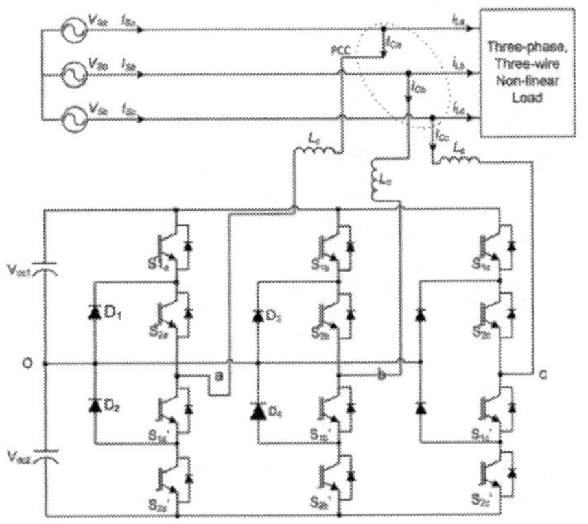

Fig. 1. DCMLI based DSTATCOM.

3. Photo voltaic cell

The approximate constant current model of the Photo voltaic cell is shown in the Fig.2.

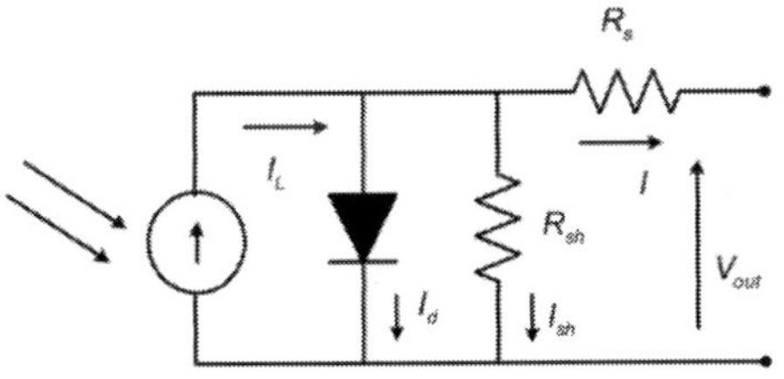

Fig. 2. Photo voltaic cell model.

The equation, which is used to define I-V characteristics photo voltaic cell is as follows

$$I = I_L - I_d - I_{sh} = I_L - I_D\left[e^{\frac{qV_{oc}}{CkT}} - 1\right] - \frac{V_{out} + R_S I}{R_{sh}} \quad (1)$$

Where I_D is the saturation current, q is the electron charge $\left(1.6\times10^{-19}\,C\right)$, C is the diode emission factor, k is the Boltzmann constant $\left(1.38\times10^{-23}\ \mathrm{J/K}\right)$ and T is the temperature.

The maximum power generated from the photo voltaic cell is at the knee point of the I-V curve as it can be seen from Fig.3.

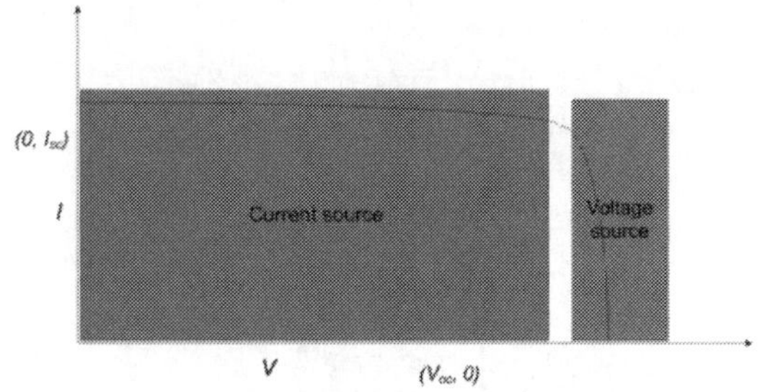

Fig. 3. V-I characteristics of photo voltaic cell.

The maximum power is the product of maximum value of voltage and current. The DSTATCOM is set to operate at optimal voltage to achieve maximum power by maximum power point tracking algorithm.

In this paper, the incremental conductance and integral regulator technique used for tracking the maximum power from the photo voltaic modules [6]. This MPPT method is based on the fact that the power slope of the PV is null at MPP point (Where $dp/dv = 0$), positive in the left, and negative in the right. In the following equations, *dv* and *di* are obtained by one-sample delayed values V.

$$\frac{dp}{dv} = \frac{d(vi)}{dv} = i + v\frac{di}{dv} = 0 \tag{2}$$

$$\frac{dv}{di} = -\frac{i}{v} \tag{3}$$

$$\begin{cases} \frac{dv}{di} > \frac{i}{v} : left \\ \frac{dv}{di} < \frac{i}{v} : right \end{cases} \tag{4}$$

I. CONTROL SCHEME

The instantaneous reactive power theory is based on the transformation of quantities and subsequent calculation of real and reactive power (*p-q* or IRP theory). It is the based on time domain scheme. The application of IRPT can be extended to the three phase or four phase with small modification in computation of reference theory.

Let the system terminal voltages be given as,

$$\begin{aligned} v_{Sa} &= v_m \cos(\omega t) \\ v_{Sb} &= v_m \cos\left(\omega t - \frac{2\pi}{3}\right) \\ v_{Sc} &= v_m \cos\left(\omega t - \frac{4\pi}{3}\right) \end{aligned} \tag{1}$$

The *p-q* theory starts with the transformation of three-phase voltage and currents into $\alpha\beta0$ stationary reference frame. The Clarke transformation maps the three-phase instantaneous voltages in the *abc*-phases, v_{Sa}, v_{Sb}, and v_{Sc}, into the instantaneous voltages on the $\alpha\beta0$-axes v_α, v_β, and v_0 as,

$$\begin{bmatrix} v_0 \\ v_\alpha \\ v_\beta \end{bmatrix} = \sqrt{\frac{2}{3}} \begin{bmatrix} \frac{1}{\sqrt{2}} & \frac{1}{\sqrt{2}} & \frac{1}{\sqrt{2}} \\ 1 & -\frac{1}{2} & -\frac{1}{2} \\ 0 & \frac{\sqrt{3}}{2} & -\frac{\sqrt{3}}{2} \end{bmatrix} \begin{bmatrix} v_{Sa} \\ v_{Sb} \\ v_{Sc} \end{bmatrix} \tag{5}$$

Similarly, three-phase instantaneous compensating currents in the *abc*-phases, i_{ca}, i_{cb}, and i_{cc}, are transformed into the instantaneous currents on the $\alpha\beta0$-axes i_α, i_β, and i_0 as,

$$\begin{bmatrix} i_0 \\ i_\alpha \\ i_\beta \end{bmatrix} = \sqrt{\frac{2}{3}} \begin{bmatrix} \frac{1}{\sqrt{2}} & \frac{1}{\sqrt{2}} & \frac{1}{\sqrt{2}} \\ 1 & -\frac{1}{2} & -\frac{1}{2} \\ 0 & \frac{\sqrt{3}}{2} & -\frac{\sqrt{3}}{2} \end{bmatrix} \begin{bmatrix} i_{ca} \\ i_{cb} \\ i_{cc} \end{bmatrix} \tag{6}$$

As no zero-sequence component exists in the 3P3W systems, v_0 and i_0 can be eliminated in equations and, respectively.Elimination of v_0 from equation leads to,

$$\begin{bmatrix} v_\alpha \\ v_\beta \end{bmatrix} = \sqrt{\frac{2}{3}} \begin{bmatrix} 1 & -\frac{1}{2} & -\frac{1}{2} \\ 0 & \frac{\sqrt{3}}{2} & -\frac{\sqrt{3}}{2} \end{bmatrix} \begin{bmatrix} v_{Sa} \\ v_{Sb} \\ v_{Sc} \end{bmatrix} \tag{7}$$

Similarly, elimination of i_0 from equation (4) leads to,

$$\begin{bmatrix} i_\alpha \\ i_\beta \end{bmatrix} = \sqrt{\frac{2}{3}} \begin{bmatrix} 1 & -\frac{1}{2} & -\frac{1}{2} \\ 0 & \frac{\sqrt{3}}{2} & -\frac{\sqrt{3}}{2} \end{bmatrix} \begin{bmatrix} i_{ca} \\ i_{cb} \\ i_{cc} \end{bmatrix} \tag{8}$$

The instantaneous value of active and reactive expressed as follows

$$p = v_\alpha i_\alpha + v_\beta i_\beta \tag{9}$$

$$q = -v_\beta i_\alpha + v_\alpha i_\beta \tag{10}$$

$$\begin{bmatrix} p \\ q \end{bmatrix} = \begin{bmatrix} v_\alpha & v_\beta \\ -v_\beta & v_\alpha \end{bmatrix} \begin{bmatrix} i_\alpha \\ i_\beta \end{bmatrix} \tag{11}$$

The instantaneous value of active and reactive powers *p* and *q* can be computed from IRPT scheme. Further decomposed as DC and AC power and as follows

$$\begin{bmatrix} p \\ q \end{bmatrix} = \begin{bmatrix} \bar{p} + \tilde{p} \\ \bar{q} + \tilde{q} \end{bmatrix} \tag{12}$$

The DSTATCOM is used to compensate the alternating component of the active power. The oscillatory component of active power obtained with low pass filter $\tilde{p} = \bar{p} - p$. In addition to this, DSTATCOM has to compensate reactive power $q = \bar{q} + \tilde{q}$. This can be calculated directly from the transformed quantities. The compensating currents of the DSTATCOM are calculated in $\alpha\beta 0$-reference frame as

$$\begin{bmatrix} i_{C\alpha}^* \\ i_{C\beta}^* \end{bmatrix} = \frac{1}{v_\alpha^2 + v_\beta^2} \begin{bmatrix} v_\alpha & v_\beta \\ v_\beta & -v_\alpha \end{bmatrix} \begin{bmatrix} -\tilde{p} \\ -(\bar{q} + \tilde{q}) \end{bmatrix} \quad (13)$$

The significance of the negative sign is used to specify the action of DSTATCOM. This sign indicate the DSTATCOM should draw the compensating current, which is in phase opposition to the load current. Let P_{loss} be the power required meet to temporary load demand and to overcome the capacitive power leakage losses. The compensating power P_{loss} can be obtain by comparing the reference voltage with measured value and processed in proportional integral controller. As a result, equation becomes,

$$\begin{bmatrix} i_{C\alpha}^* \\ i_{C\beta}^* \end{bmatrix} = \frac{1}{v_\alpha^2 + v_\beta^2} \begin{bmatrix} v_\alpha & v_\beta \\ v_\beta & -v_\alpha \end{bmatrix} \begin{bmatrix} -p_C + p_{Loss} \\ -q_C \end{bmatrix} \quad (14)$$

These compensating currents in $\alpha\beta o$ frame can be used to compute the *abc* quantities using inverse Clarke's transformation as,

$$\begin{bmatrix} i_{Ca}^* \\ i_{Cb}^* \\ i_{Cc}^* \end{bmatrix} = \sqrt{\frac{2}{3}} \begin{bmatrix} 1 & 0 \\ -\frac{1}{2} & \frac{\sqrt{3}}{2} \\ -\frac{1}{2} & -\frac{\sqrt{3}}{2} \end{bmatrix} \begin{bmatrix} i_{C\alpha}^* \\ i_{C\beta}^* \end{bmatrix} \quad (15)$$

The control scheme based on IRPT is shown in Fig.2.

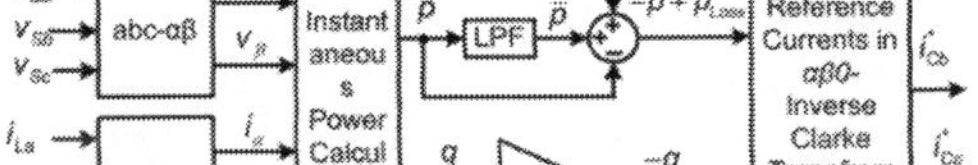

Fig. 4. Instantaneous reactive power theory.

4. Modified PQ theory based on LMS algorithm

The load current can be expressed as

$$i_L = I_{m1}\cos(\omega t - \phi) + \sum_{n=2}^{\infty} I_{mn}\cos(n\omega t - \phi_n) \quad (16)$$

The source voltage harmonics and expressed as

$$v_s = V_m \cos(\omega t) \quad (17)$$

This can be further written as

$$v_s = V_m \cos(-\omega t) \quad (18)$$

The instantaneous active power can be calculated from frequency analysis as

$$p(t) = V_m I_{m1}\cos(\phi_1) + \sum_{n=2}^{\infty} V_m I_{mn}\cos(\omega t(n-1) - \phi_n) \quad (19)$$

The value of instantaneous active power can be expressed as

$$p(t) = p_{m1}\cos(\phi_1) + \sum_{n=2}^{\infty} p_{mn}\cos(\omega t(n-1) - \phi_n) \quad (15)$$

$$p(t) = \bar{p}(t) + \tilde{p} \quad (20)$$

The oscillating component of the active power can be expressed as

$$\tilde{p} = \bar{p}(t) - p(t) \quad (21)$$

Where

$$W = [\cos(\phi_1)\sin(\phi_2)\cos(\phi_2)\ldots\sin(\phi_n)\cos(\phi_n)]$$
$$X^T = [1\cos(\omega t)\sin(\omega t)\ldots\cos(N\omega t)\sin(N\omega t)]^T$$

The LMS algorithm used for estimation of the harmonics component of active power is expressed as

$$W_{(n+1)} = W_n + \eta(d - Y)X \quad (22)$$

https://doi.org/10.1051/e3sconf/20198701013

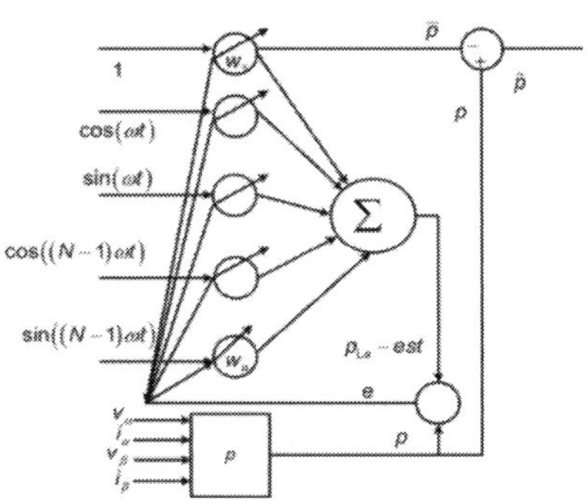

Fig. 5. Estimation harmonics component of active power.

Where, η is the constant also known as learning rate. The convergence value of learning rate is from 0.1 to 1.0. The control diagram is shown in Fig.2. The LMS algorithm eliminates the requirement of low pass filter in improved IRPT. The LMS based estimation of oscillatory component of the current is depicted in Fig.3. And control scheme based on LMS integrated with IRPT is shown in Fig.4.

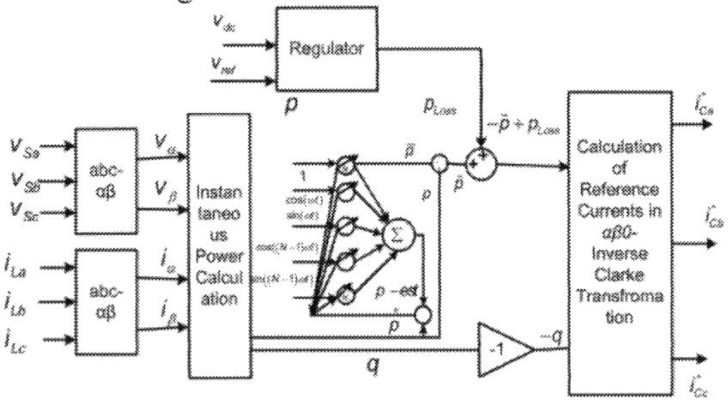

Fig. 6. LMS based IRPT control scheme.

5. Simulation results and discussion

The simulation model of the DSTATCOM and its control scheme is implemented with MATLAB/Simulink power system block sets. The nonlinear load is implemented with diode bridge rectifiers. The diode bridge rectifier generates the waveforms which is seems to be in stepped waveform in shape. The stepped waveform contains triplen harmonics current. The instantaneous value of the load voltage and currents are used to estimate the compensating currents with IRPT scheme.

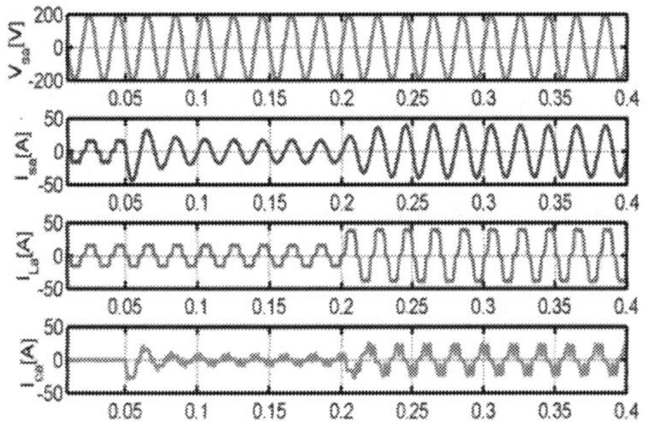

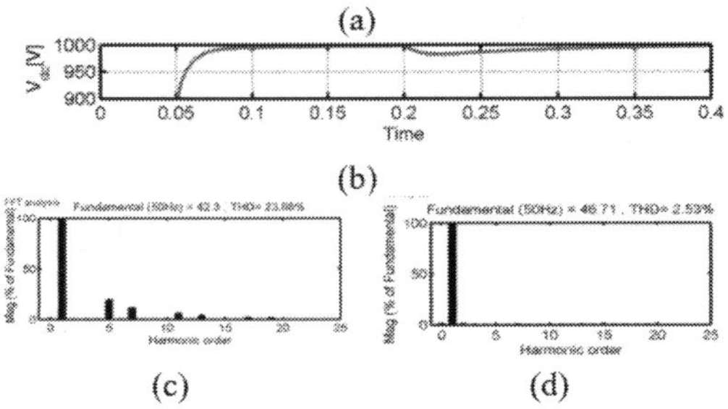

Fig. 7. Performance of the DSTATCOM with LMS based IRPT.

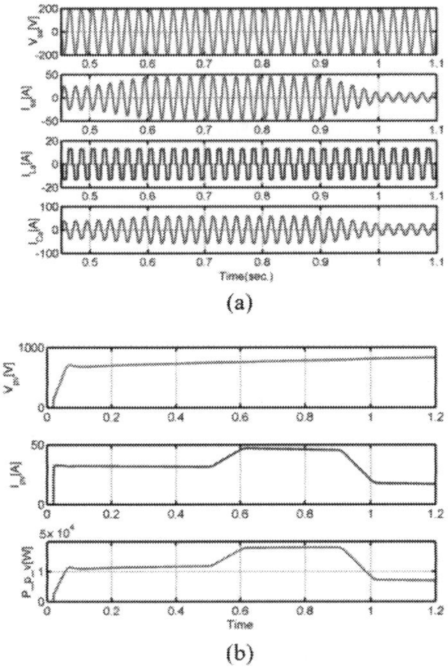

Fig. 8. DSTATCOM waveform with photo voltaic module

The diode bridge rectifier used as nonlinear load. The rectifier circuit consists of commutation inductance on input side and resistive-inductive elements on load side as non-linear load. The characteristics DSTATCOM while feeding the non-linear load is depicted in Fig. 5 (a)-(d). The waveforms of DSTATCOM characteristics are obtained with MATLAB simulations. The waveforms are identified as follows: source-voltage, source-current, load-current, compensating-current, dc link voltage across the capacitors. The control scheme under consideration is the LMS-IRPT for DSTATCOM. After the compensation with DSTATCOM current-waveform is tends to sinusoidal from stepped waveform. The power factor of source is almost unity.

After compensation with DSTATCOM, Source phase current is tends to sinusoidal. The source power factor is almost unity and source phase current is in phase with source voltage. At the instant DSTATCOM is ON, the source current THD reduced to the value 3.41% as depicted in Fig.5 (d) with LMS-IRPT scheme. But the load current THD remains at value 26.7%. The DC link side voltage of DSTATCOM is controlled at the reference levels with LMS-IRPT based schemes. The transient study shows that the DSTATCOM have settling time such as 0.05 for LMS-IRPT scheme. As the load changes at t=0.3s, from 20.2A to 40A.As load current changes, then the equivalent change in source current and dc link voltage is detected. This leads to the increase in current injected by the DSTATCOM, which are results in decrease of DC link capacitor voltages. This decrease in voltage is used to compensate the load demand. The change in voltage is detected from the Fig 5. (b) Conversely, the increase of current injected by the DSTATCOM is gradually in case of the LMS-IRPT.

The current, voltage and power waveform shown in Fig.6. (a) and Fig.6.(b) respectively. At the instant when DSTATCOM is switched ON, the active power generated in photo voltaic cells is injected into the grid. These are observed from variation in source currents and compensating current of the DSTATCOM. The maximum power also tracked from the using incremental conductance method. and variation in the power also shown in Fig.6(b).

6. Conclusion

The LMS based IRPT control scheme is derived for control of DSTATCOM to mitigate the abnormalities caused by nonlinear load. The LMS algorithm eliminates the uses of the low pass filter in the IRPT theory. The LMS algorithm is used for estimation of the oscillatory component of the active power and subsequent computation of reference current. The simulated response shows that the LMS-IRPT exhibits the better compensation characteristics and active power injection from the photo voltaic cells

Acknowledgments

Thanks to the Science and Engineering board and DST government of India for providing grant under ECR/2016/000813.

Reference

1. B. Singh, K. Al-Haddad and A. Chandra, "A review of active filters for power quality improvement," in *IEEE Tran., on Ind. Electron.*, vol. 46, no. 5, pp. 960-971, Oct 1999.
2. D. M. Divan, S. Bhattacharya, and B. Banerjee, "Synchronous frame harmonic isolator using active series filter," in Proc. Eur. Power Electron. Conf., , pp. 3030–3035, 1991.
3. A. Chandra, B. Singh, B. N. Singh and K. Al-Haddad, "An improved control algorithm of shunt active filter for voltage regulation, harmonic elimination, power-factor correction, and balancing of nonlinear loads," in *IEEE Transactions on Power Electronics*, vol. 15, no. 3, pp. 495-507, May 2000.
4. B. Singh and J. Solanki, "An Implementation of an Adaptive Control Algorithm for a Three-Phase Shunt Active Filter," in IEEE Transactions on Industrial Electronics, vol. 56, no. 8, pp. 2811-2820, Aug. 2009.
5. H. Li, Z. Wu and F. Liu, "A Novel Variable Step size Adaptive Harmonic Detecting Algorithm Applied to Active Power Filter," 2006 IEEE International Conference on Industrial Technology, Mumbai, 2006, pp. 574-578.
6. B. Singh, Sunil Kumar Dube, Sabha Raj Arya, A. Chandra and K. Al-Haddad, "A comparative study of adaptive control algorithms in Distribution Static Compensator," IECON 2013 - 39th Annual Conference of the IEEE Industrial Electronics Society, Vienna, , pp. 145-150, 2013.
7. Dhanavath Suresh, S.P. Singh, "design of single input fuzzy logic controller for shunt DSTATCOM," IETE journal of research, vol.61, no.5, 2016.
8. H. Akagi, Y. Kanazawa, A. Nabae, "Generalized theory of the instantaneous reactive power in three-phase circuits", Proc. IEEE JIEE IPEC, pp. 821-827, 1983.
9. B. Singh, J. Solanki, "An implementation of an adaptive control algorithm for a three-phase shunt active filter", IEEE Trans. Ind. Electron., vol. 56, no. 8, pp. 2811-2820, Aug. 2009.
10. Malesani, L., Rossetto, L., and Tenti, P.: 'Active filter for reactive power and harmonics compensation'. IEEE Power Electronics Specialist Conf., Vancouver British Columbia, Canada, , pp.321–330, 1986.

E3S Web of Conferences **87**, 01014 (2019) https://doi.org/10.1051/e3sconf/20198701014
SeFet 2019

Power Savings with all SiC Inverter in Electric Traction applications

B.K. Chakravarthy[1], and *G.* Sree Lakshmi[2]

[1]Research Scholar, Department of Electrical Engineering, University College of Engineering, Osmania University
[2]Member, IEEE

Abstract. The advantage of Silicon Carbide (SiC) based devices are less thermal management requirements and smaller passive components which result in higher power density. SiC devices have higher blocking voltages, lower on-state resistance and switching losses and higher thermal conductivity and operating temperatures. SiC devices can operate at higher voltages, higher frequencies and higher junction temperatures than comparable Si devices, which results in significant reduction in weight and size of the power converter and increase in system efficiency.

1 Introduction

The performance and efficiency of traction systems was improved using high efficiency induction motors and has been further enhanced by advances in power devices. Inverters and other traction systems become smaller by using low-loss power devices. Low-loss power devices use Silicon Carbide (SiC) instead of Silicon (Si). SiC has approximately 10 times the critical breakdown strength of Silicon. The thickness of drift layer in SiC is approximately one-tenth of that in Si [6]. This allows for dramatic reduction in conductivity loss and switching loss in SiC devices [9]. SiC has three times the band-gap width of Si which prevents the flow of leakage current and enables operation at high temperatures.

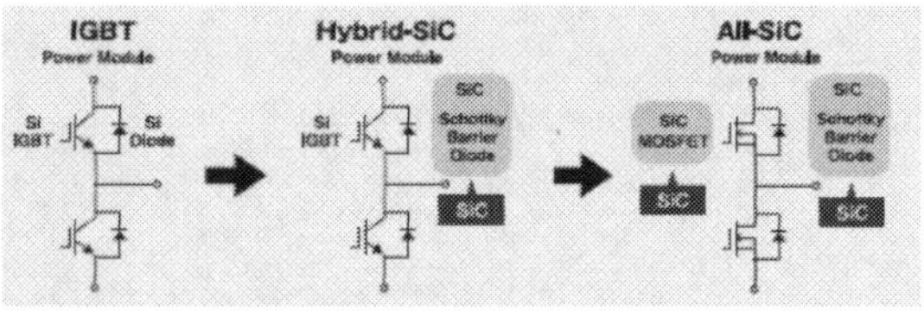

Fig. 1. Evolution in devices

1.1 Hybrid SiC and All SiC

1.1.1 High speed switching operation

A Hybrid – SiC module is shown in figure 1 It uses Schottky Barrier Diodes (SBDs). High speed switching can be realized with SBDs as they don't have accumulation carriers. The all SiC module is very compact and has high speed switching capability [1]. It is possible to reduce the size of additional components like reactors due to high speed switching.

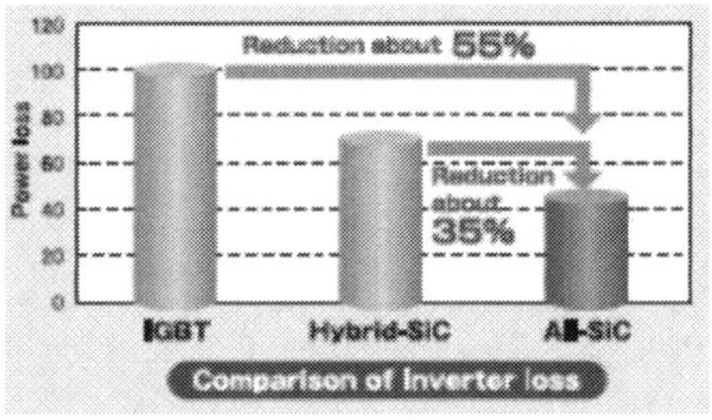

Fig. 2. Power savings in all SiC device.

1.1.2 Reduction in power loss

SiC IGBT modules with SBDs when used with active gate control techniques can reduce the inverter losses by about 35% [2]. To increase the efficiency of drive system in traction applications, the area of regenerative braking is to be expanded by minimising the use of pneumatic brake. Losses due to harmonics in induction motors can be reduced by optimizing induction motor design and PWM control techniques.

Figures 3 and 4 show the braking force and power fed back during regenerative braking of a locomotive using a conventional IGBT based inverter and those for a two-level inverter using all SiC power modules. Table 1 compares the losses in conventional IGBT modules and all SiC based modules. Figures 5 compares the turn-on characteristics and Figure 6 compares the turn-off characteristics of SiC and IGBT. The reduction in losses leads to reduction in the size of cooling systems. Traction inverters with all SiC modules are therefore smaller and lighter.

* Corresponding author: *B.K.* Chakravarthy

 https://doi.org/10.1051/e3sconf/20198701014

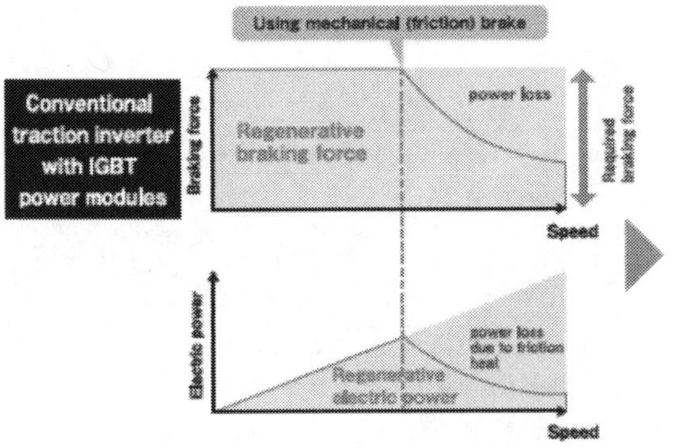

Fig. 3. Regenerative braking with IGBT modules

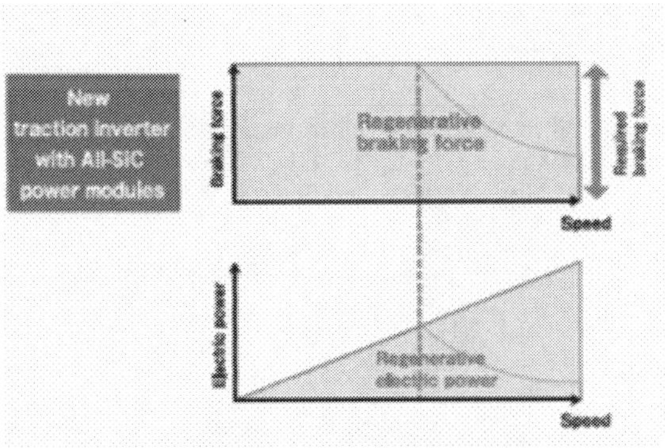

Fig. 4. Regenerative braking with SiC module

Table 1. Difference in switching of SiC and IGBT

	SiC	IGBT
Turn -on delay time (ns)	45	120
Turn – on rise time (ns)	50	80
Turn - off delay time (ns)	170	450
Turn – off fall time (ns)	60	350
Reverse recovery time (ns)	30	150
Turn - on loss (mJ)	2.7	4.5
Turn -off loss (mJ)	1.7	10
Recovery loss (mJ)	0.1	8

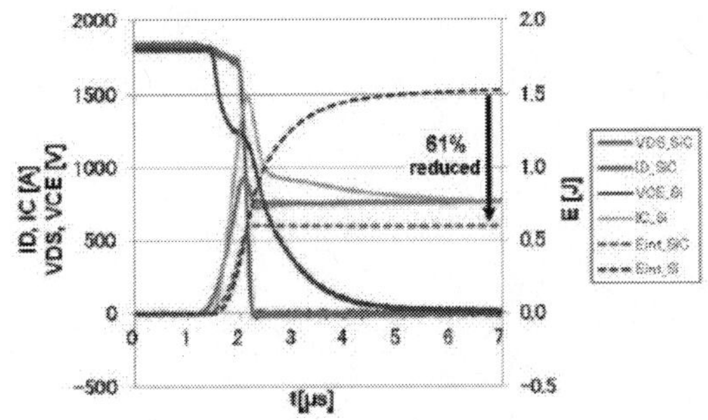

Fig. 5. Turn-on characteristics

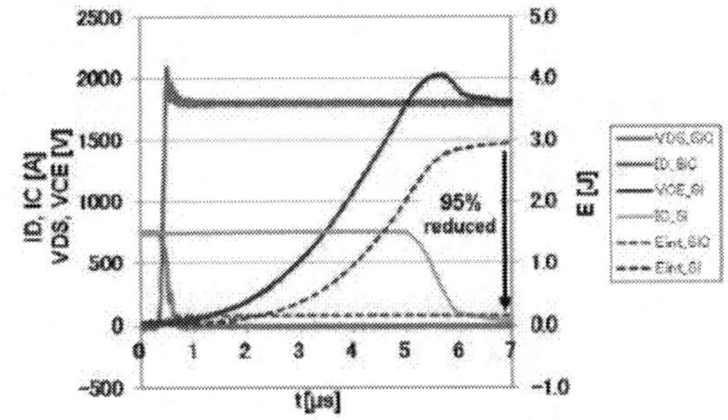

Fig. 6. Turn-off characteristics

2 Reduction in inverter losses

Losses can be reduced by as much as 35% when using all SiC based modules. The reduction in losses is more predominant [3] as the switching frequency increases as shown in figure 7.

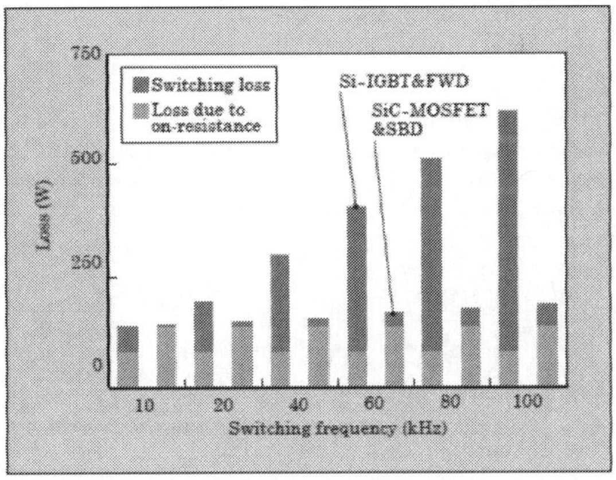

Fig. 7. Variation of losses with switching frequency

The inverter usually switches between three techniques of PWM depending on the speed. A combination of synchronous and asynchronous PWM can be used in the low and medium speed range. The control range for asynchronous PWM can be expanded for inhibiting lower order harmonics through optimization of carrier frequency. Low distortion synchronous PWM control can be used to reduce current harmonics in the upper end of medium speed range. In the high-speed range, pulse traction system can be used to reduce lower order harmonics. These techniques are shown in figure 8.

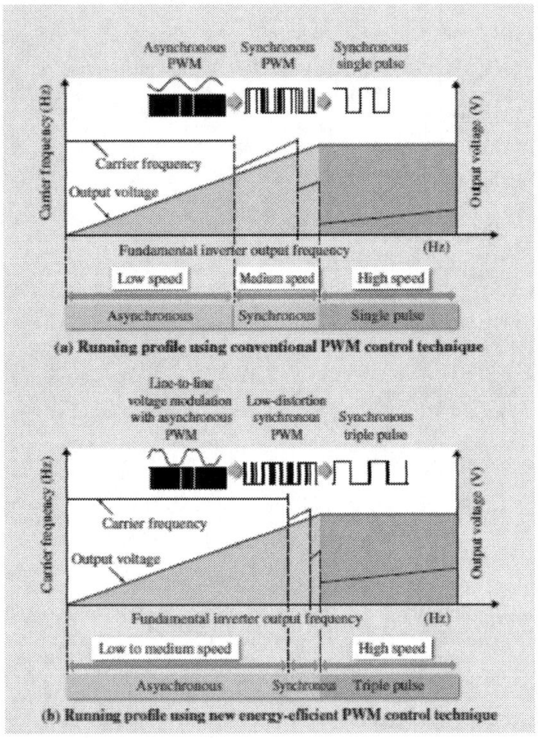

Fig. 8. Modulation techniques used with traction converter

2.1 Application in Metro trains

Metro train in an Electric Multiple Unit (EMU). EMU is an example of distributed traction system. Each unit typically has three cars. Multiple units can be added or removed depending on operating requirements as shown in figure 9.

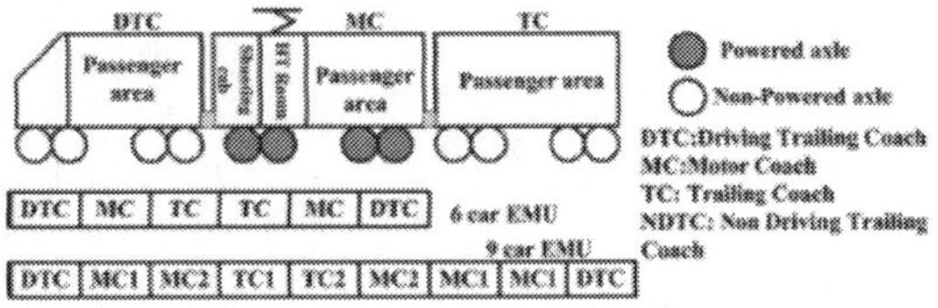

Fig. 9. Electric Multiple Unit (EMU)

Typical power architecture of EMU [2] is shown below in figure 10.

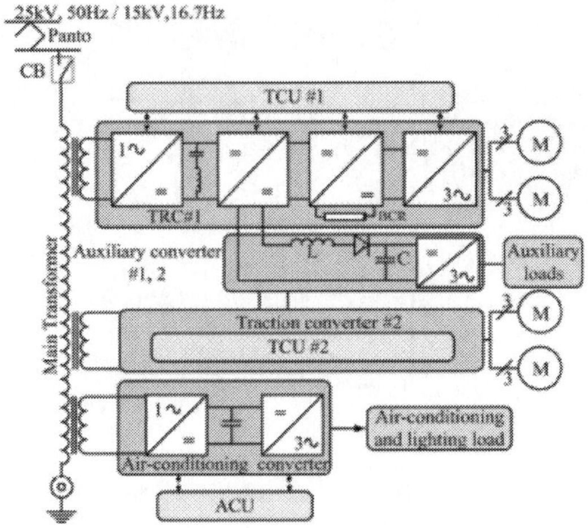

Fig. 10. EMU architecture

When SiC modules are used in inverter the reduction in the size of the inverter [4] [8] [10] is shown in the figure 11 below. There is also a significant reduction in weight.

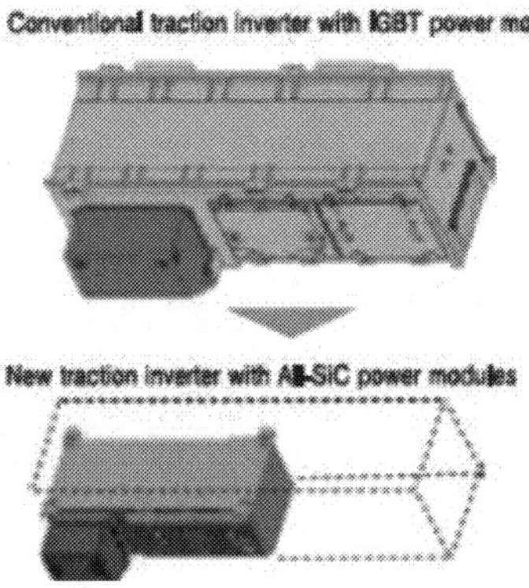

Fig. 11. Size reduction in traction converter

Losses and efficiency are compared for 80KW traction inverter: [5] [7]

Topology: Three phase inverter

Synchronous rectification (SiC version)

DC-link voltage: 400V DC

Current 480Arms (peak) 230Arms (nom)

Switching frequency: 16kHz

Vgs=+20V/-5V for SiC, Vge=±15V for IGBT

Cos(phi): 0.8

Modulation index (MI): 1

https://doi.org/10.1051/e3sconf/20198701014

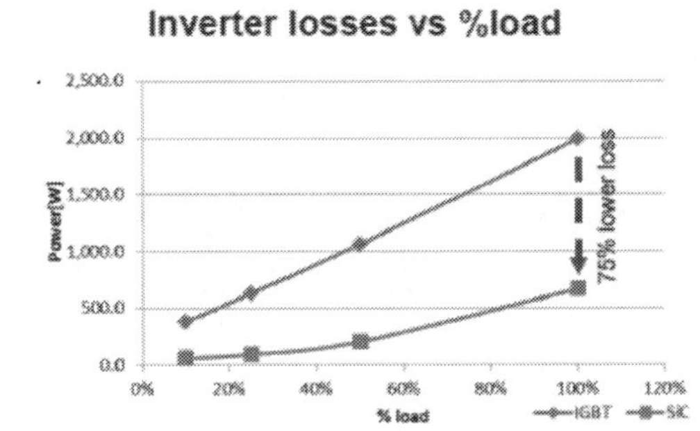

Fig. 11. Comparison of inverter losses

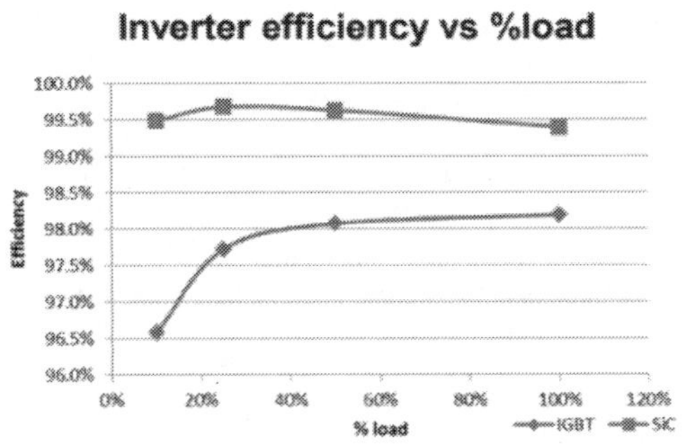

Fig. 12. Comparison of inverter efficiency

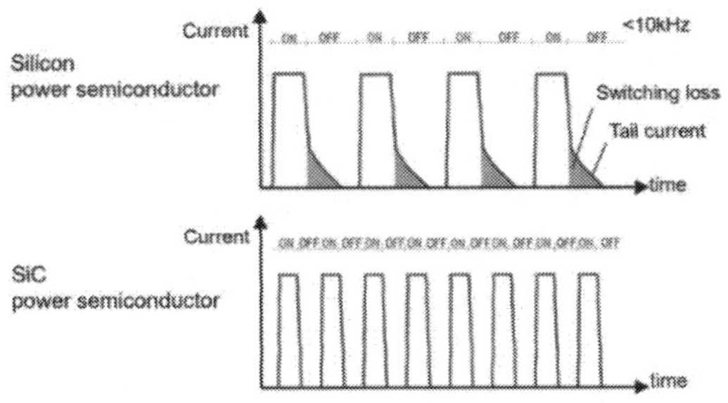

Fig. 13. Low switching loss illustration in SiC

The current waveforms in figure 13 clearly illustrate the low switching loss. As shown in Figure 14, low losses even in high frequency operation helps in reducing the size of the inverter and improves efficiency.

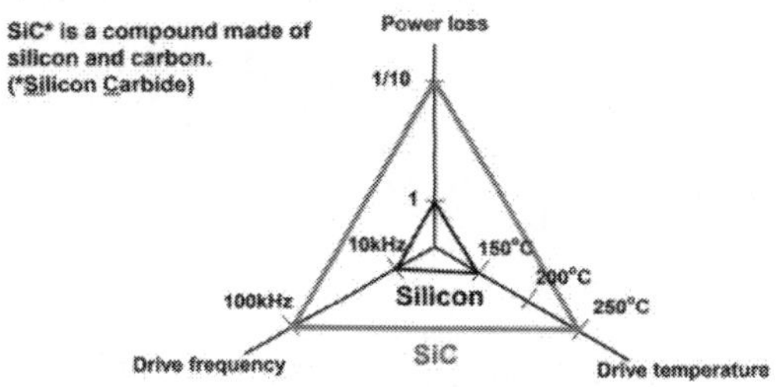

Fig. 14. Improved efficiency by using SiC

Conclusion

Silicon Carbide (SiC) based devices will have a large impact on inverter efficiency and thermal management issues in traction applications. SiC has clear operational advantage over conventional devices. SiC based inverters are also being developed for use in Electric Vehicle and Hybrid Electric Vehicle (HEV) applications. These inverters help in increasing the range of the vehicle and in reducing the weight of the vehicle.

References

1. K. Ishikawa, K. Terasawa, T. Sakai, S. Sugimoto, T. Nishino, Development of rolling stock inverters using SiC, Hitachi Review. **66**, 2 (2017)
2. D. Ronanki, S.A. Singh, S.S. Williamson, Comprehensive Topological Overview of Rolling Stock Architecture and Recent Trends in Electric Railway traction Systems. IEEE Trans. on Transportation Electrification, **3**, 3 (2017)
3. N. Hayato, H. Yuichiro, *H.* Masafumi. Ultra-compact, High-reliability, All-SiC Module, Fuji Electric Review, **59**, 4 (2013)
4. www.mitsubishielectric.com
5. www.st.com
6. P. Shamsi, M.McDonough and B. Fahimi, Wide-Bandgap Semiconductor Technology: Its impact on the electrification of the transportation industry., in IEEE Electrification Magazine,**1**, 2 (2013).
7. M. M. Swamy, Jun-Koo Kang, K. Shirabe, Power Loss, System Efficiency, and Leakage Current Comparison Between Si IGBT VFD and SiC FET VFD With Various Filtering Options, in IEEE Transactions on Industry Applications, **51**, 5 (2015).
8. Z. Zhang, F. Wang, L. M. Tolbert, B. J. Blalock and D. J. Costinett, Evaluation of Switching Performance of SiC Devices in PWM Inverter-Fed Induction Motor Drives, in IEEE Transactions on PowerElectronics, **30**, 10 (2015).
9. I. Josifović, J. Popović-Gerber and J. A. Ferreira, Improving SiC JFET Switching Behavior Under Influence of Circuit Parasitics, in IEEE Transactions on Power Electronics, **27**, 8 (2012).
10. A. Merkert, T. Krone and A. Mertens, Characterization and Scalable Modeling of Power Semiconductors for Optimized Design of Traction Inverters with Si- and SiC-Devices, in IEEE Transactions on Power Electronics, **29**, 5 (2014).

A Hybrid Cascaded Multilevel Converter for Battery Vitality Administration Connected In Electric Transport for Battery Energy Management

Srinivas Gadde[1], *Bhanu Ganesh* Lukka[2] and *Shankar Mahesh Kumar* Gali[3]

[123]Department of Electrical Engineering, Vignan Institute Of Technology and Science (VITS) Deshmukhi, Yadadri,

Abstract. In the (EV) electrical vehicles energy storage system plays a crucial role. Normally, in electrical vehicles large number of cells connected in series to increase the output voltage for driving motor. Battery cells will have different electrochemical characteristics which cause the difference in terminal voltage (or) (SOC) state of charge imbalances between the each cell. In this paper cascaded multilevel converter which involves in both management of energy and motor drives proposed for electric vehicles and each battery cell can be controlled to the circuit or bypassed by a half-bridge converter. All these half-bridges are cascaded to the output staircase shape dc voltage. And H-bridge converter will be used to change the direction of dc bus voltage to convert it ac voltage. The advantage of the converter will have multilevel voltages with small dv/dt and lesser harmonics. So, it is helpful to make better performance of motor drives. Through separate control according to state of charge of each cell the energy utilization of batteries can be improved drastically. By using the fuzzy logic controller for current measurement the THD can be reduced drastically. In order to verify the performance of proposed converter simulation done in matlab.

1 Introduction

In electrical vehicles, energy storage system plays a crucial role. Lead–acid batteries or lithium batteries are most popular one because of their suitable energy density and cost. Due to low voltage of these kind of batteries need to be connected in series to get appropriate voltage requirement for driving motor [1-2]. Normally, there are two kinds of equalization circuits are used. First one which consumes the unused energy on parallel resistance, to keep the voltage of all cells equal. Second kind of equalization circuit, it composed of a group of transformers or inductances and converters, which can realize transfer of energy between cells. The energy in cells with more terminal voltage or state of charge can transferred to other cells to realize state of charge and voltage equalization [1]. The demerit is that it require lot of inductances and isolated multiple winding transformers in this topology and it leads to complexity in operation of converter [5]-[12]. So, some studies have been performed to simplify the circuit and also to improve balance speed by multiphase equalization [9]-[13]. Such as zero current and zero voltage switching also performed to reduce the loss of equalization circuit [13].

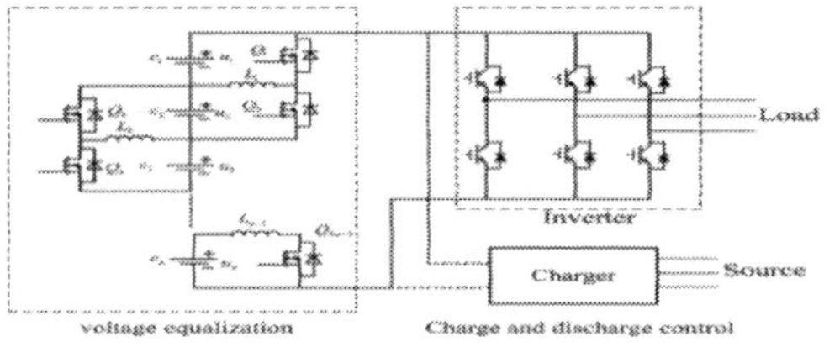

Fig.1. Obsolete power storage system with inverter and voltage equalization circuit.

Now-a-days multi converters are mostly used in the medium voltage and high voltage motor drives [5]-[9]. If their isolated direct current sources or flying capacitors are replaced by battery cells and these battery cells are cascaded in series combining with converters instead of connecting directly in series. [13]-[14].

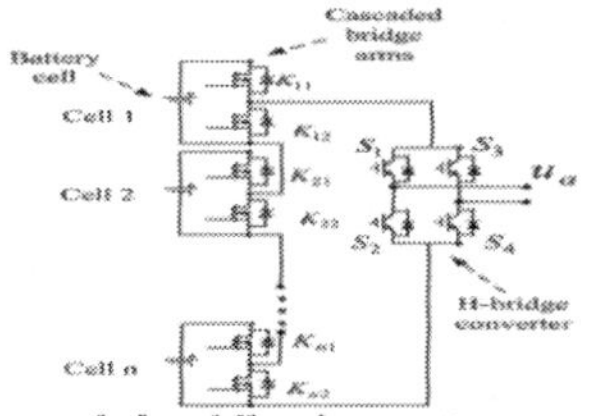

Fig.2. Hybrid cascaded multilevel converter.

* Corresponding author: gadde.cnu@gmail.com

https://doi.org/10.1051/e3sconf/20198701015

The proposed Hybrid cascaded multilevel converter in this paper, it can realize the SOC or terminal voltage between the cells. The proposed converter can realize the discharging and charging control of the battery cells. And we get the desired AC voltage at the output of H-bridge converter So, it does not require additional battery chargers or inverters in any circumstances. The obtained ac output voltage of the converter is multilevel voltage and the number of voltage levels is directly proportional to number of battery cells connected in cascaded.[1]

2 Hybrid Cascaded Multilevel Converter Topology

It is one of the most popular methods for voltage balancing circuit through energy transfer as shown in the figure.1 [5], [19] it consist of an Half bridge arm and inductance between the two cells. So, number of inductance is n-1 and number of switching devices are 2*n-1.where n stands for number of battery cells. And moreover, this circuit requires an additional inverter for the motor drive and a charger is required for recharging the battery. [1]

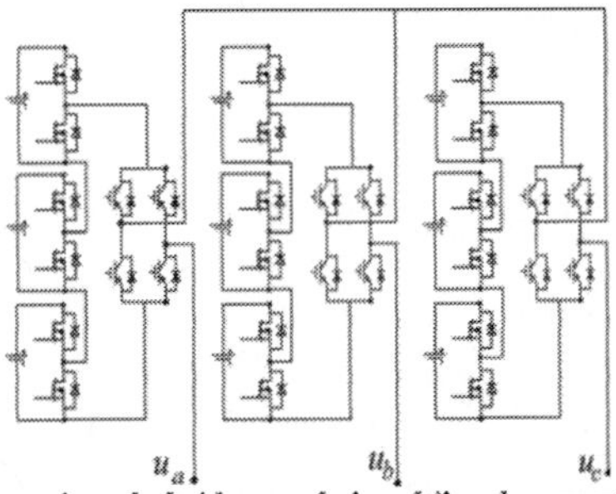

Fig.3. Three phase hybrid cascaded multilevel converter.

From the figure it is clear that output of the cascaded Half bridges is the dc bus is connected to the dc input of H-bridge. In the above circuit, each half bridge makes the battery cell involvement in voltage produced or bypassed to the circuit. H-bridge converter is equals to base frequency of desired alternating current voltage.[1]

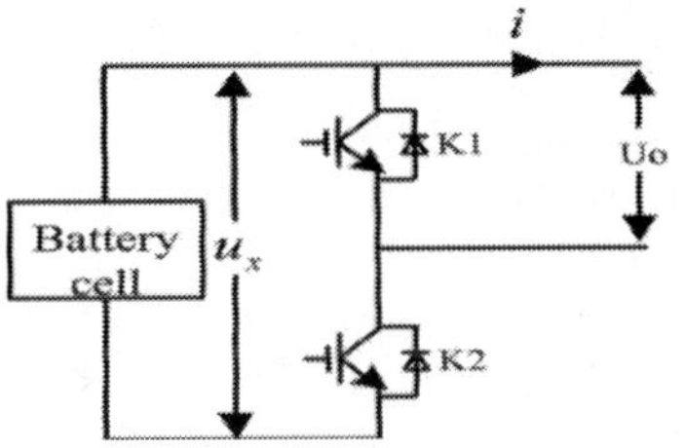

Fig.4. Output current and voltage of the battery.

There are two kinds of power electronics devices are proposed in this circuit. MOSFETs and IGBTs [1].The proposed topology of three phase converter is shown in the figure.3. if 'n' number of battery cells used in each phase, then the number of devices used in H-bridge each phase is 2*n. when compared to explained in [1].

3 Control Method of the Converter

In the cascaded half bridges are defined by S_x=1 upper switch is conducted and lower switch will be in off state.S_x=0 lower switch is conducted and upper switch will e in off state. The m_x is the modulation ratio of each half bridge defined as average value of the switching state in a pulse width modulation. The half bridge is shown in figure4. When S_x=1, battery is connected to the circuit and is charged or discharged it is determined by direction of external current. When Sx=0, battery is bypassed from the circuit, it is neither charged nor discharged. Half bridge switching state working principle is between 0<mx<1.Instantaneous power discharge is

$$p = Sx.ux.i \qquad (1)$$

Here I is the charging current of the bus and u_x is voltage of the battery cell.H-bridge is simply used to change the direction of dc bus voltage, so the reference voltage obtained at the dc bus is absolute value of alternating current reference voltage , it s similar to half sinusoidal wave at the steady state. From this we can say that not all the battery cells are needed to supply to the load at sometime

In the cascaded multilevel converter there are two modulation methods are used: they are carrier-cascaded PWM and phase shift PWM. Since, the SOC or terminal voltage balance control is realized by PWM. Hence, carrier-cascaded PWM is best suitable since modulation ratio of difference between different battery cells is used for balance control [16-17].

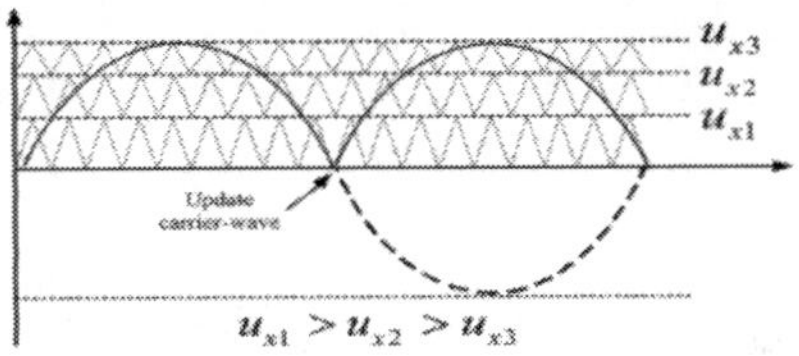

Fig 5. Carrier wave while discharging

In order to reduce EMI and dv/dt here only one half bridge allowed to change the switching state at a same time continuous reference voltage. Hence, carrier wave is rearranged when the modulation carrier wave is arranged when the modulation carrier wave is zero and rearranged carrier wave is zero. Here carrier wave is rearranged only two times during one reference voltage cycle as shown in the figure5. The cells SOC and voltage vary slowly during normal usage soothe carrier wave updated by the base frequency is enough for SOC and voltage balance.

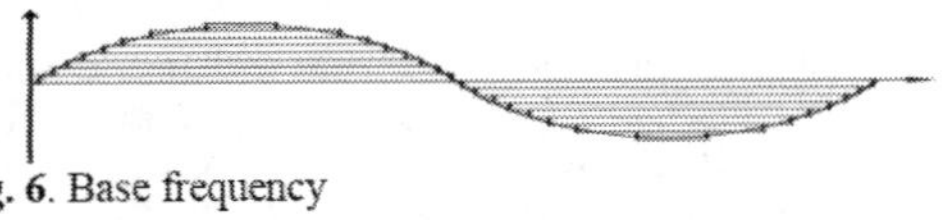

Fig. 6. Base frequency

Here all the half bridges are worked at the base frequency as shown in the figure 6, where the alternating current output voltage is approaches to sinusoidal wave it is similar to the multilevel converter [15].

 https://doi.org/10.1051/e3sconf/20198701015

If one of the battery cell got damaged, the half bridge is bypassed, and it will be no influence on the other battery cells. The output voltage of the bypassed battery cell is reduced. And we need to reduce the three phase reference voltage to fit to the output voltage ability. In order to improve the output voltage, the neutral shift three phase is adopted. Which is explained in [17-18].

4 Losses Analysis and Comparision

The circuit topology and voltage balance of the proposed circuit is quite different from the traditional circuit shown in the figure.4. in the traditional circuit method the energy transfer circuit used for the voltage balance and three phase two level DC-AC converter used for discharging control. In this proposed Hybrid cascaded circuit, discharging control is associated with the H-bridge converter and cascaded Half-bridges are used for voltage control. For this two circuits, the conduction losses and switching losses are quite different. In order to get clear thought, the conduction and switching losses are analyzed as follows:

$$J_{Loss}=J_{C\_B}+J_{C\_H}+J_{S\_H}+J_{S\_B} \quad (2)$$

Where J_{C_B} and J_{C_H} are conduction losses of cascaded half bridges and H-bridge converters and J_{S_B} and J_{S_H} are switching losses of Half bridge and H-bridge converter .

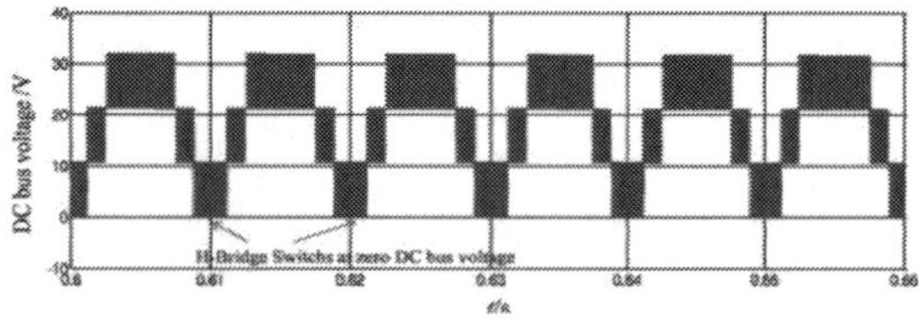

Fig 7. Dc output voltage of the cascaded half bridges

Table 1. Switching and conduction losses comparision of traditional and proposed circuit

		loss in a single switching course	Conduction loss power
Traditional circuit	Energy transfer circuit	determined by imbalance	determined by imbalance
	3-phase inverter legs	J_{s_I}	$P_{c\_I}=I^2R_{c\_I}$
Proposed novel circuit	Cascaded half-bridges	Much less than J_{s_I}/n	$P_{c\_B}=I^2R_{c\_B}\cdot n$
	H-bridges	Near zero	$P_{c\_H}=I^2R_{c\_H}\cdot 2$

5 Charging Method

In the proposed converter circuit, a dc voltage source is needed for charging the battery. The charging voltage and current can be controlled by the proposed circuit itself according to necessity of the battery cells. Circuit diagram of charging circuit is shown in the figure.8. in order to switch the dc bus from H-bridge the dc voltage source, a circuit breaker is used. Further, a filter inductor is also used in order to realize the current control, which is connected in the series. The dc voltage source can also be realized by using capacitor and H-bridge as shown in figure.9. Here H-bridge which is worked as a AC-DC converter i.e., rectifier by using diodes and a steady state dc voltage is produced by using capacitors. While charging the battery, the charging current needs to be controlled. The current state equation is

$$R_f + L_f\frac{dx}{dt} = u_{dc} - u_{char} \quad (3)$$

Where u_{charge} the voltage of dc source, u_{dc} is the dc bus output voltage of cascaded Half bridge converter and L_f, R_f are the inductance and resistance of the inductive filter between the dc source and cascaded Half bridges. In this charging method, the bus voltage must be smaller than that of the possible maximum value of the dc bus voltage.

$$u_{charge} \leq u_{dc} \leq n.u_0 \quad (4)$$

Where u_0 is the discharging cut-off voltage of the battery cell and n is the number of cascaded half bridges in each phase.

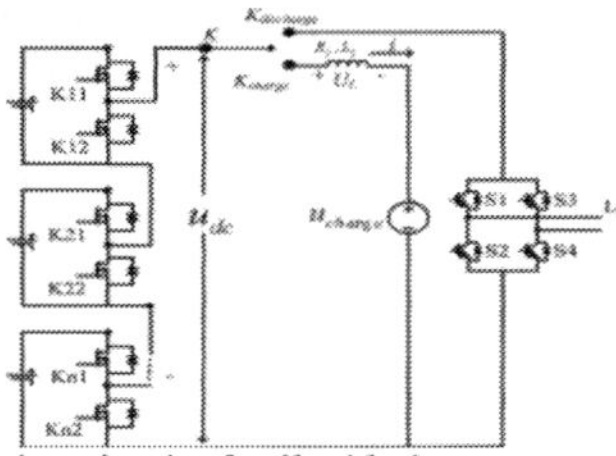

Fig 8. Charging circuit of cell with dc source

While charging cycle their might be variation in the voltage of battery cells and the dc source voltage, to make the charging current constant and the switching states of cascaded half bridges are switched. Circuit of current control scheme is shown in the above figure.10

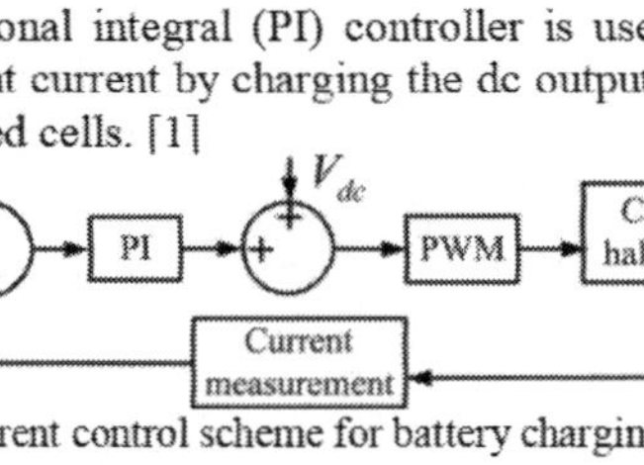

Fig. 9. Charging circuit of cell with ac source

A proportional integral (PI) controller is used to make the constant current by charging the dc output voltage of the cascaded cells. [1]

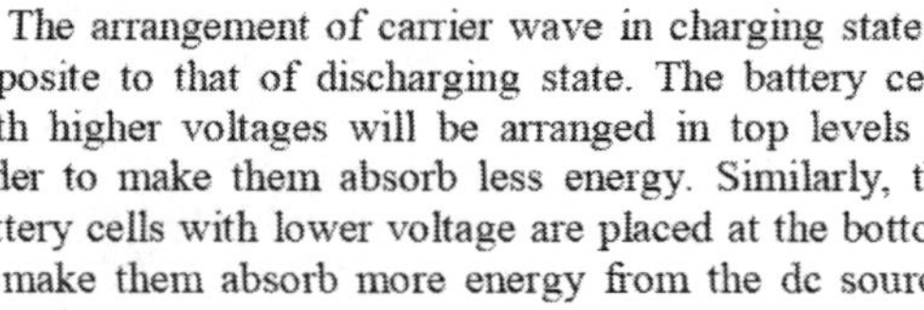

Fig 10. Current control scheme for battery charging

The arrangement of carrier wave in charging state is opposite to that of discharging state. The battery cells with higher voltages will be arranged in top levels in order to make them absorb less energy. Similarly, the battery cells with lower voltage are placed at the bottom to make them absorb more energy from the dc source.

 https://doi.org/10.1051/e3sconf/20198701015

During the discharging state similar analysis is performed and the positions of the carrier waves is shown in the figure.9. During the regenerative breaking mode of the motor drives i.e., when electric vehicles EV is breaking, the battery cells are charged, hence the modulation will be changed to charging state/mode as shown in the figure.11

Energy charged in the cell battery is same as the traditional one i.e.,

$$P_{Charge} = U_X . i\ S_X \qquad (5)$$

Here i is the charging current controlled by the scheme and S_x is the switching state of the bridge arms.

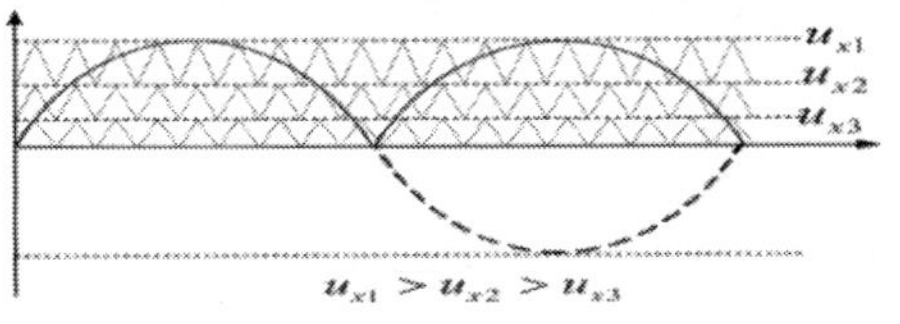

Fig. 11. Carrier wave during charging

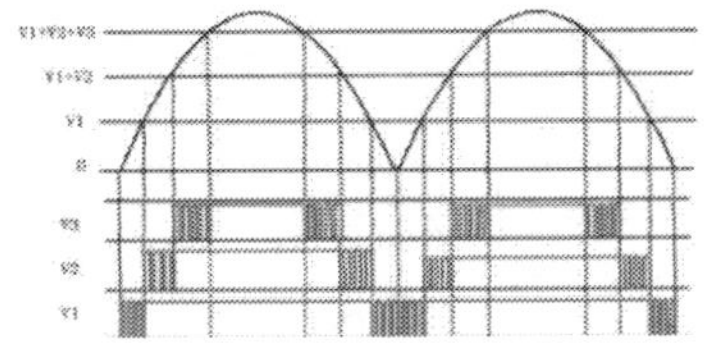

Fig. 12. Intermediate charging current of cells

Whenever, there is variation in the reference dc bus voltage, like half wave sinusoidal waveform, because of the pulse width modulation all the battery cells are intermittently charged. This method is similar to the fast charging method of the battery cells and to improve the charging speed. Charging current proposed in this converter will be more than the ordinary charging principle. Whenever steady state dc source is used to change the battery stack, we get referenced bus voltage nearly constant. In order to realize this kind of fast charging method of battery cell, the carrier wave should be forced to exchange its position from time to time which is not arranged by terminal voltages of state of charge any more. In order to get the intermittently charging current the switching state of half bridges will be charged.

6 Results of Experiment

In order to verify the performance of proposed converter and its control method, an three phase 3-cells cascaded circuit simulation done in matlab. It is shown in the figure 2. In this we used lead-acid battery module with 16.67v. Since, it is difficult to estimate the SOC of battery cells, the terminal voltages are used for the PWM carrier wave arrangement. Cascaded half bridges are designed with MOSFETs and H-Bridge designed with IGBTs.

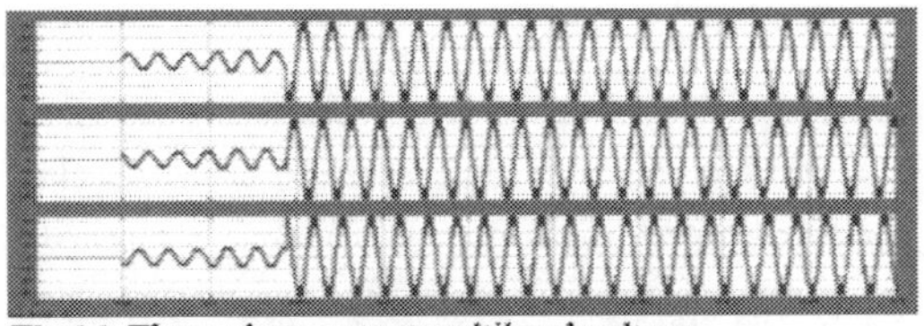

Fig 14. Three phase output multilevel voltage

Table. 2. Induction motor parameters

Paramter	Value
Rated power	0.55kW
Rated line voltage	380V
Rated line current	1.5A
Rated frequency	50Hz
Number of pole paires	2
Rated speed	1390rpm

Three phase output voltage is shown in the figure 14. In each phase there are nine levels and the waveform is more similar to ideal sinusoidal waveform than the traditional two-level inverter.

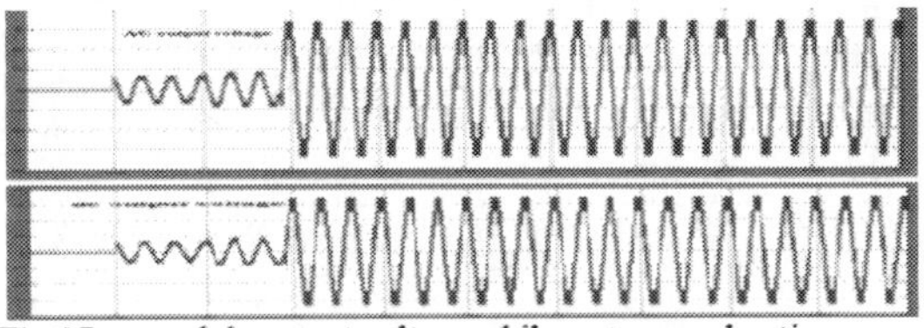

Fig 15. ac and dc output voltage while motor acceleration

An induction motor driven with (VVVF) variable-voltage-variable-frequency control method is applied. The parameters of above induction motor is shown in given table.2

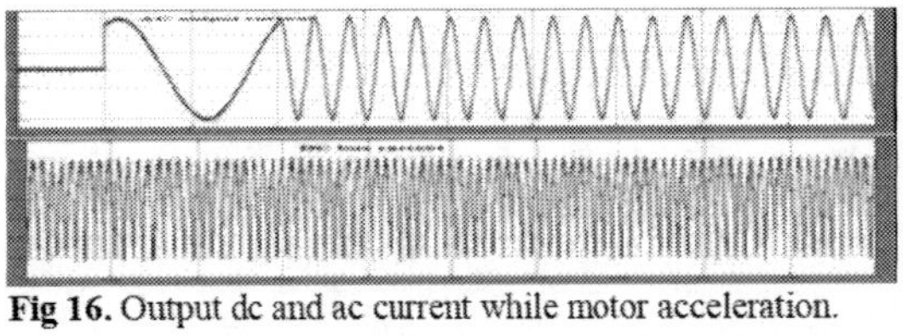

Fig 16. Output dc and ac current while motor acceleration.

The ac voltage, dc bus voltage, dc bus current, output are shown in figure15 and 16, it indicate that whole process from starting to stable state of induction motor. From the figure.15, it is clear that motor speed, its voltage levels are increasing. The stator current in the motor is shown in the figure.16 is a improved sinusoidal waveform, it will reflects the control performance of the motor.[1]. But, here the phase ac current will change its direction after certain period of time, hence the direction of dc bus is reversed when the current and phase voltage direction are different. [1]

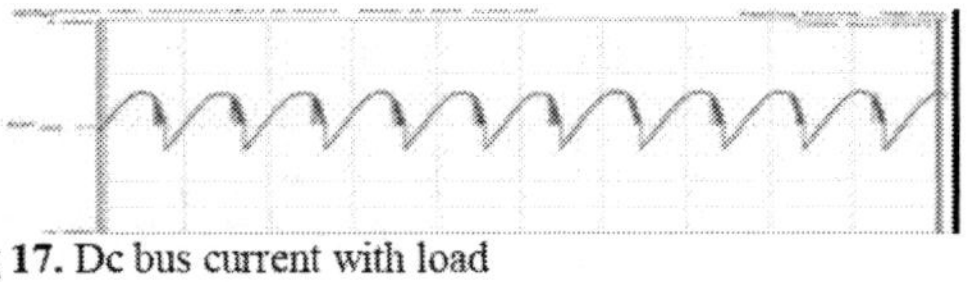

Fig 17. Dc bus current with load

The dc bus current is shown in figure.17 if there is some loads and it has some reverse current. This reverse current is greatly reduced by means of average power flowed from converter to the motor.

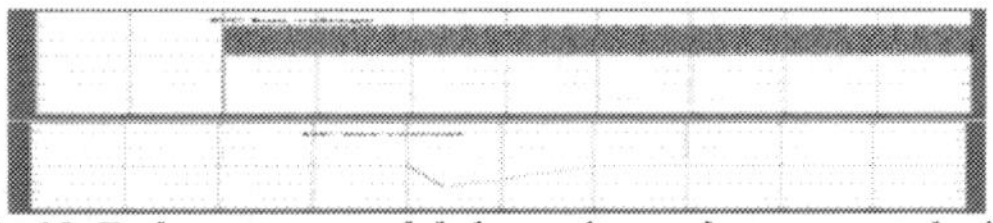

Fig 18. Dc bus current and dc bus voltage when connected with dc source

Figure 18 shows the dc bus current and dc bus voltage when the dc source was connected. It shows when dc source was connected. In the figure 17, there are two half bridges working at Sx=1 and third one working in switching state.

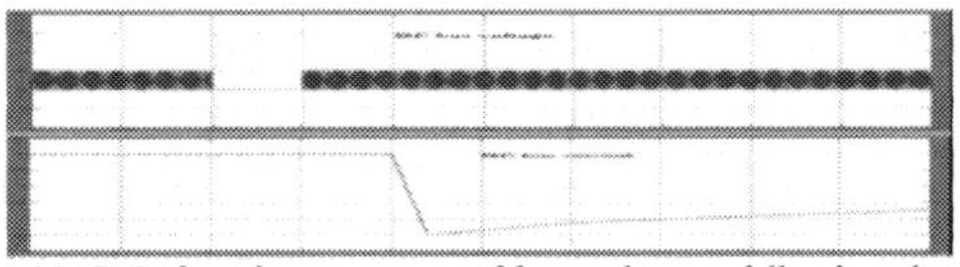

Fig. 19. DC charging current and bus voltage while charging current being givc PI controller

When the charging current reference given, the dc voltage reduce first to establish with dc current is shown in the figure 19. The resistance of the filter inductance is very small, the voltage drop on the filtered inductance through dc current is almost near to zero, and the dc bus voltage becomes same as non-charging state.

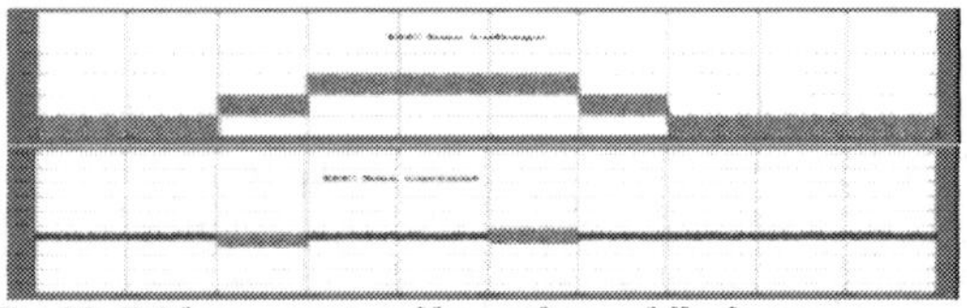

Fig. 20. DC bus current and bus voltage while dc source voltage change

At the time of battery charging, if dc source voltage is charged, the charging current control results is shown in figure 20. As we did not used feed forward compensation in our system the output has some ripples during the course of time.

7 Proposed Work

7.1 FUZZY LOGIC CONTROLLER

A fuzzy control system it is based on fuzzy logic i.e., a mathematical system which takes input analog values in the form of logical variations, that takes continuous values between 0 and 1 compared to digital logic which can operate on discrete value either 0 or 1 i.e., false or true respectively.

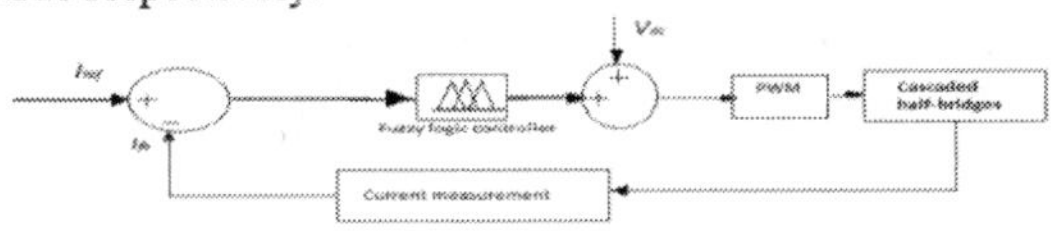

Fig 22. Current controlling scheme for battery charging using fuzzy logic controller

Total Harmonic Distortion (THD) in the current measurement scheme is reduced to 1.39% by using fuzzy logic controller and THD present PI controller is 6.27%

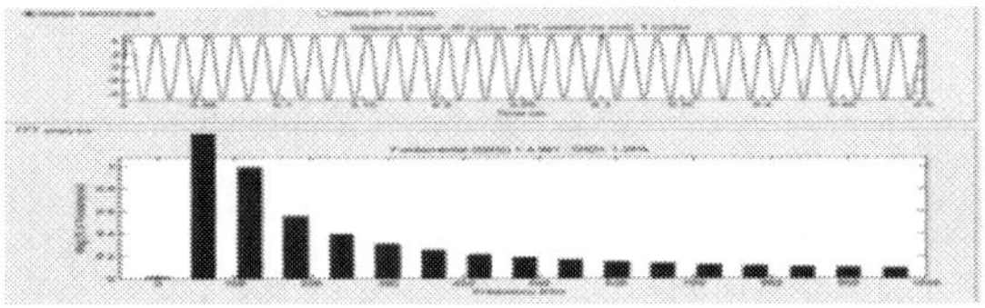

Fig. 21. THD of ac output current by FFT analysis for fuzzy logic controller

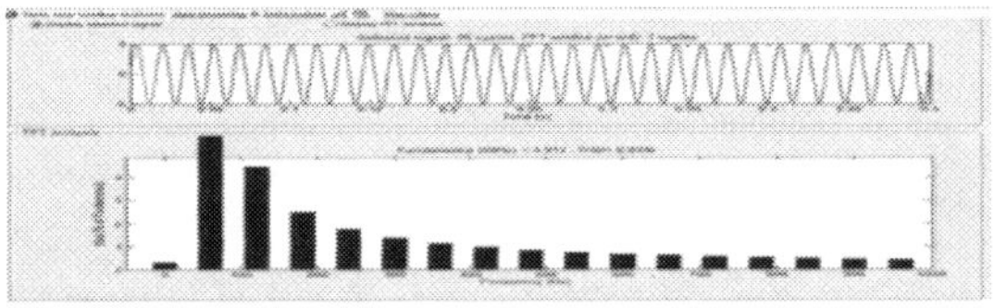

Fig. 22. THD of ac output current by FFT analysis for PI controller

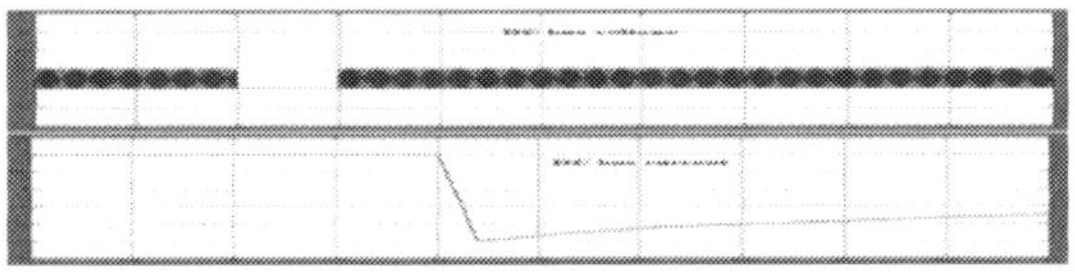

Fig. 23. DC charging current and bus voltage while charging current being given –fuzzy controller

8 Conclusion

In the proposed paper, hybrid cascaded multilevel converters, can realize the discharging and charging by battery cells, while SOC or terminal voltage balance can be realized at same time. By its modular structure it is suitable for n number of cascaded levels and it also suitable for energy storage system with lower voltage battery modules or cells. If there is any fault in battery module it can be bypassed without affecting the other which are running, hence this converter has good fault controlling capability, so it can improve the reliability of the system. Here, we used dc current control method using fuzzy logic controller for battery cells charging and discharging with external using ac or dc is studied, where additional charger is not required for constant current control. Simulation results are verified.

References

1. Zedong zheng, member, IEEE, kui wag, member, 1EEE, lie xu, member IEEE and yongdong li, member "A hybrid cascaded multilevel converter for battery energy management applied in electric vehicle" IEEE 2014.
2. S.M.Luick, J.Cao, R.C.Bnasal, F.Rodriguez and A.Emadi, "energy storage systems for automotive applications" *IEEE Trans.Ind.Electron. vol.55 no.6 pp, Jul 2008.*
3. H.M.Zhang and S.P.Ding "Application of synergic electric power supply in HEV" in *8th world congr. Intelligent control action, 2010.*
4. A.Emadi, Y.J.Lee and K.R.Rajashekara, "power electronics and motor drives in electric, hybrid electric and plug-in hybrid electric vehicles" *IEEE Trans. Ind. Electron., vol.55, no.6- June 2008*
5. K.Joghoon, S.Jhongwon, C.Changyoon, and B.H. Cho, "Stable configuration of a Li-ion series battery pack based on a screening process for improved voltage/SOC balancing," *IEEE Trans. Power Electron., vol.27, no.1, pp.411-424, Jan 2012*
6. L.Yuang-Shung, T.Cheng-En, k.Yi-Pin, and C.Ming-Wing, " Charge eualistion using quasi-resonant converters in battery string for medical power operated vehicle application," in *Proc. Int.Power Eletron. Conf., 2010 pp. 2722-2728.*
7. Y.C. Hsieh, C.S.Moo and W.Y.Ou-Yang, "A bi-directional charge equalization circuit for series-connected batteries ." in *Proc. IEEE Power Electron. Drives Syst.,2005,pp.1578-1583.*
8. S.Yarlagadda, T.T.Hartley, and T.Hussain, "A battery management system using an active charge equalization technique based on DC/DC converter topology, *" in Proc. Energy Convers.Congr.Expo.,2011,pp.1188-1195*
9. K.Chol-Ho, k.Young-Do, and M. Gun-Woo, "Individual cell voltage equalizer using selective two current paths for series connected Li-ion battery strings." In Proc. Energy Convers. Congr.Expo.2009.pp.1812-1817.
10. H.Shen, W.Zhu, and W.Chen, "Charge equalization using magnetic winding model for lithium ion battery string." In *Proc. Asia-pacefic power Energy Eng.Conf.2010,pp.1-4.*
11. P.Sang-Hyun, P.Ki-BUM, K.Hyoung-suk, M.Gun-woo, and Y. Myung-Joong, "Single-magnetic cell-to-cell charge equalization coverter with reduced number of transformer windings." *IEEE Trans. Power Electron., vol.27 no.6,pp, 2900-2911, Jun 2012.*
12. K.Chol-Ho, K.Moon-Ypumg and M.Gun-woo,"A modularized charge equalizer using a battery monitoring IC for series-connected Li-Ion battery strings in electric vehicles" *IEEE Trans. Power Electron., vol.28.no.8, pp.3779-3789, Aug.2013.*
13. L.Maharjan, S. Inoue, H. Akagi, and J.Asakura, "state-of-charge(soc)- balancing control of a battery energy storage based on a cascaded PWM converter" *IEEE Trans. Power Electron., vol.24,no.6,pp.1628-1636, Jun.2009.*
14. L.Maharjan, T.Yamagishi, and H.Akagi. "Active-power control of individual converter cells for a battery energy storage system based on a multilevel cascade PWM controller, " *IEEE Trans. Power Electron., vol.27,no.3, pp. 1099-1107, Mar.2012.*
15. K. Ilves, A.Antonopoulos, S.Norrga, and H.Nee, "A new modulation method for the modular multilevel converter allowing fundamental switching frequency," *IEEE Trans. Power Electron., vol. 27, no.8, pp.3482-3494, Aug.2012.*
16. B.P.McGrath and D.G.Holmes, "enhanced voltage balancing of a flying capacitor multilevel converter using phase disposition (PD) modulation," *IEEE Trans. Power Electron., vol.26, no.7, pp.1933-1942, Jun.2011.*
17. L.Maharjan, T. Yamagishi, H.Akagi, and J.Asakura, "fault tolerant operation of a battery –energy-storage system based on a multilevel cascaded PWM converter with star configuration," *IEEE Trans. Power Electron. vol.25, no.9, pp.2386-2386, Sep.2010.*
18. M.Ma, L.Hu, A.Chen, and X.He, "Reconfiguration of carrier-based modulation strategy for fault tolerant multilevel inverters," *IEEE Trans. Electron., vol.22.no.5,pp.2050-2060,Sep.2007.*
19. S.Gui-Jia, "Multilevel DC-link inverter," *IEEE Trans. Ind. Appl., vol.41,no.1,pp.848-854, jan-2005.*

E3S Web of Conferences **87**, 01016 (2019)
SeFet 2019
https://doi.org/10.1051/e3sconf/20198701016

IoT Based Displacement Detection Using Wireless Sensor System

Bajarangbali[1], *Vinay Kumar* Jadhav[1], *Nagarur* Druvitha[1] and *Pavan Kumar* S[1]

[1]Electronics and Communication Engineering Department, PESIT – South Campus, Bangalore

Abstract. The main objective of this work is to monitor the base position of evenly or unevenly distributed containers by analyzing and establishing a reference point for the object. The displacement of the container is measured with the help of ultrasonic sensors each connected to nodemcu. Using nodemcu, sensor values are sent wirelessly to microcontroller that is specifically designed for IoT applications. Microcontroller will then compare the received sensor values with reference values. If there is any displacement of the container, warning is given to truck driver by buzzer and SMS, E-mail is sent to intended recipient.

1 Introduction

Trade is vital to the success of a civilization. The original form of trade was barter, but as the world progressed so did the form and ways of trading and shipping. In present day most of the bulk shipping happens via ships or trucks.
Our aim is to monitor various containers that are shipped across the globe via road using trucks.
This specific case is taken because the constructions of flat beds are low cost than other kinds of trailers. Strapping loads down is easy and oversize loads can be carried with ease.
The trucks have certain disadvantages associated with it, a truck carrying a heavy container on it has a chance of meeting with accidents in case the driver is driving across sharp bends or curves where the weight of the container concentrates towards one side and brings down the entire truck. There are various situations wherein the container slips from the trailer and is damaged. So the objective is to design a set-up that will monitor the position of the container on the trailer through out its journey so as to ensure that the container reaches its destination safely.
The work proposed in this paper is an extension of the work published by Bajarangbali et al. [1].

1.1. General Description

The objective of the work proposed in this paper is to monitor various containers that are shipped across the globe via road using trucks. There are various situations wherein the container slips from the trailer and is damaged. This set-up will monitor the position of the container on the trailer throughout its journey so as to ensure that the container reaches its destination safely. In-order to monitor the position of the container the displacement sensors have been placed at the appropriate locations. The displacement of the container is measured with the help of these sensors. The readings from these sensors are then given to a micro-controller that is specifically designed for IoT applications to send the data to intended recipient.

1.2 Problem Definition

Fig. 1. Problem Definition

- There are various situations wherein the container slips from the trailer and is damaged.
- This prototype is used to monitor the position of the container on the trailer throughout its journey so as to ensure that the container reaches its destination safely.

1.3 IoT And It's Role In This work

The Internet of Things is nothing but a network of physical objects which are embedded along with electronic devices which are connected to internet for

* Corresponding author: *B.K.* Chakravarthy

 https://doi.org/10.1051/e3sconf/20198701016

communication purpose such as data or information exchange. The objects can be controlled remotely with the help of IoT. IoT enables increased connectivity of objects and people.

The IoT vision is a massively instrumented world of intelligent sensors and actuators improving performance and efficiency. Streamlining, collapsing and creating system architectures which are more effective, affordable and responsive are done by IoT and also it eliminates the need for expensive and difficult to maintain middle-level automation software.
The role of IoT in this project is to send the data to the client that is received on the server by the micro controller. There are various ways in which the data that is received by the server can be communicated to the client. Few such methods are,

- Sending an email
- Sending a text message
- Sending a voice call

But the method used to communicate the information to client in this work is using a text message. This text message is auto generated using IoT based applications which gets triggered when the cargo is displaced beyond tolerable limits.

2 Design And Implementation

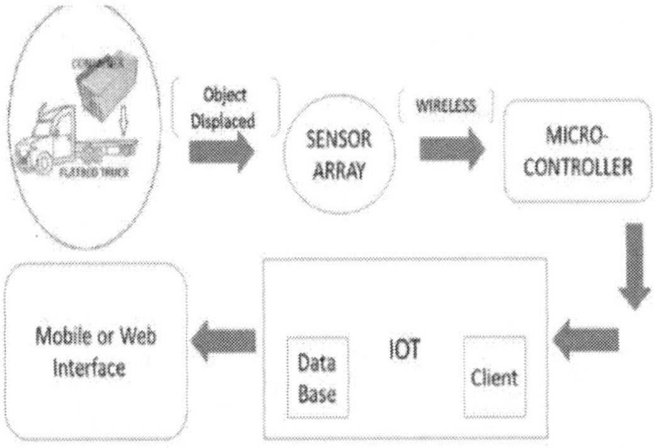

Fig. 2.1. Block Diagram

- In this displacement sensors will be used to continuously monitor the position of the container under consideration.
- Any movement or displacement of the object from its reference point has to be measured and analyzed.
- For smart devices and sensors, each event is capable of creating data for further analysis. This information is sent over the network back to the intended recipient. This part is implemented using IoT.

A brief description of various devices used in this work is mentioned below:

A. HCSR-04 Ultrasonic Sensor

The Ultrasonic sensor used here is the HC-SR04 ultrasonic ranging sensor which provides 2cm to 400cm of non-contact measurement functionality. Also it has a ranging accuracy which can reach up to 3mm. This device consists of a receiver, an ultrasonic transmitter and a control circuit.

B. NodeMCU

NODEMCU is an open source platform used for IoT programming. This NODEMCU includes hardware based on the ESP – 12 module and firmware operating on ESP8266 Wi – Fi system on chip (Soc). The NODEMCU term refers to the firmware more than the development kits.

C. LCD

LCD stands for Liquid Crystal Display. The LCD display used here is 16X2. Here information like cargo on the truck is safe or there is any displacement is displayed in a matrix of 5x7 or 5x8. With the help of this information appropriate action can be taken to avoid accidents.

D. Buzzer Interface

A buzzer or beeper is used to generate an audio signal. This device may be piezoelectric, mechanical or electromechanical. A Piezo buzzer has a Piezo disc and an oscillator inside. Any displacement of the cargo beyond the reference point is detected by the system which sends signal to this buzzer. This sound can be used as an alarm or warning which alerts the truck driver.

E. Energia

The software being used to code CC3200 is Energia, which is an open source community-driven integrated development environment (IDE) software framework. This software provides an intuitive coding environment as well as a robust framework of easy to use functional APIs and libraries for programming a particular microcontroller.

F. Blynk

It is a platform used for IoT devices with a drag-n-drop mobile application builder which helps to visualize sensor data used to control remotely placed electronics in short time. It is a used to control devices like Arduino, Raspberry Pi and Android apps over the Internet.

With the help of its digital dashboard a graphic interface can be built by simple drag and drop facility for different widgets. It can be noted that any specific shield or board is not tied to Blynk. Whether ESP8266 chip, Raspberry Pi or Arduino is linked to the Internet over Wi-Fi, Ethernet Blynk will be connected online and IoT will be ready.

G. IFTTT

It is short form of if this, then that, which is a free web-based service used to create chains of simple conditional statements, called applets. Whenever any changes occur within other web services like Telegram, Gmail, Instagram, or Pinterest an applet is triggered. This is used to receive short message service (SMS) and email messages on your device and send messages to other phone numbers.

H. Ultrasonic Sensor Interface

Here the sensor measures the difference time between sent and received waves to find the distance between the object and sensor. The ultrasonic sensor has 4pins which are Vcc, Ground, Trigger and Echo. Four ultrasonic sensors are used in this application.

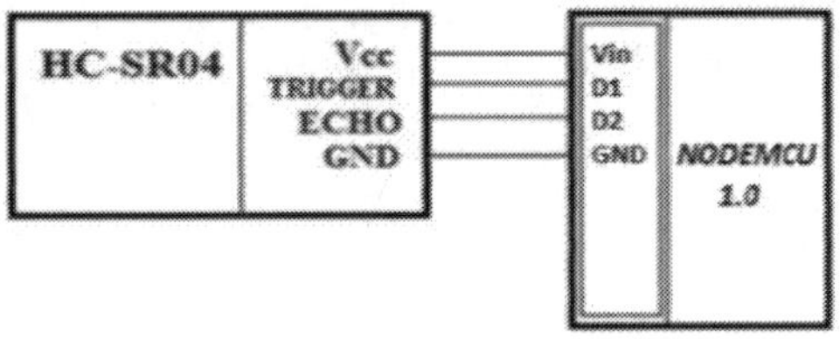

Fig. 2.3. HC-SR04 Interface

I. LCD Interface

It uses liquid crystal to produce a visible image. The LCD display has eight digital pins out of which D4, D5, D6 and D7 pins are connected to the microcontroller digital pins with pin number p1.6, p1.7, p1.8 and p1.9. The logical low (0) on register select pin selects command register, and data register when it is logical high (1). This pin is interfaced with microcontroller at pin p1.4.

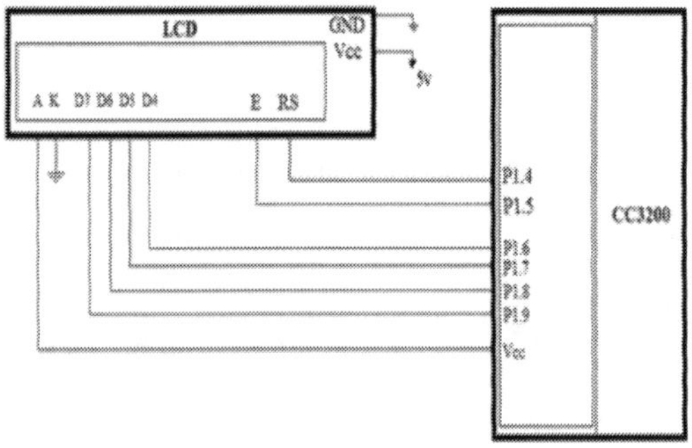

Fig. 2.4. LCD Interface

J. Buzzer Interface

The buzzer has two pins Vcc and Ground. Vcc is connected to pin number 19 of microcontroller. When the cargo/container is not in the safe zone the buzzer is made high and it beeps and when cargo/container is in safe zone the buzzer does not beep.

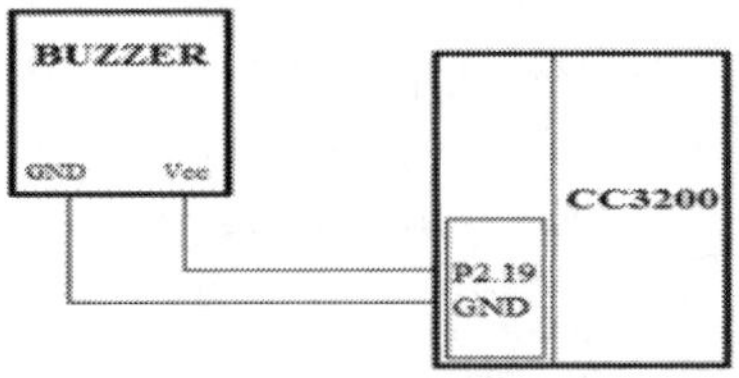

Fig. 2.5. Buzzer Interface

K. Complete Interface

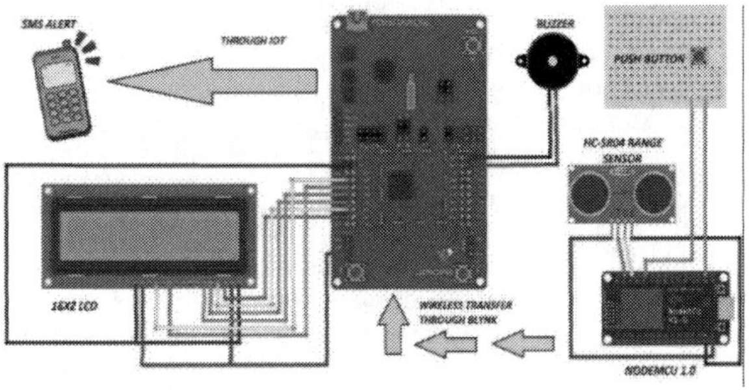

Fig. 2.6. Complete Interface

The CC3200 Microcontroller is interfaced with Buzzer and LCD Display and it is connected to ultrasonic sensor and NODEMCU wirelessly.

L. Setup

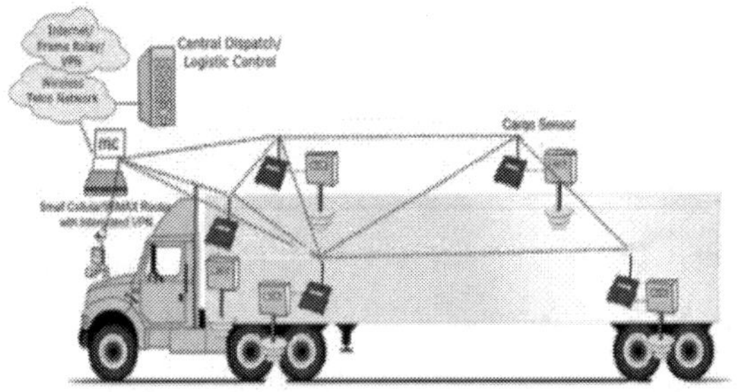

Fig. 2.7. Setup with Ultrasonic Sensor

Assumptions made in this setup

1.The cargo/container under consideration is evenly/unevenly distributed in space.

2. The horizontal displacement, vertical displacement and angular displacement across flatbed trailer is considered.

3. Both displacement at the base and height are taken into consideration.

3 Results

A. Final Prototype

Fig. 3.1. Final Prototype

B. Prototype results

Fig. 3.2. Final Prototype Readings

From the results that were obtained on the hardware setup following details could be inferred,

- Uses ultrasonic sensors for range measurements.
- Operates in soft real time (with few seconds delay).
- This setup is valid for evenly and unevenly distributed loads (container).
- Both horizontal and Vertical displacements are being measured to determine if the load (container) is safe of not.

4 Conclusion

This work is a set-up that is designed to measure any displacements that occurs during the safe shipping of load (container) on flatbed truck that could cause any casualties due to slipping of the container. The set-up was successfully designed and tested for evenly and unevenly distributed containers using a small prototype.

Even though this set-up was successfully tested using a small prototype, there are various aspects that could be considered for improving the work.

One can also experiment with various other displacement sensors for improved accuracy.

This concept can also be used to build a setup to monitor the alignment of railway tracks.

References

1. Iot based displacement sensor; Bajarangbali, Nisha S., Sujith R., Raju Singh and Nithu Singh, 5th International Conference on Advancement in Engineering, Applied Science and Management (ICAEASM-2017), Bengaluru, India.
2. Sensors and Actuators A: Physical Volume 107, Issue 1, 1 October 2003, Pages 2125; T.W. Ng.
3. Exploiting accelerometers to estimate displacement: Embedded Computing (MECO), 2016 5th Mediterranean Conference; Renato Ferrero; Filippo Gandino; Masoud Hemmatpour.
4. Precision positioning using a microfabricated electrostatic actuator: IEEE Transactions on Magnetics (Volume: 35, Issue: 2, Mar 1999); D.A. Horsley ; N. Wongkomet ; R. Horowitz ; A.P. Pisano.

 https://doi.org/10.1051/e3sconf/20198701017

A case Study on wired and Wireless charger standards in India for Electric Vehicle Application

Shivanand M N[1], Y. Maruthi[1] *Phaneendra Babu Bobba*[2], Sandeep Vuddanti[1]

[1]GRIET, EEE Department, Hyderabad, Telangana, India
[2]CUK, Electrical Engineering Department, Kalburgi, Karnataka, India

Abstract. India has taken major step in adopting the electric vehicle by means of FAME Scheme (Fast Adoption and Manufacturing of Electric Vehicles), a government initiative. ARAI (Automotive Research Authority of India) and DHI (Department of Heavy Industry) have published standardization protocol for both EV charging infrastructure. Many of those standards are derived from the SAE (Society of Automotive Engineers) Internationals and IEC (International Electrotechnical Commission). USA, Europe and China are also following the same standards to build the EV (Electric Vehicle) infrastructure. This paper provides the Indian standards to build EV charging infrastructure and comparing it with other countries. Glimpses on energy demand for electric vehicles in Indian market. It also provides the demanding wireless power transfer technology in EV's. Status of Standards provided by the industry on wireless power transfer. Factors that are necessary to be considered before drafting the standards for WPT.

1 India's Charging Infrastructure

India aims to have at least 15 percent of the vehicles on its roads to be electric in five years, signalling the government's wish to join a long list of countries around the world that are already seeking to cut fossil fuels aggressively. While cumulative global sales of passenger electric vehicles likely surpassed 4 million in August, with China accounting for more than a third since 2011, India sold an estimated 2,000 EVs 2017. EVs may account for about 7 percent of sales in India by 2030 by [1]. The government of India (Department of Heavy Industries – DHI) is already providing incentives through FAME scheme (Faster Adoption and Manufacturing of Electric Vehicles) from the year 2015 in order to reduce the price of EV's. The government has also approved pilot projects, charging infrastructure projects and technological development projects. Energy Efficiency Services Limited (EESL), under the administration of Ministry of Power, Government of India (GoI), has ordered 10,000 EVs [2]. With all the initiation that India has been taking, there is serious issue of charging infrastructure for both wired and wireless charging system. Standardization to the technical aspects of the EVs charging, range and price are the major barrier for deployment. Currently in India, there are very small scale of AC and DC charging stations available, which are installed by EV manufacturers. Even though the AC slow charging infrastructure at residences, workplaces and public places requires very low investment when compared to fuel stations. Still, the EV charging infrastructure growth in India is not up to the mark. To grow large scale of charging infrastructure in India it requires continuous support from the government, utility grid authorities and automobiles manufacturers.

1.1 Indian EV Charging Standards

Departemnt of Heavy Industry (DHI) & Automotive Research Association of India (ARAI) drafted industry standard for Electric Vehicle Charging infrastructure on December 2015. The first draft of AC-charging configurations are Type-I and Type-II. Type-III standards are published in the second draft on May 2017. [3]

1.1.1 Type-I: AC001(3.3KW,15A,230V)

- Derived from IEC 60309 Standard
- AC Slow charging
- Each outlet will have up to three independent charging sockets.
- Input: 3 phase AC Supply, 5 wire (3 phase+ Neutral +PE- Protective Earth). Nominal Input Voltage is 415V (+6% and -10%) as per IS 12360
- Frequency is 50Hz ± 1.5 Hz
- Output: Single phase, two wire system 230V (+6% or -10%) and 15 Amps as per IS12360
- No Communication protocol is used between EV and EVSE (EV Supply Equipment)

1.1.2 Type-II: AC001(>3.3KW)

* Corresponding author: shivanandnaduvin@gmail.com

• Charger power is greater than 3.3KW.
• This AC Fast charging.
• Outlet is derived from the Standard IEC 62196 and IEC 61851.
• Charging with 415V, three phases, 63A.
• Control Pilot (communication protocol) extends to control EV charging Equipment system (EVSE).

1.1.3 Type-III: DC001(48V/2V,10KW/15KW)

• Input: 3 phase AC Supply, 5 wire (3phases+N+PE). Nominal Input Voltage is 415V & frequency is 50Hz ± 1.5 Hz (+6% and -10%) Maximum 200 Amps as per IS 12360.
• Output- 48V or 72V DC, based on suitable charger Configuration for vehicle battery.
• Charger Configuration Types
• Single Vehicle charging at 48V/72V with a maximum of 10kW, or a 2 Wheel vehicle charging at 48V with maximum 3.3kW.
• Single Vehicle charging at 48V with a maximum of 10KW, or 72V with maximum 15kW or a 2 Wheel vehicle charging at 48V with maximum 3.3kW.

1.2 Charging System Available in The World

1.2.1 The International Electrotechnical Commission (IEC 62196) modes definition:

Mode-1: Slow charging from a regular electrical socket (single or three phase)
Mode-2: Slow charging from a regular socket but which equipped with some EV specific protection arrangement
Mode-3: Slow or Fast charger using a specific EV multi-Pin Socket with control and protection functions
Mode-4: Fast charging using some specific charger technology such as CHAdeMO. [4]

1.2.2 European EV Standards

• Normal power or Slow Charging with rated power inferior to 3.7kW is used for domestic application or for long-time EV parking.
• Medium power or quick charging with a rated power from 3.7-22kW is used for private and public EV
• High power or fast charging with a rated power superior to 22kW is used for public EV [4]

1.2.3 American EV Standards

TYPE-1: The charger is on-board and provides an AC voltage at 120 or 240 volts with a maximum current of 15A and a maximum power of 3.3kW
TYPE-2: The charger is on-board and provides an AC voltage at 240V with a maximum current of 60A and a maximum power of 14.4kW.
TYPE-3: The charger is off-board, so the charging station provides DC voltage directly to the battery via DC connector with maximum power of 240kW. [4]

Table 1. Electrical Rating of Different Charge Method in North-America

Charging - Method	AC Level 1	AC Level 2	DC Level 3
Nominal Supply Voltage(V)	120	240	207-600
Max. Current (A)	12	32	400
Max. Power (KW)	1.44	7.7	240
Charger Location	On-Board 1-Phase	On-Board 1-Phase	Off-Board 3-Phase
Communica tion Protocol	Control Pilot (CP)	CP	Power Line Carrier (PLC)

1.2.4 EV charging based on Power and Usage Location

Charging ratings of different electric vehicles based on their usage are classified from normal power rating of 3.7kW to >22kW.

Table 2. Classification of EV charging based on Power and Usage Location.

Charging Method	Connecti on Type	Power (kW)	Max. Current (A)	Location
Normal Power	I-Phase AC	3.7	10-16	Domestic
Medium Power	I or 3 Phase	3.7-22	16-32	Semi-Domestic
High Power	3- Phase AC	>22	>32	Public
High Power DC	DC	>22	>32	Public

2 Electric Vehicles in India

2.1 Energy Demand for EV

Energy required by utility grid increases as demand in the electric vehicle market. The automobiles production in India is at a CAGR of 5.81% from 2007 to 2018. The vehicles include 80.9% of two wheelers, 13.2% of passenger cars, 2.5% of three wheelers & 3.4% of commercial vehicles [5]. Out of these, close to 99.9% are conventional ICE vehicles which are burning petrol&

 https://doi.org/10.1051/e3sconf/20198701017

diesel for traction. The figure 1 shows the historical vehicle production data along with the estimated production volumes by 2030 which is indicated with blue line and also it shows the historical & estimated fuel (both petrol & diesel) consumption in MMT (Million Metric Tonnes) which is indicated with red line.

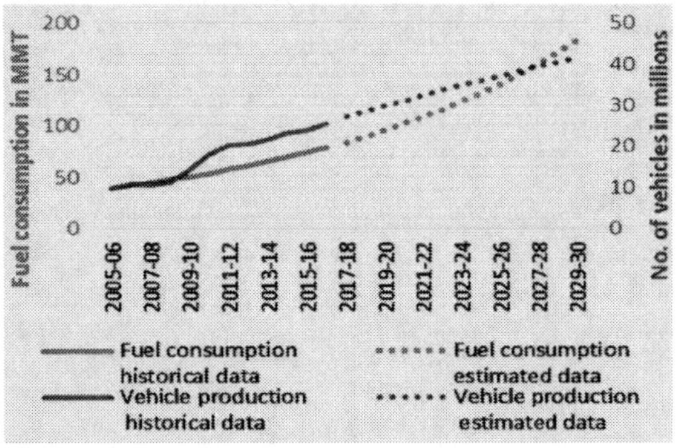

Fig. 1 Fuel Consumption & Vehicle Production Data

The total electricity generation in India during the year 2016-17 is 1160TWh which is used in domestic, commercial, industrial & agriculture sectors. The total electricity required, if all the existing ICE vehicles are converted into BEV's during the year 2016-17 is 439TWh which is 34.26% additional to the existing electricity generation. By 2030, the electricity generation (Egen) in India will grow along with the demand in the domestic, commercial, industrial & agriculture sectors which is shown in the figure 3 (indicated with blue). Apart from this, there will be a growth in the production of automobiles in India by 2030 and if all these vehicles are 100% BEV's the required electricity generation (Egen_tr) demand on the utility grid increases which is also shown in the figure 3 (indicated with red).

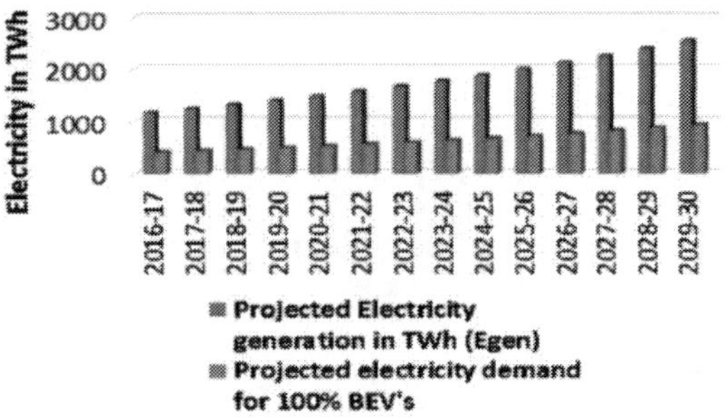

Fig. 2 Projected Electricity demand upto 2030

Hence by 2030 in India, 100% of BEV's would require an energy of 929.3 TWh which is about 37% additional electricity to be produced apart from estimated electricity generation of 2500 TWh (includes domestic, commercial, agriculture & industrial sectors etc.) [6].

2.2 Indian Automobile Industry – An Overview

The Indian Automobile Industry is currently ranked 5th largest in the world and is set to be the 3rd largest by 2030.The requirement of mobility in India is set to change dramatically in the near future to cater to the requirement of 1.30 billion population While there is a vision for 100% electric vehicles by 2030, most industry experts indicate that around 40-45% EV conversion by 2030 is a realistic expectation. A major push towards EVs will be led by the public transportation requirements in India – Fleet cars, E-Buses, 3 wheelers and 2 wheelers. Personal vehicle options for EVs will still be a relatively smaller element in the whole pie. The Government plans to work towards creating a demand for EVs by buying in bulk, which could provide for large orders for automakers. A tender for 10,000 cars is already issued and now a major tender for electric buses in 11 cities is likely to be released soon. [7]

2.3 The outlook for 2018

-Passenger cars to see a higher increase in new model launches compared to utility vehicles.
-Production capacity will also be added at car makers to reduce waiting periods and to boost demand. (Passenger vehicle segment growing between 7-9%).
-In the two-wheeler segment, motorcycles are expected to grow moderately while scooters will continue to grow in double digits with two-wheelers growing between 9-11% in FY'18.
-Industry moving towards a March 2020 launch of BS-6 and most OEM/Auto component firms have made investments to meet this deadline.

2.4 EV Chargers

2.4.1 EV Chargers in India

Post FAME and the Niti Ayog Plan announcement, we have seen an increased trust on setting up EV chargers in India. Some names we could confirm, who have joined interest in EV Chargers manufacturing are Raychem RPG India, Analogic India, Deltron, EOS Power, AdorPowertron, Kraft Power Con, Elind etc.

2.4.2 Global EV Charger firms on Indian Market

More than six key global firms eyeing the EV Chargers market closely. Firms like ABB India, Delta India, Schneider India, Siemens India etc are looking at the Indian market closely. These firms have their global designs and products. They are studying the technical/specifications, business models and potential for their products. All these firms are only looking at the 4 Wheelers' (Cars) EV Chargers.

Table 3: Total number of EV Charging Stations in India

Title	Installed base 2016-17	Additions in the last 4 months	Total installed base as of August 2017
EV Charging stations	100	30	130
Average Norms of EV Chargers	1.5	4	-
Total EV Chargers installed	150	120	270
% of AC Slow chargers	100%	80%	x
% of DC Fast Chargers	0%	20%	x
No of AC Slow chargers installed	150	96	246
No of DC Fast Chargers installed	0	24	24

Table 4: Likely future market for EV Chargers in India

Title	2017-18	2018-19	2019-21	2021-25	Potential up to 2026
No of EV Charging stations likely to be set up	1000	5,000	50,000	350,000	406,000
Total EV Chargers likely to be installed	4000	20,000	300,000	2,100,000	2,424,000
% of AC Slow chargers likely	90%	80%	80%	70%	x
% of DC Fast Chargers likely	10%	20%	20%	30%	x
No of AC Slow chargers likely to be installed	3600	16000	240,000	1,470,000	1,729,600
No of Fast Dc chargers	400	4000	60,000	630,000	694,400

3 EV Wireless Charging Standards

As the world is advancing towards the wireless charging from smaller biomedical application to the electric vehicle charging system. Standardization is the main barrier holding the commercialization of high-voltage and high-power WPT for EV charging. It includes safety criteria, efficiency, EM(Electromagnetic) limits, and interoperability targets, along with test setup for getting wireless charging. It should provide the compatible charging station to all the EV owners. IEC-61980-1 standard contains the total system of WPT from supply network to EVs charging the battery or any equipment of the same at the standard supply of 1000-Vac or 1500-Vdc. These all are addressed by SAE in its standard SAE TIR J2954 (TIR-Technical Information Report). This is the first standard developed by SAE in WPT for an EV charging application. This standard is developed specifically for SWC. The frequency band, interoperability, safety, coil definitions, and EMC/EMF limits from SAE TIR J2954 allow any attuned vehicle to charge wirelessly from its wireless home charger, office or a commercial charger with the same charging ability. Table 5 shows key standards for wireless charging. [8] In the near term, vehicles that are capable to be charged wirelessly under recommended practice should also be able to be charged by SAE J1772 plug in chargers. SAE recommended practice J2954 is intended to be used for stationary applications (charging while vehicle is not in motion). Dynamic applications may be considered in the future based on industry feedback. SAE Recommended Practice J2954 is meant to be used for interoperability, performance and emissions testing, where a single standard coil-set has been chosen for the WPT Power Class 1 and 2 to 7.7kW, per Z-Classes (1 through 3) which is circular topology. However, there are two reference options for WPT 3 to 11kW per Z-classes (1 through 3) with two topologies. The next revision of the Recommended Practice in 2018 is slated to have one standard coil set for WPT 3. SAE TIR J2954 establishes a common frequency band using 85 kHz (81.39 - 90 kHz) for all light duty vehicle systems. [9].

Table 5: Wireless Charging standards

Standard Developer	Standard Name	Published/Update Date	Description	Status
IEC	IEC 6180-1Ed. 1.0-New Addition	2015-07-01	Electric Vehicle wireless power transfer (WPT) System-Part 1: General Requirements	Active
IEC	IEC 61980-1Ed2.0	2020-03-30	Electric Vehicle wireless power transfer (WPT) System-Part 1: General Requirements	ACD
IEC	IEC 61980-1/IAMD I Ed1.0	2015-07-24	Electric Vehicle wireless power transfer (WPT) System-Part 1: General Requirements	ACD
IEC	IEC 61980-1/COR1A	2017-01-01	Electric Vehicle wireless power transfer (WPT) System-Part 1: General Requirements	VALID
IEC	DRAFT IEC/TS 61980-2Ed.1.0	Will be published in 2017	Electric Vehicle wireless power transfer (WPT) Systems-part2: Specific Requirements for communication between Electric Road Vehicle (EV) and Infrastructure concerning Wireless power transfer (WPT)	Active
IEC	Draft IEC/TS61980-3Ed1.0	2015-08-28	Electric vehicle wireless power transfer (WPT) systems-part3: Specific Requirements for the magnetic field power transfer system	Draft, Valid
SAE	J2954SAE	WIP	Wireless charging Electric and Plug in Hybrid vehicles	WIP
SAE	J2954_201605	2016-05-26	Wireless power transfer for light duty plug in /Electric vehicles and Alignment methodology	Issued
SAE	J1773_201406	2014-06-05	SAE electric Vehicles inductively coupled charging	Stabilized
SAE	J2847-6	2015-08-05	Communication between Wireless charger charged vehicles and wireless EV Chargers	Issued
SAE	J2831	2015-8-27	Signalling Communication for wireless charged Electric vehicle	Issued
UL	Subject 2750	NA	Outline of Investigation for Electric Vehicle wireless charging	Draft Released

4 CONCLUSION AND FUTURE WORK

The major obstacle to the adaptation wireless EV's charging is the standardization. The inductive coupled power transfer has given greater results than the other types of system. The factors that needs standardization are geometry of coils, volume and weight, position of coil, alignment, frequency, compensation topologies, distance and Efficiency. Including the human and environmental consideration due to electromagnetic radiation. DHI and ARAI has provided standards for conductive charging infrastructure. The standards provided are very limited to the charging equipment and its features. But lacks in charging time, driving range, lack of awareness and efficiency.

References

1 Bloomberg Technology https://www.bloomberg.com/news/articles/2018-09-06/india-proposes-a-goal-of-15-electric-vehicles-in-five-years

2 Energy Efficiency Services Ltd http://www.eeslindia.org/content/dam/doitassets/eesl/pdf/Coffee%20Table%20Book/EESL_Book_final.pdf

3 Department of Heavy Industry https://dhi.nic.in/UserView/index?mid=2418

4 M. C. Falvo, D. Sbordone, I. S. Bayram and M. Devetsikiotis, "EV charging stations and modes: International standards," Ischia, 1134-1139, 2014.

5 http://www.siamindia.com/statistics.aspx?mpgid=8&pgidtrail=14

6 S. Nair, N. Rao, S. Mishra and A. Patil, "India's charging infrastructure — biggest single point impediment in EV adaptation in India," Pune, 1-6, 2017

7 India EV Story- https://www.innovasjonnorge.no/contentassets/815ebd0568d4490aa91d0b2d5505abe4/india-ev-story.pdf

8 A. Ahmad, M. S. Alam and R. Chabaan, "A Comprehensive Review of Wireless Charging Technologies for Electric Vehicles," **4**, 38-63, 2018.

9 SAE International "Wireless Power Transfer for Light-Duty Plug-In/Electric Vehicles and Alignment Methodology" ,2017

10 Green Car Congress - Energy, Technology, issues and policies for sustainable mobility https://www.greencarcongress.com/2016/05/20160518-j2954.html

https://doi.org/10.1051/e3sconf/20198701018

Communication between PHEV's and Smart Grid using Zigbee Protocol

Karthik Rao R[1,*], *Sai kiran P*[2,*], and *Phaneendra Babu* Bobba[3]

[1,2,3]Department of Electrical Engineering, GRIET, Hyderabad, India

Abstract. Plug-in-hybrid electric vehicles commonly known as PHEV's are hybrid electric vehicles that use rechargeable batteries for operation. Since PHEV's run on electric batteries, they require charging after the charge reaches a certain minimum level. The batteries can be charged using external sources usually a smart grid. This requires a wireless technology that can be used to send the information of the battery charge to the smart grid so that it can be charged. This paper is a detailed description of how this communication can be achieved using the ZigBee wireless technology. The battery level information can be sent to the smart grid using this technology and the smart grid operator can then decide whether the PHEV needs charging or not. If not, the battery can be used to provide Vehicle-to-grid (V2G) services i.e. the charge from the vehicle can be sent back to the grid depending on the will of the vehicle owner. Thus, in this way a system can be developed where in both the PHEV driver and the grid operator can benefit.

1 Introduction

Plug-in-hybrid electric vehicles commonly known as PHEV's are hybrid electric vehicles that use rechargeable batteries for operation that can be charged using an external power source. This external power source can be a smart grid. PHEV's draw energy from the smart grid and use it to supply propulsive energy to the vehicle. While the conventional hybrid vehicles used petroleum/diesel for propulsive energy, PHEV's are a step forward and use electric current to achieve the necessary propulsion. PHEV's are seen as an important alternative to the conventional hybrid vehicle because they will reduce dependence on petrol and diesel as well as reduce carbon dioxide emissions associated with the burning of fossil fuels. Therefore, PHEV'S are not only sustainable but also environmental friendly.

PHEV's consist of a gasoline or diesel engine with an electric motor and a large rechargeable battery which can be charged from an outlet. When the battery of the PHEV is emptied, the conventional gasoline engine turns on and the vehicle starts operating as a non-plug-in hybrid. PHEV's produce significantly less carbon dioxide emissions than their gas only counterparts. The efficiency of the PHEV's is extremely high due to the presence of an electric motor and a battery. The cost of a PHEV's is less due to use of less gas as compared to their gas only counterparts.The PHEV's work in two modes-: 1) All-Electric mode and 2) Hybrid mode.In all-electric mode, the electric motor and the rechargeable battery provide all the propulsive needs of the vehicle. The PHEV's continue to run in the all-electric mode until their battery charge is depleted. Usually, the batteries are able to provide propulsive energy up to 40 miles after which they require to be charged.

In the hybrid mode, both the motor and the gasoline are used.The PHEV battery needs to be continuously recharged in order for it to fulfil the propulsive needs of the vehicle. While the smart grid can essentially supply energy to the car battery to charge it, there is a possibility of reverse flow of energy i.e. from the battery to the grid when the electric vehicle is not in use such as during the night time or early mornings. This gives rise to the possibility of two types of power interactions when the electric vehicle battery comes in contact with the smart grid. These are Grid-to-vehicle (G2V) and Vehicle-to-grid (V2G) types of power interactions. The V2G is a new concept which is currently practiced in a selected few countries. Its proper implementation requires the development of communication architecture as well as an efficient technology (wired or wireless) that can transmit the battery information to the smart grid operator.

Two types of V2G architectures have been developed. These are the Direct-deterministic model and the Aggregative model. [1-4]

2 V2G Architectures:overview

2.1 Direct deterministic architecture

* Corresponding author: r.karthikrao427@gmail.com,saikiran170897@gmail.com

 https://doi.org/10.1051/e3sconf/20198701018

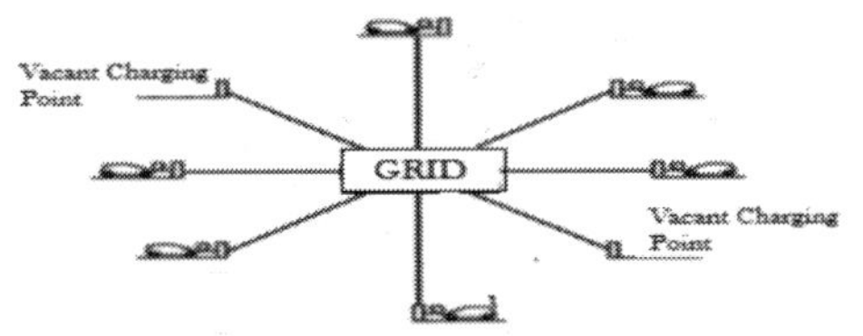

Fig 1. Vehicles under Direct deterministic architecture[4]

In this type of architecture, a direct line of communication exists between the grid system operator and the vehicle owner such that the vehicle acts as a deterministic resource which acts according to the commands of the grid operator.The vehicle can provide ancillary services till the time it is at the charging station i.e. till the time it is connected to the charging outlet.

2.2 Aggregative architecture

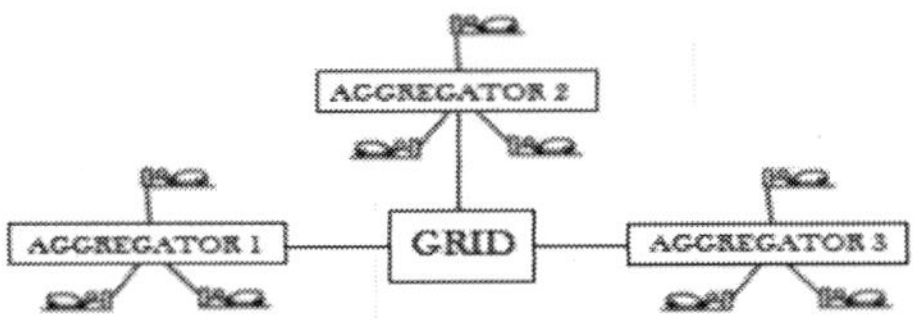

Fig. 2. Vehicles under Aggregative architecture [4]

Aggregative architecture aggregates the vehicles to make a single controllable power source. In this architecture, an intermediary is inserted between the smart grid operator and the vehicles performing ancillary services. The aggregator receives ancillary service requests from the grid system operator which issues the power commands to the vehicles that are available at the station. The aggregator can modify the power commands to the vehicles based on the number of vehicles present at the charging station. Thus even if the number of vehicles at the station is less, the aggregator can still provide the ancillary services to the grid by modifying the power commands issued to the available vehicles.
During V2G power interactions, two primary stakeholders are involved- the vehicle owner and the grid system operator. The V2G architecture under consideration should satisfy the demands of both. Due to this reason, the two architectures mentioned above are compared on the basis of three factors viz Availability, Reliability and Compensation.

3 V2G Architectures: a study availability of v2g architecture

The availability of the architecture is calculated using the Availability factor which is defined as the ability of an individual generation resource to enter into a contract with the grid system operator. It can be calculated as:

$$AF = \frac{\text{Number of hours of service}}{\text{Total number of hours of operation}} \qquad (1)$$

The availability factor is then compared with the availability of the gas turbine power plant which is the conventional ancillary service provider to the grid. Its availability is found to be 92.91%.

3.1.1 Availability of Direct deterministic architecture

The availability of the direct deterministic architecture can be calculated by considering two scenarios during which the vehicles can take part in V2G services. These are V2G from home scenario i.e. vehicles providing services when they are parked at home and V2G from home and work scenario i.e. vehicles providing services when they are parked at either home or workplace. A graph is plotted between the minute-by-minute availability of an average vehicle (A*vehicle*) using the NHTS dataset and the time for both the scenarios.

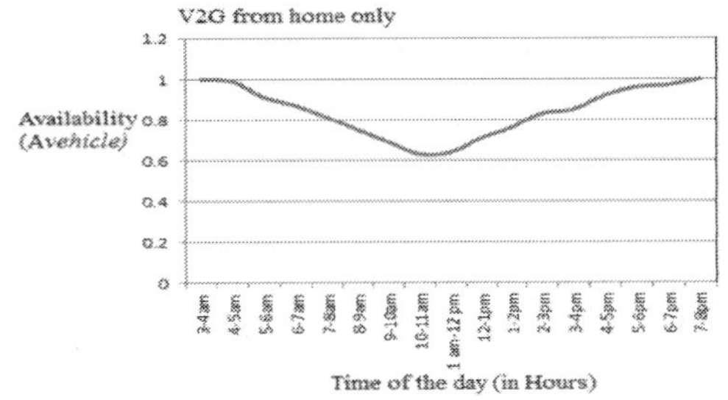

3(a)

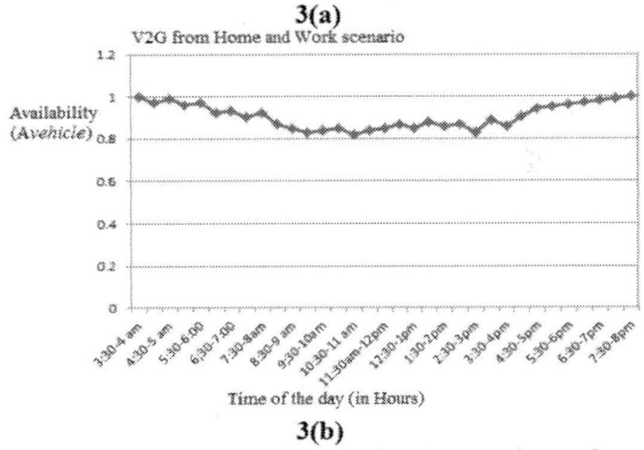

3(b)

Fig 3(a) and 3(b). Graphs for V2G from home and V2G from work and home scenarios.

It is observed that the availability of the vehicles is Maximum during the early morning hours for both the scenarios but decreases gradually as the dayprogresses. However, in case of the work and home scenario the availability is improved due to the ability of the vehicle to take part in V2G services from the workspace also which increases the charge penetration and therefore increases its availability.

 https://doi.org/10.1051/e3sconf/20198701018

3.1.2 Availability of vehicles under Aggregative architecture

In case of the aggregative architecture, the ability of the grid operator to enter into a contract with the vehicle owner is independent of the availability of the vehicle at the charging station due to the presence of an aggregator in between. The aggregator receives power commands from the grid operator. It modifies the power commands for each vehicle depending upon their number. Thus, even if the number of vehicles at the charging station is low, the aggregator is still capable of sending some energy back to the grid due to which the availability of aggregative architecture approaches 100% which is greater than that of the conventional gas turbine power plant.Therefore, the aggregative architecture provides greater benefits to the grid operator due to the larger availability of the vehicles.

3.2 Reliability of V2G architecture

The reliability of a power plant is calculated by its ability to fulfil a V2G ancillary service contract even during its forced down time. The reliability of a V2G service provider is calculated by a factor called the Forced Derated Hours Ratio (FDHR). The FDHR is defined as the ratio of the Equivalent Forced Derated Hours (EFDH) to the Service Hours (SH) i.e.

FDHR= EFDH/SH------------------ (2)

The reliability of a system to provide the contracted and commanded ancillary services is calculated as:-
Reliability,

$$R = (1 - FDHR) \text{------------- (3)}$$

The reliability of the V2G service provider under consideration is compared with the reliability of the gas turbine power plants which are the conventional ancillary service provider to the grid which is found to be:-
FDHR=1.11%
Therefore, R=1-0.011=98.9%
Thus the reliability of the gas turbine power plant is found to be 98.89%.

3.2.1 Reliability of vehicle under Direct deterministic architecture

Under Direct deterministic architecture, the FDHR of the vehicles is found to be 4.65% for the home connection scenario and 5.13% for the home and work connection scenario. The reliability thus becomes 95.35% for the home connection scenario and 94.87% for the home and work connection scenario. In both the cases, the direct deterministic architecture is unable to meet the industry standards reliability of 98.89%.

3.2.1.1 Reliability *of* vehicles under Aggregative architecture

The reliability of the aggregative architecture is found by calculating the total fleet size (Nvehicles) which allows the aggregator to fulfil an hourly contract for a certain power. The fleet scaling factor *x*fleet is determined as follows:-

$$\text{xfleet} = \frac{\ln(1-R)}{\ln(1-AF)} \qquad (4)$$

Where R is the reliability of the baseline gas turbine generator and AF is the availability factor.

- For the home connection scenario, Availability Factor=83.6, R=98.89%. Therefore, xfleet=2.49.
- For the home and work connection scenario, Availability Factor=91.7%, R=98.89%. Therefore, xfleet=1.81.

Using the fleet scaling factor the amount of power contracted by the aggregator (PNvehicles) can be calculated as follows:

$$\frac{\text{PNVEHICLES}}{\text{XFLEET}} \text{-----------------}(5)$$

By increasing the size of the aggregator's vehicle fleet, the reliability can be increased to match the industry standards.

3.3 Compensation from V2G architectures

Compensation is the potential revenue earned by the vehicle owner for taking part in the V2G ancillary services. The vehicle owners will take part in V2G services only if they are given a good compensation for their investment.
The revenue earned is calculated as the
Average Annual Gross Profit=

$$AAGP = \left[rReg_{up} + rReg_{down} - cReg_{up} - cReg_{down}\right] \text{------------------}(6)$$

Where,
$rReg_{up}$ = Revenue earned for regulation up service
$rReg_{down}$ = Revenue earned for regulation down service
$cReg_{up}$ = Cost for a single regulation up contract
$cReg_{down}$ = C ost for a single regulation down contract

The compensation received by the vehicle owners for taking part in V2G ancillary services for the two architectures is compared.

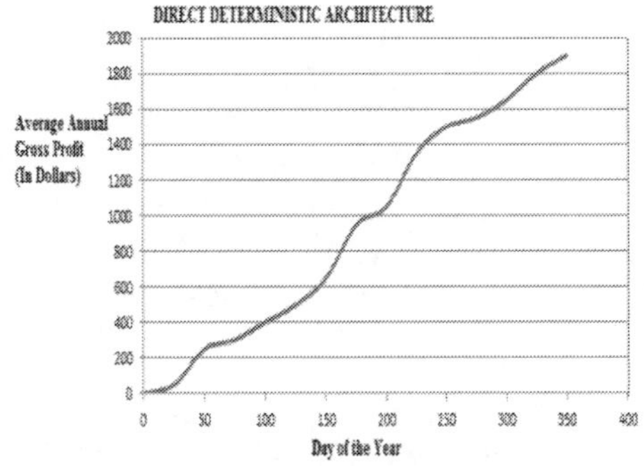

Fig.4. Revenue earned by vehicles for taking part in V2G services under direct deterministic services

It is observed from the graphs that for the same day of the year, the revenue earned under direct deterministic architecture is more than the earned revenue under aggregative architecture. This is because the fleet size increases in an aggregative architecture as more vehicles come to the charging station but the revenue earned by the aggregator (intermediary) for its services to the grid system remains the same which is to be divided equally among all the vehicles connected to the aggregator thereby reducing the amount received by the individual vehicle.

Thus, it can be concluded that from the point of the view of the vehicle owner, the direct deterministic architecture is more beneficial than the aggregative architecture.

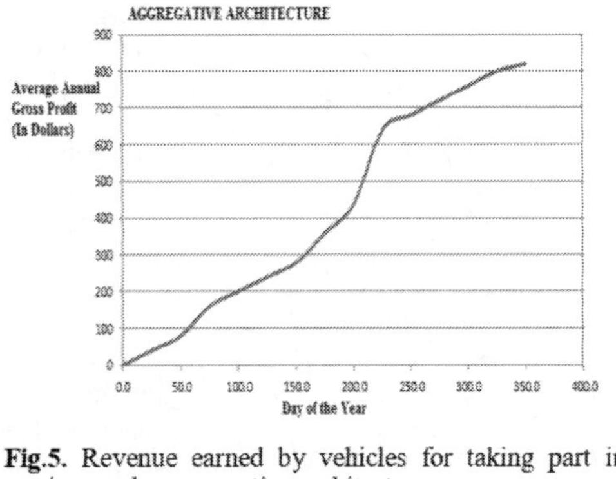

Fig.5. Revenue earned by vehicles for taking part in V2G services under aggregative architecture

4 Establishment of communication using ZIGBEE

Communication plays an important role in sending information to and from the vehicle. The type of communication technology that can be used to transmit information (battery level) can be either wired or wireless. Here, the ZigBee wireless technology which belongs to the IEEE 802.15.4 standard has been used.The following block diagram shows the communication between the vehicle battery and the server.

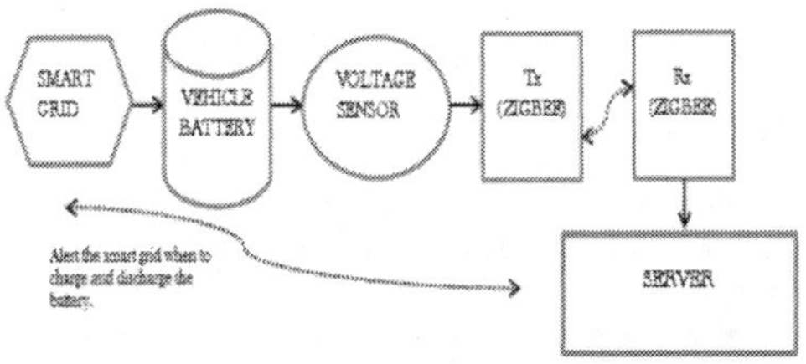

Fig.6. Communication from the vehicle battery to the server.

4.1 Working

The battery of the vehicle is connected to a charging outlet which is connected to the smart grid. A voltage sensor is connected to the battery to sense the amount of charge present in it. The voltage sensor measures the voltage value of the battery and sends the information to the servers wirelessly. The wireless communication is achieved using the Zigbee transceivers. The voltage value is received at the receiving station which is the smart grid operator where its value is analysed and it is decided whether it can take part in V2G ancillary services or not.The various steps involved in the one-way communication as shown in the block diagram are briefed as follows:-
The voltage of the battery is measured by a voltage sensor. The output of the voltage sensor is sent to the QPSK modulator after converting it in the form of bits. After modulation, the modulated information is sent to the demodulator through a ZigBee transceiver and the information is received by another transceiver which sends it to the QPSK demodulator. The demodulator then sends the information to the server where it is analysed. If the value of the charge is less than a certain minimum value (which is predefined in the smart grid's computer system) then the battery is put on the charging mode and if the value is more, the battery can take part in V2G ancillary services depending on the will of the vehicle owner.[5-7]

5 Software implementation using simulink

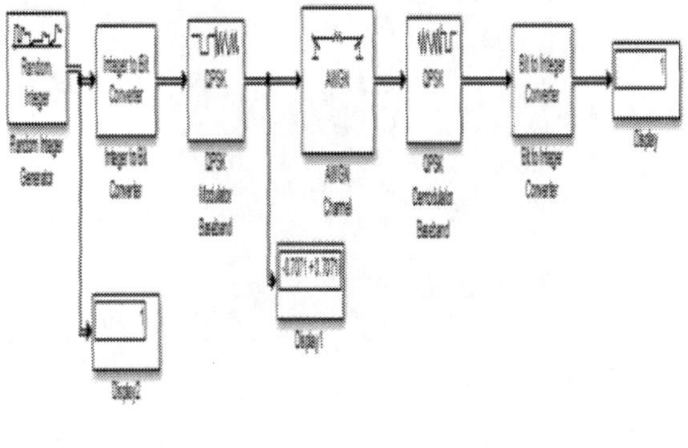

https://doi.org/10.1051/e3sconf/20198701018

Fig.7. Matlab simulation of one way communication between the vehicle battery and server.

5.1 Working

The Random Integer Generator represents the voltage levels in the battery, Integer to Bit converter is Analog to Digital Converter because the input to the QPSK Modulater and Demodulater block should be in the form of bits, AWGN (Additive White Gaussian Noise) is the wireless channel through which the data is transmitted, the display block is used to show the received voltage levels of the battery. In the model, the Random Integer Generator block represents the battery of a car which has voltage and display block represents the server which gets the information about the battery of the car. As the Random Integer Generator takes the integer valueas the input from time to time, the same value is being displayed on the Display block by passing through the various blocks in the Simulink. To be more precise, Random Integer Generator takes one integer at a time and transmits it to the integer to bit converter block which acts as Analog to Digital Converter from which the converted data is being given to QPSK modulator block in the form of bits where it is wirelessly transmitted over a wireless channel to the QPSK demodulator block where it is transmitted to Bit to Integer Converter block which acts a Digital to Analog Converter before finally getting displayed on the Display block which is the sever. The same is being implemented using hardware but in a slightly different manner.

6 Hardware implementation

The model is implemented using the following hardware components.

1. Arduino boards
2. ZigBee transceiver modules
3. A voltage sensor
4. Jumper wires
5. X-CTU software
6. Arduino software
7. DC power supply

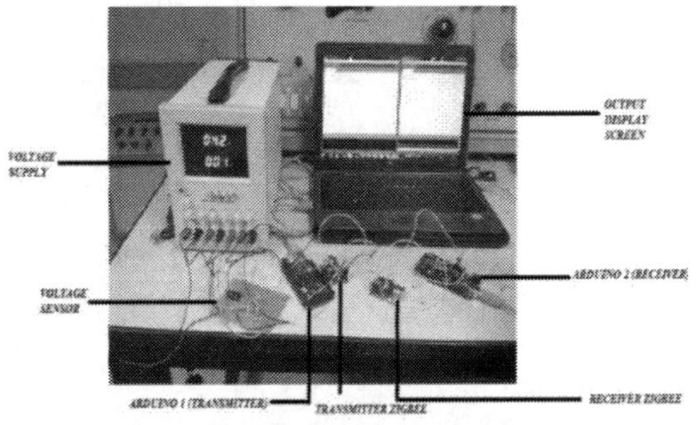

Fig 8. Experimental Setup using Hardware components

6.1 Construction of voltage sensor circuit

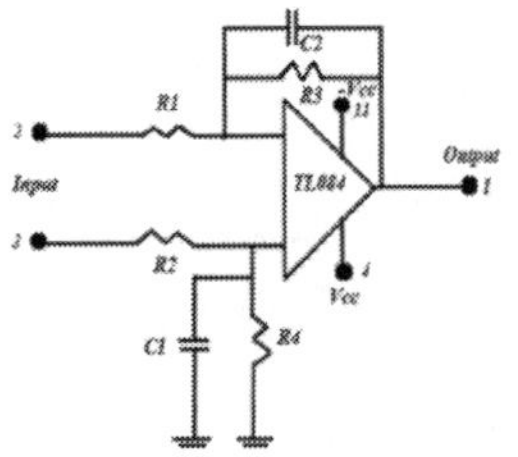

Fig.9. Voltage sensor using TL084 IC

Where R1=R2=10 kΩ, R3=2.2kΩ, R4=2.2kΩ, C1=C2=103pF

The first step to achieving the one-way communication is the development of a voltage sensor circuit whose output is given as an input to the transmitter. TL084 IC is used in the voltage sensor circuit. The voltage senor circuit de-amplifies the magnitude of the input voltage. The output voltage at pin 1 after de-amplification is given by the formula:-

$$Vo=\frac{Rf}{Ri}*Vin=\frac{R3}{R1}*Vin\text{----------} \quad (7)$$

Where Vo=Output voltage at pin 1, Vin=Input voltage. Thus, if the input voltage is 5V, it gets de-amplified to 1.1V and is shown as the output at pin no 1. During the hardware implementation, the output of the voltage sensor is connected to the 'A0' pin of the transmitting Arduino.

6.2 Testing of ZigBee transceiver modules

Before using the ZigBee transceiver modules, it is imperative to check whether they work properly or not. This can be done by checking whether the two Zigbees communicate with each other or not. For this, the XCTU software can be usedwhere the communication between modules is tested by assigning different modes to each module also called as the configuration of modules. The ZigBee module acting as the sender is assigned the router mode and ZigBee module acting as the receiver is assigned the coordinator mode. One of the ZigBee transceiver modules is connected to computer through USB cable and assigned the router mode after which the other Zigbee is connected to the same computer and assigned the coordinator mode. The result of the communication test can be seen on the console log. In the console window of the transmitter, we write something (let's say 'hello world'). If the Zigbees are working and communicating, then this 'hello world' would appear on the console log screen of the receiver screen as well.

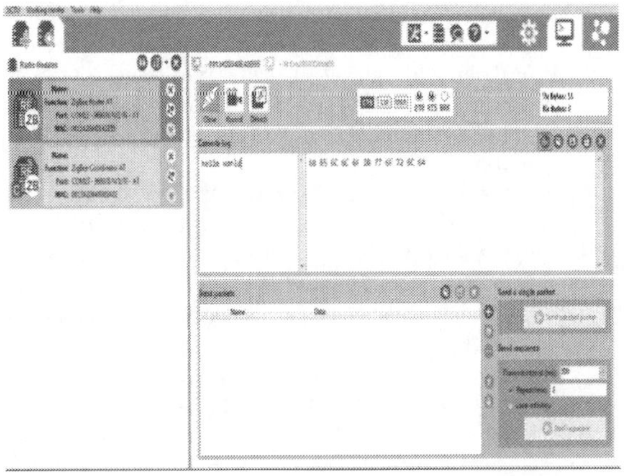

Fig.10. 'Hello World' writtenon Transmitter Console log screen.

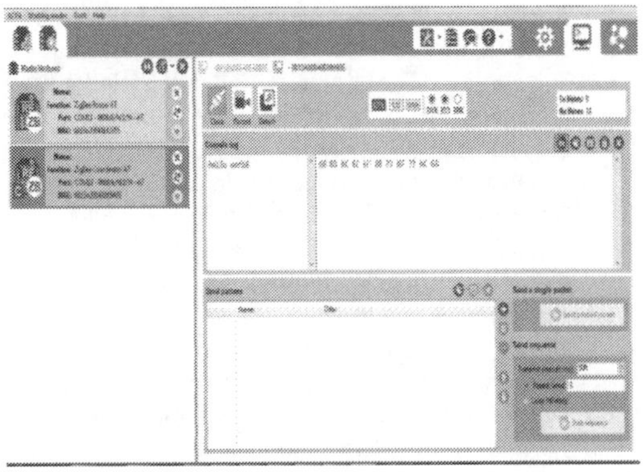

Fig 11. 'Hello World'appears on Receiver Console log screen.

6.3 Connecting the ZigBees to Arduino boards

Once the ZigBee transceivers are tested, they are disconnected from the computer and connected to the Arduino boards.The Arduino boards are then connected to computer through the USB cables. The connections between Arduino board and ZigBee modules differ for the transmitter and receiver. For the transmitter, 5 V of Arduino board is connected to 3.3 V of ZigBee, GNDs of Arduino board and ZigBee are connected, T_0 of Arduino board is connected to Din of ZigBee module and R_0 of Arduino board is connected to Dout of the ZigBee module. For the receiver, 5 V of Arduino board is connected to 3.3 V of ZigBee, GNDs of Arduino board and ZigBee are connected, R_0 of Arduino board is connected to Din of the Zigbee and T_0 of the Arduino board is connected to the Dout of the Zigbee module. After making the connections, the Arduino is connected to the laptop. The transmitter and the receiver programs are uploaded on the respective Arduinos (it is to be noted that while uploading the program on the Arduino, it should not be connected to the Zigbee).For compiling and uploading the programs on the Arduinos, the 'Arduino' software is used.Once the transmitter and the receiver programs are uploaded, the output can be observed on the receiving Arduino's serial monitor screen.

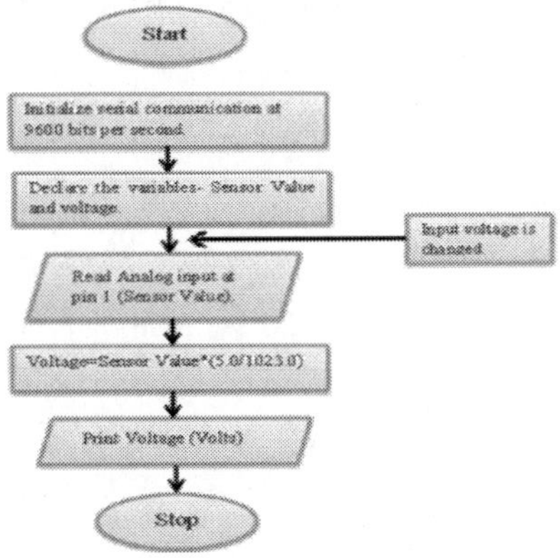

Fig 12. Flowchart showing the transmitter side program

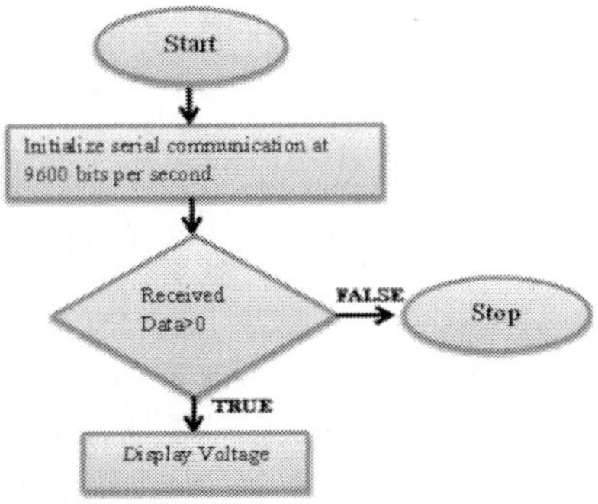

Fig 13. Flowchart showing the receiver side program.

6.4 Transmitter and Receiver Programs

The output of the voltage sensor (which carries the information about the battery level) which is transmitted wirelessly needs to be displayed on the receiving Arduino's serial monitor screen. For this, certain programs have to be uploaded on the transmitting and the receiving Arduinos. The flow chart describing the functions of the programs uploaded on the transmitter and the receiver are shown below.

6.5 Output

For the battery input=4.3V, the voltage sensor output becomes:

$$Vo = \frac{2.2x4.3}{10} = 0.946V$$

This value is wirelessly transmitted from the transmitter to the receiver which can be viewed on the serial monitor screen of the receiving Arduino.

Fig 14. Output of the voltage sensor as seen on the Serial monitor screen of the transmitting Arduino.

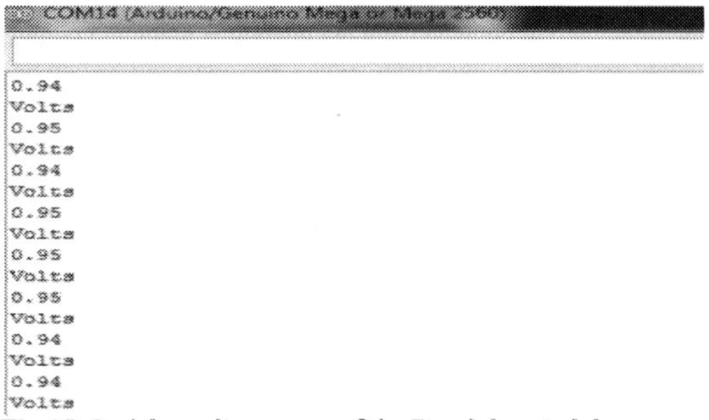

Fig 15. Serial monitor screen of the Receiving Arduino showing the wirelessly received value.

7 Conclusion

In this paper, various architectures to make Vehicle-to-grid type of power interaction a reality are discussed. The architectures should be such that they meet the interests of both the stakeholders involved in this process i.e. the vehicle owner and the grid system operator. For this, the architectures are compared on the basis of three factors namely-Availability, Reliability and Compensation to find out which of the two is able to meet the demands of both the stakeholders and provides greater benefits. In the later sections, the Zigbee wireless technology is used to demonstrate one-way communication between the vehicle and the server using a voltage sensor as the battery's voltage sensor, the Zigbee transceivers and the Arduinos.

References

1. Colak, Ilhami, et al. "A survey on the critical issues in smart grid technologies." Renewable and Sustainable Energy Reviews 54 (2016): 396-405.
2. Palak P Parikh, Mitalkumar G Kanabar, and Tarlochan S. Sidhu, "Opportunities and challenges of wireless communication technologies for smart grid applications", Power and Energy Society General Meeting, 2010 IEEE.
3. Kabisch, Sebastian, et al. "Interconnections and communications of electric vehicles and smart grids", Smart Grid Communications (SmartGridComm), 2010 First IEEE International Conference on IEEE, 2010.
4. Yan, Ye, et al. "A survey on smart grid communication infrastructures: Motivations, requirements and challenges." Communications Surveys & Tutorials, IEEE 15.1 (2013): 5-20.
5. Casey Quinn, Daniel Zimmerle and Thomas H. Bradley, "The effect of communication architecture on the availability, reliability, and economics of plug-in hybrid electric vehicle-to-grid ancillary services", Journal of Power Sources 195.5 (2010): 1500-1509.
6. SaidaElyengui, RiadhBouhouchi and TaharEzzedine, "The Enhancement of Communication Technologies and Networks for Smart Grid Applications", International Journal of Emerging Trends & Technology in Computer Science (IJETTCS, November – December 2013
7. Piyush Ghune, Ruchita N Ghune, Pawan Pandey and Pushpendra Mishra, "Applications of Wireless Sensor Networks and Smart Grid-Opportunities, Challenges and Technologies Available (A Survey)".

https://doi.org/10.1051/e3sconf/20198701019

V2G Transfer of Energy to Various Applications

Dr. G. Sree lakshmi[1], *G.* Divya[2], *G.* Sravani[3]

[1]Professor, CVR College of Engineering, Hyderabad, India
[2] Assistant Professor, CVR College of Engineering, Hyderabad, India
[3]Assistant Professor, CVR College of Engineering, Hyderabad, India

Abstract: In today's world, there is a need of verge of significantant transformation in Electrical Power System. The Vehicle-to-Gird (V2G) concept optimizes this transformation. The PEV typically has a higher capacity Energy Storage System (ESS). Each PEV stores approximately 5-40kWh of energy. This energy can be transferred to the Vehicle-to-Grid (V2G), Vehicle-to-Home (V2H) and Vehicle-to-Building (V2B) as most of the time the vehicle is kept in parking as idle. This paper presents the concept of V2G technology, their classifications, battery storages and types of batteries for V2G.

1. Introduction

Hybrid Electric Vehicle to Grid is a technology of bidirectional flow of electrical energy from vehicle to grid. The system which integrates the electric vehicle into the grid is called Vehicle to Grid system and its interesting feature is to store, sell, or to distribute power [1]-[6]. Hybrid electric vehicles are under development in the recent market by various manufacturers because it also reduces pollution level which is very much useful for future generation to live in healthy environment. In the developing market these vehicles helps in improving the transportation system and helps in reducing our dependence on consumption of petroleum and helps grid to overcome its problems during peak demand [7]-[12]. Because one properly designed vehicle can generate 10kW of power which is equal to power used by 10 houses. PHEV main aim is to shift energy demand for the individual transportation sector from crude oil to electricity. PHEV rely mostly on a battery and on electric motor. Once the battery is discharged, it is again charged by plugging PHEV to power systems. From records it is noted that the average travelling time of a vehicle is about 2 to 4 hours only, rest of the time it is in the parking slot. So, when the vehicle is in parking the battery can be used as power storage and the existing battery power can be fed back into the grid if there is any deficit of power in grid [13]-[19].

In this paper, an electric utility grid-connected energy management strategy and the stages involved in PEVs to transfer energy to V2G are demonstrated. Typically, there are multiple designs that may be used in designing the hybrid electric vehicle as per the customer. To apply this concept, communication between Vehicle and grid is needed, unidirectional power flow is best option because it reduces hardware requirement, interconnection problem and degradation of battery cycles usage. In this paper we will start with basic concept of EHV, next we will study about batteries system, technical details of V2G, stages, applications, advantages, disadvantages, finally about the regulations and standards and its conflicts are explained. Fig.1 shows the transfer of energy from V2G, V2H and V2B. Also, Renewable energy generation is taking place. Fig.2 shows the block diagram of power flow from G2V. It is expected more than 20 million of PEVs will be on road as on 2020.

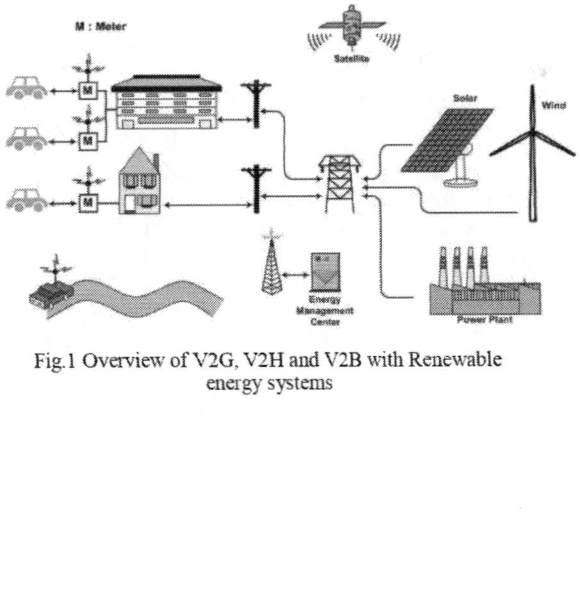

Fig.1 Overview of V2G, V2H and V2B with Renewable energy systems

* Corresponding author: *Dr. G.* Sree lakshmi

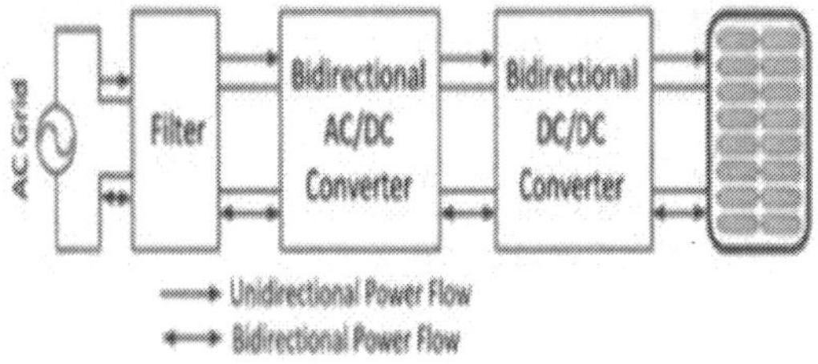

Fig.2 Block Diagram of power flow from G2V

2 Types of Batteries for V2G

The durable, safe, high energy and cost effective batteries have come from lead-acid batteries, batteries nickel-metal hydride (NiMH) to lithium-ion (Li-ion) batteries. Battery cost is the responsible of 20-40% of PEV price. The energy density of battery has been improved from 60Wh/L to 150 Wh/L. Also, the cost of the battery is going down with the improvement in the battery technologies. The present market is using Li-ion batteries as they have long range, higher density, lower cost and non-toxic behavior. They require less time to get charged. Nisan Leaf, Mitsubishi, Tesla Model and Chevrolet Volt are using Li-ion battery for energy storage system. Fig.3 shows the history of batteries for PEVs.

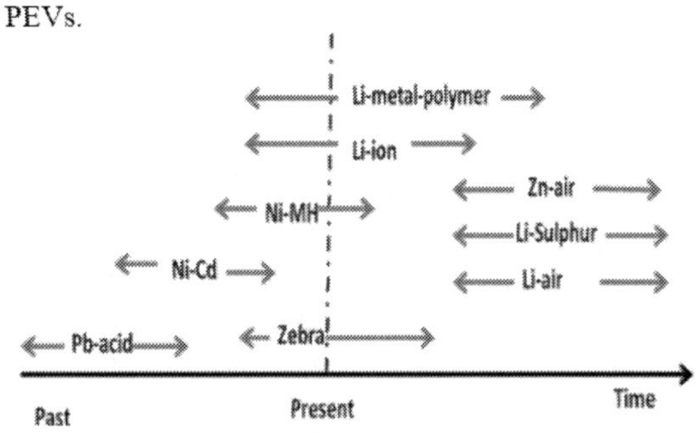

Fig.3 History of Batteries for PEV

Battery degradation is very important if the vehicle is used to travel as well as to transfer energy. The aging of batteries are due to cycling and calendar aging. For medium-long term applications Li-ion batteries are used. Lithium Sulphur (Li-S) has high energy density of 300Wh/L but low battery life. Graphene is one which can solve this problem to some extent. Another promising and cheap technology is Lithium-air batteries which can give 1500Wh/L energy density but needed further research to address several technical challenges.

2. V2G Classifications Based on Batteries Used

3.1 Fuel Cell Vehicles

It uses fossils fuels, bio fuels and hydrogen fuels to generate power using generator which is inbuilt in vehicle during high demand times.

3.2 Rechargable Batteries

These batteries store power during low peak hours that is during night and deliver power to grid during high peak hours.

3.3 Small Renewable Energy Power Stations

Solar vehicle can store power during sun hour and deliver power to grid when battery is full or during high demand period.

4 PEVs classisfications based on Energy Transfer

The integration of PEVs is very important for electric grid operation. The integration has to be done based on charging and discharging of PEVs by using various strategies and control methods. The various applications of PEVS are V2H, V2B and V2G.

3.4 Vehicle to Home (V2H)

The main objective of Vehicle to Home (V2H) is to store energy from local distributed energy resources such as solar energy and wind energy and deliver it to home loads. This will reduce energy bills, improve efficiency and can provide backup power. Other components that can be present in a V2H system are smart loads, other static storage systems and combined heat and power generators (CHP) such as micro gas-turbines. This can be integrated into smart home concept. V2H is composed of at least one PEH, a bi-directional charger, home loads, small-scale distributed generation, a smart meter, home grid and home energy management system.

The other components are smart loads, static storage systems and combined heat and power generators such as micro gas-turbines. Smart metering plays a very important role in V2H concept. It will register data about system consumption and also provides interface to send or receive information from utilities. This concept can be integrated into larger control systems to reduce burden the distribution electric grids. Also, the possibility that PEV user receiving the stop charging request from utilities in their smart phones are made available. Fig.4 shows the Vehicle-to-Home technology energy transfer which has a smart metering in it. Which effectively transfer energy from Vehicle to House during non travelling hours of the vehicles.

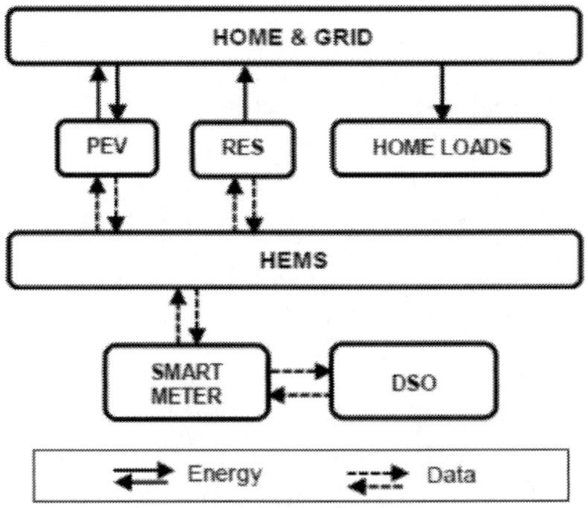

Fig.4 Vehicle-to-Home Technology

3.5 Vehicle to Building (V2B)

Vehicle-to-Building technology is attracting many people. It is an intermediate step between V2H and V2G but provides important advantages for PEVs and Building Owners. Both of them can achieve savings in energy bills as the arriving and departure timings of the workers are known. They can help to transfer energy for hospitals, hotels, universities, office buildings, shopping centers and especially during critical loads. IT consists of set of PEVs, distributed energy generators, critical loads, static storage and Building Energy Management control system. BEMS execute optimum algorithms to send the calculated set-point to different subsystems including the PEVs under its control.

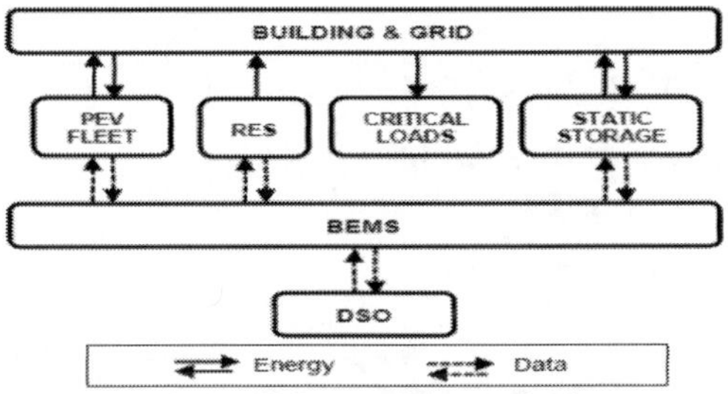

Fig.5 Vehicle to Building Technology

3.6 Vehicle to Grid (V2G)

Vehicle-to-Grid technology was introduced in the year 1997 by Willet Kempton and Steven E. Letendre. This concept explores the possibilities of using PEVs as distributed energy generators. It provides services like spinning and regulation services, peak power and even voltage regulation to distribution system operators. PEVs can provide distributed storage capacity for renewable energy sources. The batteries have very fast response and well suited for primary and secondary frequency regulation. There are two control architectures in V2G systems, centralized and decentralized. In centralized systems the charging and discharging decision in every PEV is taken by aggregator. Whereas, in decentralized controls the decision corresponds on every PEV of the system.

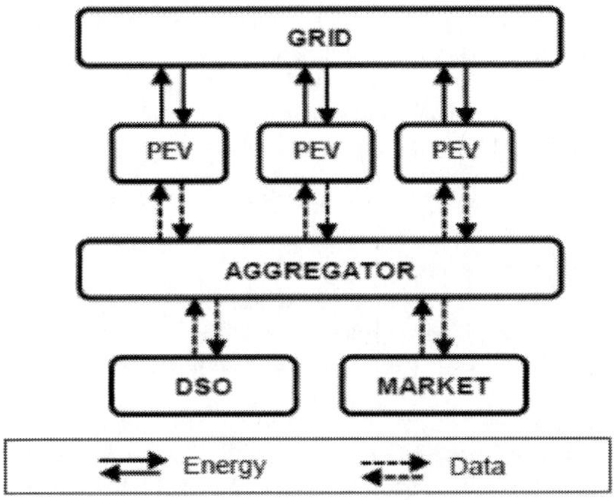

Fig.6 Vehicle to Grid Technology

4 Vehicle Energy Storage System Design

The Energy Storage Systems for PEVs is electromechanical in nature. They consist of cells, modules, packaging, a thermal management system and a battery management system. The battery pack is made up of thousands of cells that are assembled into modules which are connected in series to increase the voltage. the most common type of cell packaging are cylindrical, prismatic and pouch cells. The BMS is the control center for ESS. The BMS keep track of the state of charge of ESS and control the current flow from inside and outside of the vehicle. The TMS ensures the temperature range of ESS. There is a communication path between BMS and the pack and between BMS and TMS.

4.1 Renewable energy Integration with V2G

As the high storage capacity batteries are costly the usage of the renewable sources is not adopted quickly. Moreover the power produced by the renewable energy sources is not constant all the time, so it is becoming difficult during peak load hours from 4 to 9 pm to get the power without fluctuations. For example if wind power is taken, depending on the wind capacity the grid may be over wheeled or face shortage of power. So if PHV are connected to the grid during these peak hours, it can add advantage to the grid to overcome the peak demand. Integrating this EHV and renewable energy resources can help increasing the use of PHV.

4.2 Vehicle to Grid Communication:

The communication path should be a bi-directional. It should be common protocol, reliable, short time delay. There are multiple series connected stages in communication process, the stages is shown in block diagram.

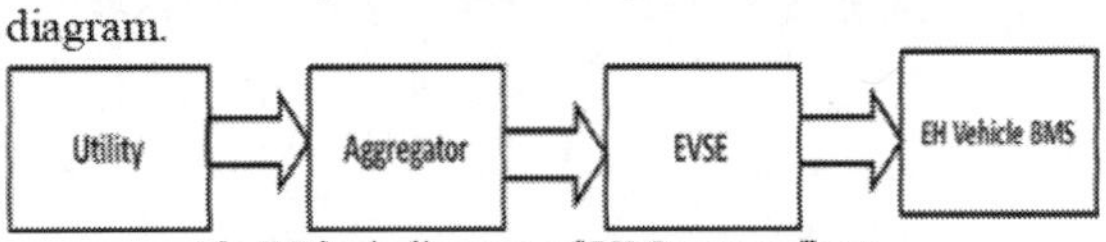

Fig.7 Block diagram of V2G power flow

V2G technology should have a software which helps the vehicle owner to set time for charging, give information about tariff of power, preferences and also should be able to manage thousands of other vehicles on the grid leading to smart grid. This leads to more synchronizing grid and integrating renewable resources as well. The notification of availability of renewable resources also should be given to consumers so that consumers will not connect their vehicles to grid.

Aggregator : It is the mediator between operator and Plug-in electric vehicle which also participate in spinning. Aggregator also gives information of market rates.

EVSE : Electric vehicle supply equipment locks charge cable to provide additional saftey. Tariff selected by the EV is sent to EVSE, with given signal switching process takes placedpending on price of electricity on grid.

BMS : Battery managemnet system controls the rechargable battery by protecting it from voltage changes, temperature e.t.c.

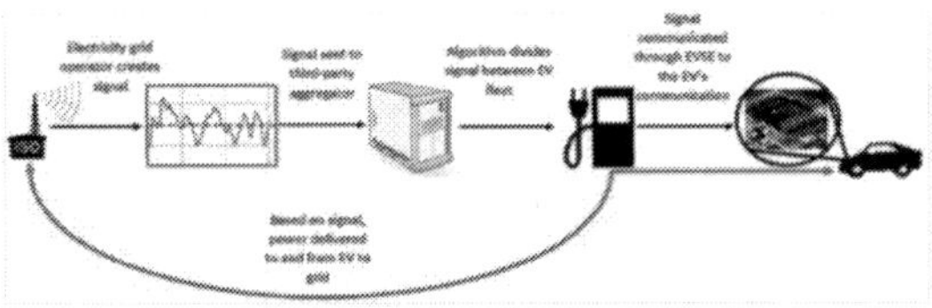

Fig :7.1 V2G communication technology

4.3 Major Components of Vehicle to Grid

- Energy resources like renewable or non-renewable resources and electric utility
- Independent system operator for monitoring the system and aggregator
- Charging point infrastructure and location for battery charging
- Unidirectional communication electric flow between PEV and aggregator(when large no of EHV vehicles are connected aggregator increases its battery size to the point it has sufficient impact) for power transmission from grid to vehicle and vice versa.
- Intelligent metering control system for identifying the capacity available
- PEV with battery charger and management

5 Charging Points for PEVs

The main divisions of charging points are based on speed of charging. They are divided as

1. Rapid chargers
2. Fast chargers
3. Slow chargers

Rapid chargers: These are again divided into 3 types

1. AC charging: In first type of charging points, normal type of charging points which is used in houses and work place. The rating of this type of charging point is 16 A and 110V to 120 V AC charging.
2. Where as in second type of charging points the maximum current is 19.2 KW and 220 to 240V AC. This type helps in fast charging and discharging of PHV.
3. DC charging points: These can be used in public and commercial places and these are similar to gas stations.

These types of chargers allow us to charge 80% of battery in 30 minutes. So the vehicle can run up to 80 to 100 miles.

Fast chargers: These types of chargers can charge PHV vehicle fully in 3 to 4hours. The power output of these devices is around 7 to 22 kW.

Slow chargers: This type of chargers time is from 2 to 4 hours and its power output is only 3 kW.

5.1 Charging Points at Home

It is the easiest way to charge a PHEV at home and it is the cheapest way. Government is also providing subsidies for this. The output power is about 3 to 7 kW. Even car manufacturers are also making deals with charger companies.

5.2 Charging points at Work place and commercial places

Many companies are planning to install sockets to plugging-in EV as most of the time the vehicle will be stationary at this place. In commercial places these charger can attract the customers to park their vehicles. In these places total output range is from 7 to 22 kW.

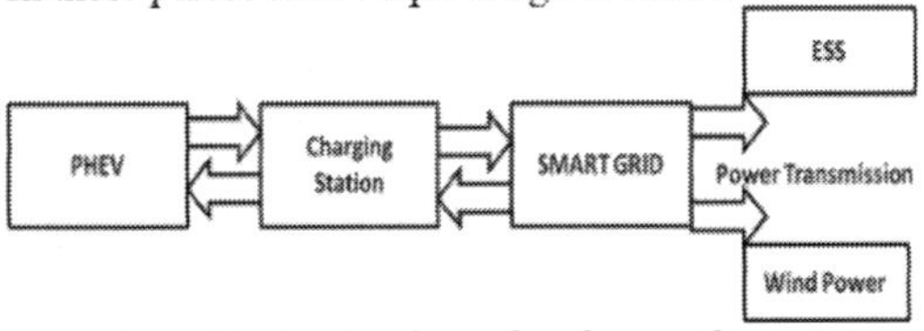

Fig.8 Bi-directional transfer of power from V2G

6 Advantages, Disadvantages & Standards

6.1 Applications of V2G

1. Power is provided to help balance the load by valley filling, peak shaving (that is charging during the low peak hours and discharging during high peak hours into the grid). This is nothing but peak leveling.
2. V2G can also provide peak power for 3- 4 hours a day. [Peak power is nothing but giving the supply during high demand time]. This helps in reducing gap between the energy demands in power system, also reduces transmission loses, investment in transmission loses, mainly the stress in power system.
3. V2G also can act as spinning reserve which means it can provide power within 10 min, when operator demands for power. But this sudden requirement of power should pay additional amount to energy delivering device. The price is divided as[a] Capacity price: Price for responding in a minute notice, [b] Energy price: Price for actual energy output given to

grid. Electric vehicle can behave as spinning reserve only for 3 to 4 hours a day.

6.2 Advantages of V2G

1. Electric vehicles are cheaper source of transportation for consumers as electricity price is lower than the fuel prices.
2. It provides additional revenue to the owners of the vehicle.
3. Electric utility system also have advantages because it will get power from the vehicle during peak period times
4. It reduces pollution by reducing the use of renewable sources.
5. The environment for V2G can be houses, parking lots, employers working place and publicity available charging stations.

6.3 Disadvantages of V2G

1. V2G is not a cheap source when compared to large power plant power generation.
2. Large scale use of these vehicles may lead to technical problems.
3. Compatibilities difficulties may arise when small scale generations are integrated into large power generating units.

6.4 Standards and Regulations for V2G

Identification of standards, regulations, codes for making V2G to be used worldwide. This project is basically divided into 3 stages

1. Stage 1: Modes of operation and functionalities of V2G
2. Stage 2: models, codes, regulations which are in the market as of today
3. Stage 3:Common set of regulations which are to be designed

The main success of this technology is achieved only when the electric utilities and the judiciary should share information to the agencies so that quick adoption of the technology can take place and fix the standards national wide. These standards of National Institute of Standards and Technology, Institute of Electrical and Electronic Engineering society, Automotive engineering proceedings should be followed by the local authorities to implement V2G.

7 Conclusion

In this paper, the concept of V2G, their classifications, batteries used, battery storage system, charging points, various V2G technologies, advantages, disadvantages and standards are clearly stated. Also, how renewable energy sources are used for PEVs has studied. It has observed that, usage of V2G reduces the use of inefficient investment in conventional generation and promotes use of renewable sources. V2G also helps grid to select best options from available renewable sources in fixing tariff at low rates. But with this cost of renewable sources increases because storage of energy is needed. As the renewable sources are not obtained all the time in a day.

References

1. J. Krumm, "How People Use Their Vehicles: Statistics from the 2009 National Household Travel Survey," Apr. 2012.
2. W. Kempton and J. Tomić, "Vehicle-to-grid power implementation: From stabilizing the grid to supporting large-scale renewable energy," *J. Power Sources*, vol. 144, no. 1, pp. 280–294, Jun. 2005.
3. E. V. Landscape, "Global EV outlook for 2020," 2013. [Online]. Available: http://www.iea.org/publications/globalevoutlook_2013.pdf. [4] "Monthly Plug-In Sales in US." [Online]. Available: http://insideevs.com/monthly-plug-in-sales-scorecard/.
4. B. K. Sovacool and R. F. Hirsh, "Beyond batteries: An examination of the benefits and barriers to plug-in hybrid electric vehicles (PHEVs) and a vehicle-to-grid (V2G) transition," *Energy Policy*, vol. 37, no. 3, pp. 1095–1103, Mar. 2009.
5. J. Tomić and W. Kempton, "Using fleets of electric-drive vehicles for grid support," *J. Power Sources*, vol. 168, no. 2, pp. 459–468, Jun. 2007.
6. B. Lunz, Z. Yan, J. B. Gerschler, and D. U. Sauer, "Influence of plug-in hybrid electric vehicle charging strategies on charging and battery degradation costs," *Energy Policy*, vol. 46, pp. 511–519, Jul. 2012.
7. J. D. K. Bishop, C. J. Axon, D. Bonilla, M. Tran, D. Banister, and M. D. McCulloch, "Evaluating the impact of V2G services on the degradation of batteries in PHEV and EV," *Appl. Energy*, vol. 111, pp. 206–218, Nov. 2013.
8. M.-K. Song, Y. Zhang, and E. J. Cairns, "A long-life, high-rate lithium/sulfur cell: a multifaceted approach to enhancing cell performance.," *Nano Lett.*, vol. 13, no. 12, pp. 5891–9, Jan. 2013.
9. X. Feng, M.-K. Song, W. C. Stolte, D. Gardenghi, D. Zhang, X. Sun, J. Zhu, E. J. Cairns, and J. Guo, "Understanding the degradation mechanism of rechargeable lithium/sulfur cells: a comprehensive study of the sulfur-graphene oxide cathode after discharge-charge cycling.," *Phys. Chem. Chem. Phys.*, vol. 16, no. 32, pp. 16931–40, Aug. 2014.
10. K. G. Gallagher, S. Goebel, T. Greszler, M. Mathias, W. Oelerich, D. Eroglu, and V. Srinivasan, "Quantifying the promise of lithium–air batteries for electric vehicles," *Energy Environ. Sci.*, vol. 7, no. 5, p. 1555, 2014.
11. D. P. Tuttle, R. L. Fares, R. Baldick, and M. E. Webber, "Plug-In Vehicle to Home (V2H) duration and power output capability," *2013 IEEE Transp. Electrif. Conf. Expo*, pp. 1–7, Jun. 2013.
12. C. Liu, K. T. Chau, D. Wu, and S. Gao, "Opportunities and Challenges of Vehicle-to-Home, Vehicle-to-Vehicle, and Vehicle-to-Grid

Technologies," *Proc. IEEE*, vol. 101, no. 11, pp. 2409–2427, Nov. 2013.

13. F. Lambert, "Norway is reaching tipping point for electric vehicles as market share reaches record breaking 37%", Electrek, 2017. [Online].
14. "Environmentally Sustainable Transportation Systems - Government Relations - SAE International", Sae.org,2017.
15. "Global ev outlook: Understanding the electric vehicle landscape to 2020", www.iea.org, 2017.
16. K. CZECHOWSKI, "Assessment of Profitability of Electric Vehicle-to-Grid Considering Battery", Graduate, KTH Royal Institute of Technology, 2015.
17. W. Kempton, and J. Tomic, Jun. 2005, "Vehicle-to-grid power fundamentals: Calculating capacity and net revenue", J. Power Sources, vol. 144, 268–279
18. S. Shao, M. Pipattanasomporn, and S. Rahman, "Grid integration of electric vehicles and demand response with customer choice," IEEE Trans. Smart Grid, vol. 3, no. 1, pp. 543–550, Mar. 2012
19. M. Pipattanasomporn, M. Kuzlu, and S. Rahman, "An algorithm for intelligent home energy management and demand response analysis," IEEE Trans. Smart Grid, vol. 3, no. 4, pp. 2166–2173, Dec. 2012.

Compact edge and end coupled wide band filter with notch at 5GHz

Raasi Chenna[1] , *Vijayadithya Doddi*[1] , *Sumanth Gurram*[1], *Vinay Sharma*[2], *Madhur Deo Upadhayay*[3]

[1] Electrical Engineering Department, B.Tech Scholar, Shiv Nadar University, Greater Noida, India, 201314.
[2] Electrical Engineering Department, Ph.D Scholar, Shiv Nadar University, Greater Noida, India, 201314.
[3] Electrical Engineering Department, Assistant Professor, Shiv Nadar University, Greater Noida, India, 201314.

Abstract. In this paper a compact Ultra Wide Band filter with notch at WLAN frequency is designed. The filter is a designed using coupled lines arranged in inter digital capacitor form. The filter is designed to allow microwave signals in the range of 2GHz to 7GHz with a notch at 5GHz. The measured results show the notch at 5GHz with transmission coefficient of -14dB. With uniform group delay in pass band and nonlinear group delay at notch frequency.

1 Introduction

Wireless systems have increased rapidly during the past generation, which has resulted in a fully covered frequency spectrum. In communication, the system can be termed as wideband when the message bandwidth considerably exceeds the coherence bandwidth of the channel. Ultra wide band (UWB) technology has great potential in the development of various modern transmission systems, for instance, through-wall imaging, medical imaging, vehicular radar, indoor, and hand-held systems, etc. Due to this UWB filter with band rejection capabilities are in great demand. Many designs of filters are proposed in recent years with size optimization and improved performance. The wideband bandwidth is forced to use for any communication link due to a high data rate. Aperture coupling integrates two bandpass filter designs for dual/tri band operation [1].Filter with multiple-mode resonators is categorized as stepped impedance resonator and is application oriented with transmitting signals in whole UWB pass band of 3.1GHz – 10.6GHz [2]. In [3] a wide bandpass filter consisting of three parallel coupled lines with the centre one connected to a radial open stub. It operates at a centre frequency of 3GHz. In [4] a filter based on transversal signal interference principle was implemented. Two transmission line section and two shunt open stubs were utilised. UWB filter with quarter wavelength spaced shunt stub transmission lines can be used in UWB communications and localization systems [5][6]. In [7] a band stop filter is designed with dual U-shaped DMS (Defected microstrip structure). DMS circuits are more immune than DGS (Defected ground structure) from crosstalk and ground plane interference. A compact filter using zeroth-order resonance of third-order coupled CRLH resonators for WiMAX applications is discussed in [8]. In [9] different designs combining a bandpass and a notch filter are developed in to operate in the receiving band from 350–470.

In this a paper a UWB filter is designed with notch at WLAN frequency of 5GHz.

2 Design of UWB

The filter is designed on a FR4 substrate with permittivity of ε=4.1 and substrate height 1.6mm. In order to create a pass band behaviour, edge coupling and end coupling are employed. Lower the end coupling gap, higher the transmission. The surface currents at the end gap between the transmission lines affect the filter behaviour. These currents are proportional to the area and thus the increase in area near the change of the direction of surface currents (the reason for higher gap in the centre) improves the design performance. The 5o ohm lines length can be varied according to need. Here, to measure the hardware results and for the probe to touch the line for measuring results, the lines on either side have been extended. In all the graphs showing S parameter results, S11 or S22 characterize reflection coefficient and S12 or S21 characterize the transmission coefficient.

2.1. Design

The layout with dimensions of proposed design are shown in fig 1. All the coupling gaps are all 0.1mm. This design exhibits UWB behaviour between 2GHz – 7GHz, with a notch stop band at 5 GHz.

* Corresponding author: vd875@snu.edu.in rc936@snu.edu.in sg864@snu.edu.in mu689@snu.edu.in

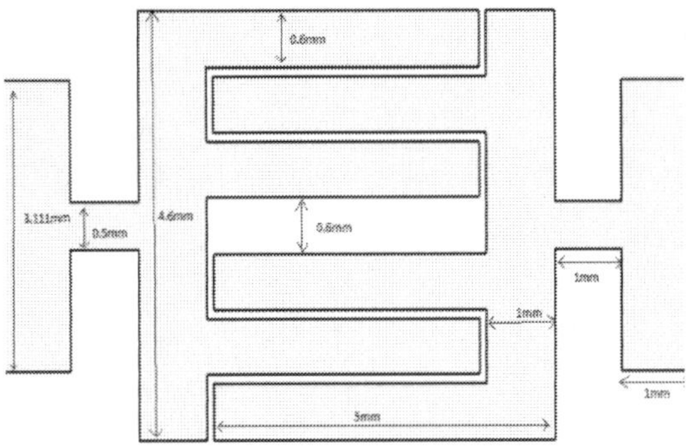

Fig. 1. Optimzed Design with dimensions

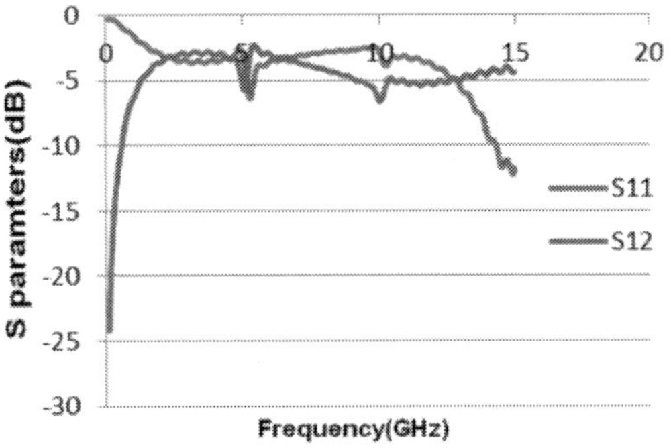

Fig. 2. S-parameters of the proposed design

2.2. Parameters

Transmission is observed to be reduced with an increase in edge gap between the horizontal impedance lines. Change in end gap between 0.1mm to 0.3mm, decreases the transmission and increases the reflection whereas increasing coupling width decreases transmission and BW, increasing reflection. Also, by increasing coupling line length, the pass band starts at an even lower frequency.

3 Results and Discussion

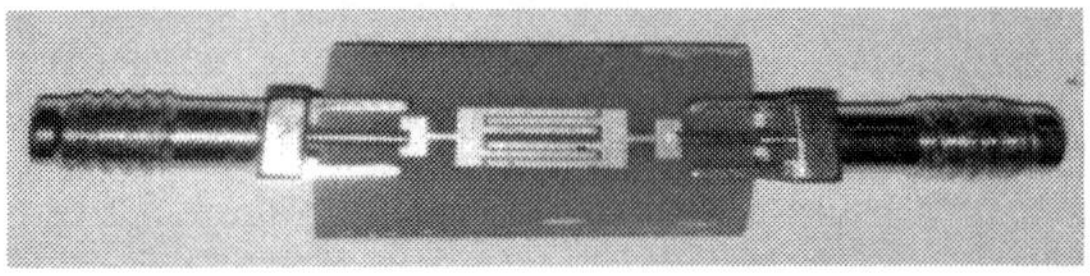

Fig. 3. Fabricated design with SAM connectors.

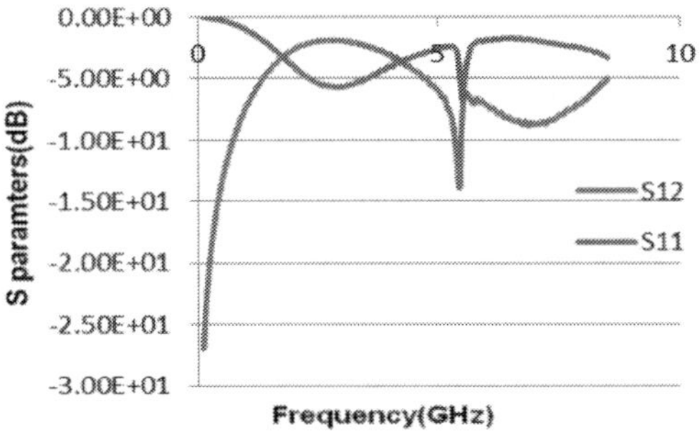

Fig. 4. Measured restum loss and transmission coefficent of proposd design.

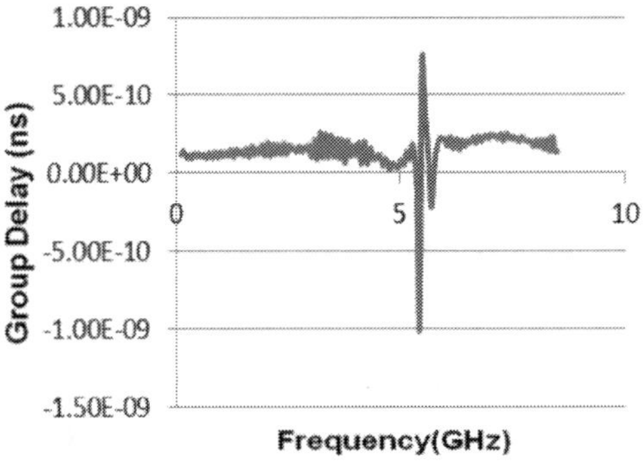

Fig. 5. Measured Group Delay of proposed design

Table 1. Size Comparison

Work	Material	Thickness	Size
Proposed work	Ɛ = 4.3 FR4	1.6mm	10mm x 4.6mm
[10]	Ɛ = 2.2	0.508mm	45mm x 5.8mm
[11]	Ɛ = 10.8	1.27mm	16mm x 1.7mm
[12]	Ɛ = 3 Rogers 3003	1.52mm	10mm x 5mm

The hardware results show the fabricated filter exhibits Ultra wide band between 2GHz to 7GHz. One notch is observed at 5GHz being the ranges of WLAN, BLE, Wi-Fi. From fig 5 it can be seen that the change in group delay is observed at notch frequency. From table I it can be seen that the desinged filter is compact in size.

4 Conclusion

The proposed design is highly size optimized with its dimensions being 10.1mm x 3mm and exhibits good UWB behaviour with less reflection in the range between 2GHz – 7GHz and with a notch at 5GHz. The notch at 5GHz avoids the interference of other communication signals travelling in the same range as the filter supports. Thus prevents extra noise and transmission of corrupted signal in the intended range. The existing UWB filter size which is currently deployed in UWB systems is around 25mm x 28mm. Hence, this a highly compact and size optimized design which behaves as an ultra-wide band filter. The fabricated design as expected, showed one notch 5GHz.

References

1. Shao Yong Zheng , Zhi Li Su, Yong Mei Pan, Zeeshan Qamar, and Derek Ho ,New Dual-/Tri-Band Bandpass Filters and Diplexer With Large Frequency Ratio, *IEEE transactions on microwave theory and techniques*, **vol.66**, no.6, (June 2018)
2. Lei Zhu, Sheng Sun, Wolfgang Menzel ,Ultra-Wideband Bandpass Filters using Multiple-Mode Resonator, *IEEE Microwave and Wireless Components Letters*, **vol.15**, no.11, (November 2005)
3. Yechou Lachen, Tribak Abdelwahed, Kacim Mohamed, Zbitou Jamal and Angel Mediavilla Sanchez, A Novel Low-cost Microstrip Bandpass Filter for Ultra Wideband RF Applications, *International Journal of Microwave and Optical Technology*, **vol.11**, no.2, (March 2016)
4. S.VinothKumar, M.Ganesh Madhan, Kaushik R A simple UWB Bandpass Filter Design based on signal Interference Approach, *CAE-ISSN:2394-4714, FCA-Newyork, International conference on computing, communication and sensor network,* (2015)
5. Subodh Kumar Singhal, Deepak Sharma, Ram Mehar Singh Dhariwal, Design of 1.3GHzMicrostrip Highpass Filter Using Optimum Disturbed Short Circuited Stubs, *First International Conference on Computational Intelligence, Communication Systems and Network*, (2005)
6. Cansever, Cem, Design of a Microstrip Bandpass Filter for 3.1-10.6 GHz Uwb Systems, *Electrical Engineering and Computer Science – Thesis.* Paper 1 (May 2013)
7. Dongsheng La, Yinghua Lu, and Siyang Sun Novel Bandstop Filter Using Dual-U Shape Defected Microstrip Structure, *Proceedings of International Symposium on Signals, Systems and Electronics,* (2010)
8. Mahmoud A. Abdalla, Ashraf Y. Hassan, Ahmed A. Ibrahim Building a Zero Order BPF with CLRH Transmission Lines, *Microwaves & RF Maganzine*, (August 2018)
9. Silas Christensen, Vitaliy Zhurbenko and Tom K. Johansen, Compact Microstrip bandpass filter with tunable notch, *International Conference on Microwave, Radar, and Wireless Communications*, (2014)
10. Jun-Dong Ye, Liu He, Xing Jian Zhong and De Xin Qu, Design of Ultra-Wideband Bandpass (UWB)Filter with Enhanced couplings by Using lumped capacitors, *International confrence on Microwave and Millimeter wave Technology,* (2018)
11. Yunxiu Wang, Yan Liu , Wenhui Zou ,Yusong Hu and weichao Yang, Design of UWB Filter with Dual Notched bands, *IEEE International confrence on RFID Technology and Application,*, (2018)
12. Nilesh Kurhe and R. P. Labade , Metamaterial based Ultra Wide Band Low Pass Filter Using New Configuration of Complementry Split Ring Resonator Cell, *International confrence on Information , Communication , Engineering and Technology,* (2018)

Development of wireless charging system along with power line communication used in Electric Vehicles

Sai Varun Alapati[1*], *Indusaiteja* Nadella[2*], *Phaneendra Babu* Bobba[3] and *Madhur Deo* Upadhayay[4]

[1]Dept of Computer Science,State University of NewYork at Buffalo, Buffalo, New York, USA
[2]Senior Software Developer Associate, NTT Data GDS PVT Ltd, Hyderabad, Telengana, India
[3]SMIEEE, Dept. of EEE, GRIET, Hyderabad, Telangana, India
[4]Electrical Engineering Department Shiv Nadar University, Greater Noida, India

Abstract. Charging an Electric Vehicle wirelessly is the latest technology being developed for the electric vehicles replacing the traditional way of plugging to the supply. In this paper, authors explained how to integrate Power Line Communication along with wireless power transfer in EV. The entire system is implemented in ADS simulation software. We are adapting to magnetic resonance coupling method for wireless power transfer in EV. The overall ideology of the project is to design an innovative system which involves higher power transfer and implement smart communication system between vehicle and the grid while following the latest magnetic resonance charging standards.,

Keywords Electric Vehicle (EV), Power Line Communication (PLC), wireless power transfer, communication protocol, Wireless Power Transfer (WPT), Power Transfer Efficiency (PTE)

1 Introduction

The cost of fuels like petrol, diesel etc. has been steadily increasing due to increased number of vehicles and proportional excess usage of fuel. Depleting sources of these fuels are also a major concern. These age old designs of vehicles are the major contributors to the problem of greenhouse gases.

The future of automotive technologies is moving toward electric vehicles which are considered as a replacement to oil-powered internal combustion engine driven vehicles, keeping in mind CO2 reduction. Alternate energy Plug-in Electric Vehicles (PEVs) have been proposed to achieve environment friendly transportation. Even though the PEV usage is currently increasing, a breakthrough would be required to overcome battery related drawbacks.

Furthermore, PEVs have not been accepted as a preferred choice by many consumers due to charging related issues. This is the major issue faced in the field currently.

Wireless power transfer (WPT) was discovered and enacted by Nikola Tesla. This was further enhanced by many other scientists. The concept has been implemented in many applications for e.g. mobile charging, laptop charging, and electric vehicles etc. However application in electric vehicles was a major finding in the field of Automobile and Transportation. Wireless power transfer methods are bounded with technologies such as Radio waves, micro waves, photoelectric, laser, inductive coupling and magnetic resonance coupling. Based on the power transfer distance WPT methods are categorized into near field and far field. Inductive coupling and magnetic resonance coupling are regarded as near field transfer methods; whereas the other methods are categorized into far field approach. The magnetic resonance coupling method is recommended for WPT of EV because of high range and higher frequencies secured by the method when in compared with other WPT methods.

In [1], Chun et al. discuss about the various methods in WPT for EV charging. Out of which the strongly recommended one is the coupled resonance method for usage in EV's. The challenges faced in any method are the coil design, as they decide upon the upper boundary limit of power capacity and also the efficiency which depends on the quality factor (Q) of the coils. The frequency range at which the coil is designed varies from 10 kHz to 150 kHz and even more, but we obtain high ac

* Corresponding author: saivarun@buffalo.edu
* Corresponding author: Indusaiteja.nadella@gmailcom

https://doi.org/10.1051/e3sconf/20198701021

resistance when there is an increase in frequency which leads to decrease of skin effect. The high ac resistance is also observed due to proximity effect as the eddy currents are flown along the strands of the wire.

In [4], Miller et al. provides an explanation to use litz wires so as to reduce the ac resistance. It reduces the effect of both skin and proximity effects by ensuring the ac current to flow uniformly throughout the wire. Smaller diameters of copper wires are used so that radius of each individual strand is less than their skin depth, so the higher the frequency the smaller the inductor. Litz wire also increases the range of frequencies which can be used, that give high efficiency and whose weight is very less when in compared with conventional wire.

In [5], Subhadeep et al. experiment in comparing coil structures among circular spiral, square spiral and helical coils. Out of which the inductance is observed to be least for circular spiral coil and highest for helical shaped coil. This paper also proposes a multilayer spiral structure which observes high Q factor when it had necessary space constraint and a better field distribution in coils.

As the coil design is important, in [3] Rikard et al. uses a circular spiral coil which at a source frequency of 85 kHz is supplied and both the coils are identical and shielding is done with a ferrite core, where the coils are places at 30 cm apart and checked the shielding effect on vehicle side and ground side both inclusive and also placing shielding only on vehicle side. The WPT's develop a lot of magnetic fields as the power is transferred wirelessly but they are ensured that they are safe to use by the general public keeping in mind about the health of humans as well as animals. The power in reference [3] is transferred at a peak value of 1.3 kW with an efficiency of 90%, air gap maintained at 0.3 m.

As WPT is hazardous for humans as well as animals, we must ensure that certain standards are to be followed while implementing WPT in the system. In [6], author discuss mainly of standards being followed in US and Europe. In US, the main standard for EV charging is the Society of Automotive Engineers (SAE) standard J1772. Whereas in Europe the standard followed is International Electro technical commission (IEC) 61851. The above mentioned standards define the requirements for "Electric Vehicle Supply Equipment" (EVSE).

In [7] the discussion is about SAE TIR J2954 standard which establishes an industry-wide specification guideline that defines acceptable criteria for interoperability, electromagnetic compatibility, minimum performance, safety and testing for wireless charging of light duty electric and plug-in electric vehicles. The standard SAE TIR J2954 is intended for stationary applications (charging while vehicle is not in motion).

Based on [3], the study about the WPT system and coil design, the circular spiral coil is taken in consideration with [3] as the required power rating for the application is near about 1.3 kW. Thus, we are implementing magnetic resonance method for WPT in our system.

The development of Li-ion battery packages show the need of a Battery Management System (BMS) which improves the performances of all cells ensuring correct charging/discharging, continuous monitoring for preventive maintenance etc. A BMS needs data transfer, which can be efficiently performed by the use of a PLC (Power Line Communications) data channel demonstrated in [8]

In [9], the Qi standard demonstrates that the communication between Transmitter and Receiver happens in the form of packets. A Packet consists of a preamble, a header byte, a message, and a checksum. A Packet is named after the kind of message that it contains.

In [10], authors proposed a magnetically coupled resonator composed by four resonant circuits, magnetically linked by coupling coefficients and simulated the WPT channel frequency response. They calculated the channel capacity for a WPT system by using data communication transfer as a function of the coupling coefficient, k and by considering its frequency response and additive white Gaussian noise (AWGN). They proposed architecture for integrating the PLC in a WPT system.

In [11], author proposed a coupled WPT-PLC as a continuation to [10] and constructed a two coil system and measured the scattering parameters used to design a four port system and implemented a simulation package in ADS. Their results confirmed the theoretical studies previously published and showing that the system guarantees power transfer together with reasonable and effective data transfer.

2 Wireless Power Transfer in Electric Vehicles

2.1. MAGNETIC RESONANCE COUPLING

Magnetic Resonance is a near field WPT technology. It works on the principle of resonance of inductance and capacitance present in the circuit. If two or more pairs of RLC resonators are used then the WPT system is used to increase the PTE and extend its range. The inductive power transfer and magnetic resonance coupling come under near field WPT technologies, but there is a small difference between these two technologies. The frequency range at which the power transfer can be done is higher in magnetic resonance coupling when in compared with inductive coupling. With the usage of more pairs of RLC's the coupling becomes stronger and is known as strongly coupled resonance circuit as mentioned in [1]. We use an equivalent circuit theory method (Thevenin equivalent) to find the circuit parameters such as currents and voltage across coils, compensation capacitors. So it is essential to use Thevenin equivalent method to estimate component ratings in designing high power WPT system. The two coil systems consists of a primary coil and secondary coil whereas the four coil system consists of two extra coils known to be drive loop and load loop which are coupled with transmitter coil and receiver coil respectively. Both the coil systems have their own advantages and disadvantages. But we consider two coil systems for our project and proceed per [2].

2.2. EV CHARGING SCHEMES

Electric vehicles are successful only when the charging stations are easily accessible to the public when required. The critical point to think is about how are the grids in support with the electric vehicle specifications.

The first place considered for charging stations should be homes and workplaces. Other potential locations with high populations include gas stations, shopping centres, restaurants, entertaining places, highway rest areas, municipal facilities, and schools.

There have been various standards regarding the energy transfer, connection interface and communication for EV charging, the following table summarizes the standards to be followed for Electric vehicle charging schemes [2]:

Table 1. STANDARDS OF EV CHARGING

Standard	Title/description
National Electric Code	Electric Vehicle Charging System
SAE J2293	Energy Transfer System for Electric Vehicles
SAE J2836	Recommended Practice for Communication between Plug-in and Vehicles and Utility Grid
SAE J1772	Electric Vehicle Conductive Charge Coupler
SAE J1773	Electric Vehicle Inductively Charge Coupler
IEC 62196	Plugs, socket outlets, vehicle couplers and vehicle inlets-Conductive charging of electric vehicles
IEEE 1547.3	Interconnecting Distributed Resources with Electric Power Systems

3 Methodology for PLC Implementation

In the scheme shown in the Fig.1 the power signal and the data signal are superimposed to form a PLC signal at the outlet. There are two tasks of this WPT system. First, transmitting the power from the power source to the load and enabling the communication to the PLC Device. Second, the high power signal at low frequency is separated from the high frequency, low power PLC signal, by using a low pass and a high pass filter respectively. By using a DC to radio frequency amplifier, the power signal is then transformed to DC and then converted to a high power sinusoid at the resonance frequency f_0 of the WPT system. In order to remove the out of band noise generated by the amplifier, the output of the RF amplifier is filtered by a band pass

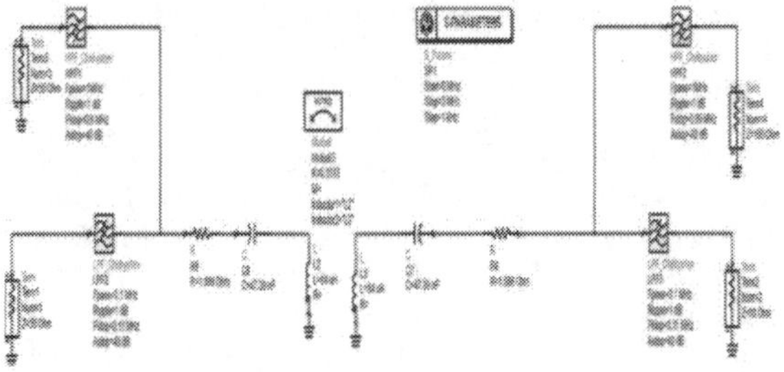

Fig.2. Proposed WPT-PLC system

filter centered at fo. The PLC signal is coupled at the exit stage of the amplifier and to prevent the propagation of high power signal generated by the DC/RF amplifier through the PLC signal path, a band stop filter with a narrow stop-band centered at fo is added. The voltage source, V_{source} at this point i.e., the input of the drive loop is represented by the sum of the high power radio-frequency signal centered in fo and the low power PLC signal. This signal then propagates to the load loop through the WPT resonators. At the output of the load loop, a band pass filter centered at f0 extracts the power signal and sends it to a RF/DC rectifier to provide the power to the DC load. A band stop filter centered at f0 is used to remove the power signal from the superimposed signal to get PLC signal, which is then coupled to a PLC device.

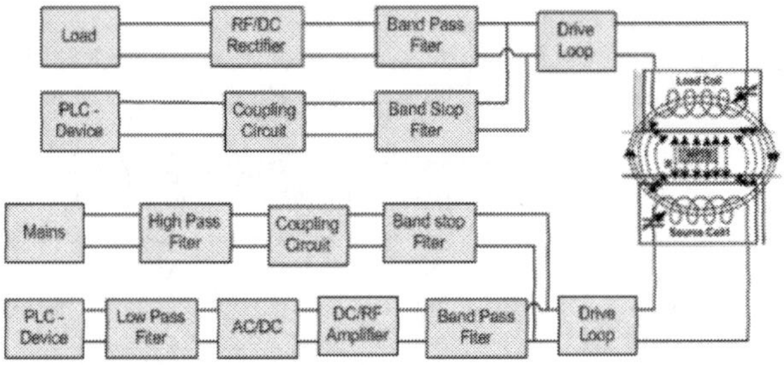

Fig.1. PLC and Implementation of system block diagram

In the above scheme, the power transmission is unidirectional where as it allows a bidirectional communication. The PLC device connected to the load loop can be enabled to transmit data that will propagate in the same method from the receiver coil to the transmitter coil as a low power broadband signal. The WPT systems are mainly designed to have two resonance frequencies one peak for power transfer. So we can use the frequencies around the other peak for the PLC communication.

4 Simulation results

4.1. Proposed WPT-PLC system

The scheme of the coupled system is shown in Fig.2, which is implemented in circuit simulator ADS. Port 1, port 2 are the input and output terminations for the power channel and port 3, port 4 are the input and output terminations for the power line data signal.

The power signal is transmitted through the coupling inductors and decided to adopt the power transmission frequency of 85kHz and the communication signal is transmitted through the band 1- 3MHz. The are two 1MHz chebyshev high-pass filters and the two 0.1 MHz chebyshev low-pass filters of band pass with a ripple of 1dB and an attenuation for all the filters in order to isolate the data from the power channels and vice-versa.

Fig.3 represents frequency response of the transmission coefficients S (2,1) (in terms of the scattering matrix) on the power channel at K =0.35 with peak value 0.82. The frequency chosen for the power link is 85 kHz.

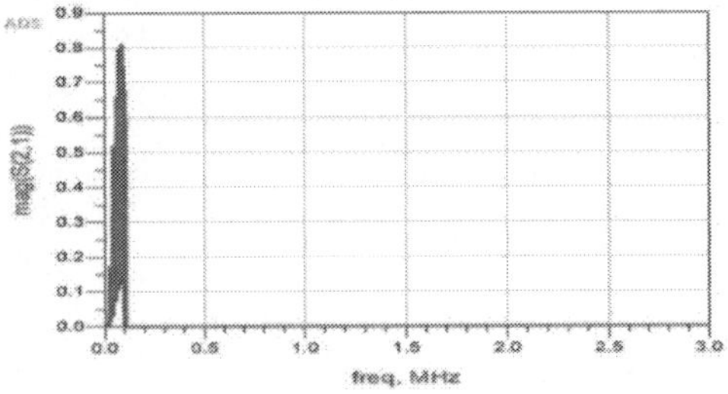

Fig.3. Transmission coefficient S(2,1) on the Power Channel at K=0.35

Fig.4, represents frequency response of the transmission coefficients S(2,1) (in terms of the scattering matrix) on the power channel at K =0.6 with peak value 0.90. The frequency chosen for the power link is 85 kHz.

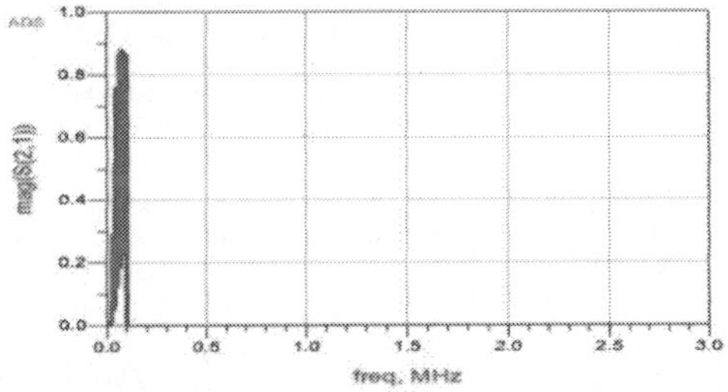

Fig.4 Transmission coefficient S(2,1) on the Power Channel at K=0.6.

Fig.5 represents frequency response of the reflection coefficients S(1,1) (in terms of the scattering matrix) on the power channel with K = 0.35. The frequency chosen for the power link is 85 kHz.

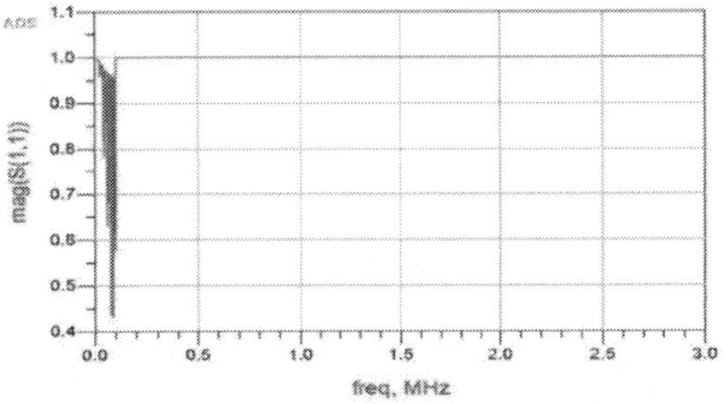

Fig.5. Reflection coefficient S (1,1) on the Power Channel

Fig.6 represents frequency response of the reflection coefficients S (1,1) (in terms of the scattering matrix) on the power channel with K =0.6. The frequency chosen for the power link is 85 kHz.

Fig.6. Reflection coefficient S (1,1) on the Power Channel at K=0.6

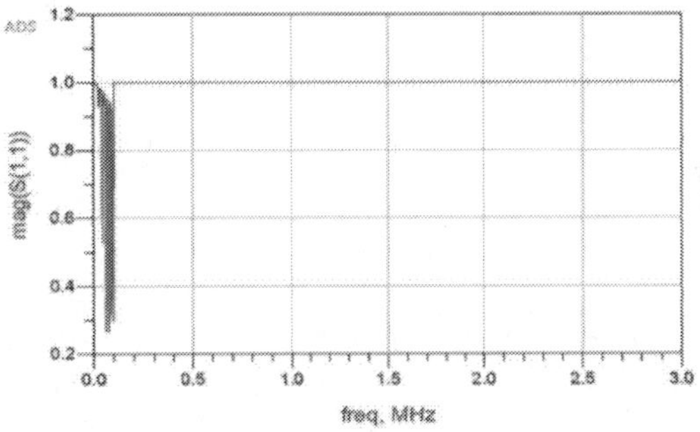

Fig.7 represents frequency response of the transmission coefficient S(4,3) (in terms of the scattering matrix) on the Power line channel in the frequency range 1-3Mhz with K =0.35. The frequency chosen for the power link is 85 kHz.

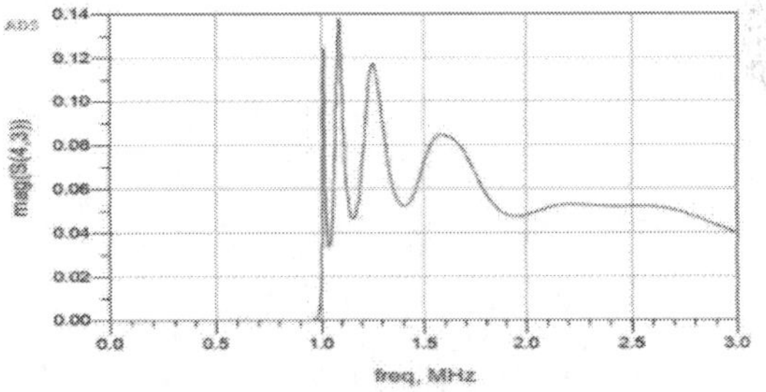

Fig.7. Transmission coefficient S(4,3) on Power Line Channel in the frequency range 1-3 MHz with k =0.35.

Fig.8 represents frequency response of the transmission coefficient S (4,3) (in terms of the scattering matrix) on the Power line channel in the frequency range 1-3MHz with K=0.6. The frequency chosen for the power link is 85 kHz.

 https://doi.org/10.1051/e3sconf/20198701021

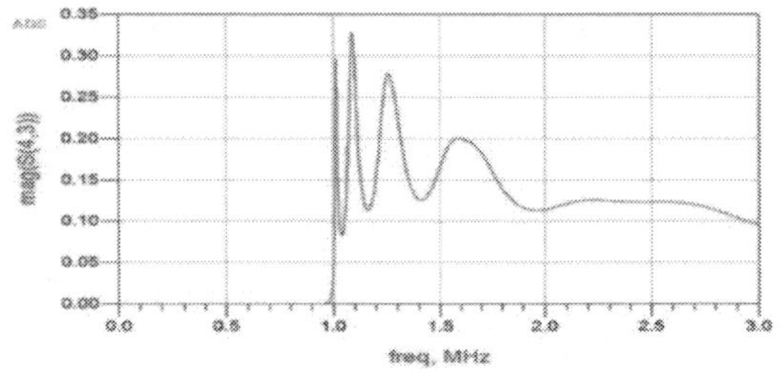

Fig.8. Transmission coefficient S (4,3) on Power Line Channel in the frequency range 1-3 MHz with K=0.6.

Fig.9 shows the isolation between the input port of the Power channel and the input/output ports of the power line data channel at K=0.35. The frequency chosen for the power link is 85 kHz.

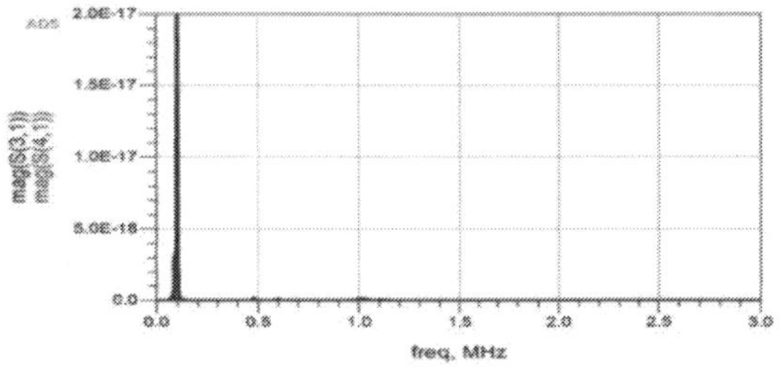

Fig.9. Isolation among the input port of the Power channel and the input (3) and output (4) ports of the Power Line Data channel at K=0.35.

Fig.10 shows the isolation between the input port of the Power channel and the input/output ports of the power line data channel at K=0.6. The frequency chosen for the power link is 85 kHz.

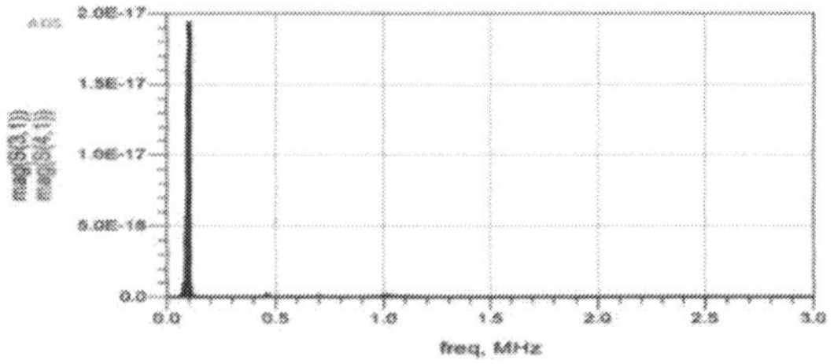

Fig.10. Isolation among the input port of the Power channel and the input (3) and output (4) ports of the Power Line Data channel at K=0.6.

5 Conclusions

In this paper authors implemented magnetic resonance coupling method for wireless power transfer in EV and also demonstrated how to integrate Power Line Communication along with wireless power transfer in EV's. The simulation results demonstrates that the transmission coefficient of the power signal is acceptable at k = 0.35 and as we increase k to 0.6 the transmission coefficient increases, while both k values guarantees power transfer together with reasonable and effective data transfer. The power signal can be improved by increasing the power transmission frequency with permissible k.

References

1. Chun Qiu, K.T. Chau, Chunhua Liu, and C.C. Chan, "Overview of Wireless Power Transfer for Electric Vehicle Charging", in EVS27 International Battery, Hybrid and Fuel Cell Electric Vehicle Symposium, Barcelona, Spain, November 17-20, 2013, pp. 1-9.

2. D.M. Vilathgamuwa and J.P.K. Sampath, "Wireless Power Transfer (WPT) for Electric Vehicles (EVs)—Present and Future Trends", in Plug In Electric Vehicles in Smart Grids, S. F. Rajakaruna S., Ghosh A., Ed., ed Springer International Publishing AG, Part of Springer Science+Business Media: Springer Singapore, 2015, pp. 33-60.

3. Rikard Vinge, "Wireless Energy Transfer by Resonant Inductive Coupling", , Master of Science Thesis, Department of Signals and systems, CHALMERS UNIVERSITY OF TECHNOLOGY, Goteborg, Sweden 2015.

4. J.M. Miller, C.P. White, O.C. Onar, and P.M. Ryan, "Grid side regulation of wireless power charging of plug-in electric vehicles," in Proc. Energy Conv. Congress Expo. Raleigh, NC, 2012, pp. 261-268.

5. Subhadeep Bhattacharya, "Design of Static Wireless Charging Coils for Integration into Electric Vehicle", IEEE ICSET 2012, Nepal, pp. 146 - 151.

6. WiTricity, WiT-3300, [Online]. Available: http://witricity.com/innovation/standards/

7. SAE International, Hybrid-Ev committee [Online]. Available: http://standards.sae.org/wip/j2954/

8. I. Ouannes, P. Nickel, and K. Dostert, "Cell-wise monitoring of lithiumion batteries for automotive traction applications by using power line communication: battery modeling and channel characterization," in Proc. 18th IEEE Int. Symp. on Power Line Commun. and its Appl., Mar. 2014, pp. 24–29

9. "http://www.qiwireless.com/category/wireless-standards/" QIWireless

10. S. Barmada, M. Tucci, M. Raugi "Power Line Communication Integrated in a Wireless Power Transfer System: a Feasibility Study",Proceedings of International Symposium on Power-Line Communications and Its Applications, ISPLC 2014, 30 March - 2 April, 2014, Glasgow, Scotland, pp. 116 – 120.

11. S. Barmada, M. Tucci,M. Raugi M. Dionigi, P. Mezzanotte "Experimental Validation of a Hybrid Wireless Power Transfer - Power Line Communication System" International Symposium on Power Line Communications and its Applications (ISPLC), 20

E3S Web of Conferences **87**, 01022 (2019) https://doi.org/10.1051/e3sconf/20198701022
SeFet 2019

Analysis and Design of PMBLDC Motor for Three Wheeler Electric Vehicle Application

V. Sandeep[1,*], and *Sharankumar* Shastri[2]

[1]Department of Electrical Engineering, Central University of Karnataka, Gulbarga 585367, India
[2]Department of Electrical Engineering, Central University of Karnataka, Gulbarga 585367, India

Abstract. This paper deals with analysis and design of permanent magnet brushless dc machine (PMBLDCM), primarily aimed for three wheeler applications. The motor sizing accounts for the forces acting on the motor and the design variables such as number of stator and rotor slots, stator and rotor dimensioning, air-gap approximation, slot sizing, flux per pole and permanent magnet sizing has been explained using simplified equations. The designed motor rated at 1.5 kW, 3000 rpm, 120 V radial flux surface mounted permanent magnet rotor, is then assessed using analytical tools for design such as ANSYS's RMXprt to verify the analytically obtained results. These results are then verified using the computer aided analysis tool, finite element analysis, using ANSYS Maxwell, to obtain the electromagnetic characteristics of the motor for further modification of design.

1 Introduction

OF late, due to issues like increasing global oil demand and automobile emissions due to various factors [1], there has never been a higher demand for research and development of environmentally safer technologies to prevent this global resource depletion. One of the fastest technologies to develop among these has been in the field of electric vehicles (EVs). It has shown prominent advancements in the past 10-15 years. EVs can prove to be a very viable alternative to petroleum powered automobiles due to their features like zero tailpipe emissions, higher power efficiencies and lower costs compared to their gasoline counterparts [2].

Past developments in EV technologies, can be generally classified into these three subsystems: Battery technology/ energy storage, electric motors and their drives. As a key element of the EV system, electric motors have to offer high efficiency, wide-speed range, high power density and maintenance-free operation [3].

Fig. 1 shows the different types of motors being developed / used for EV applications.

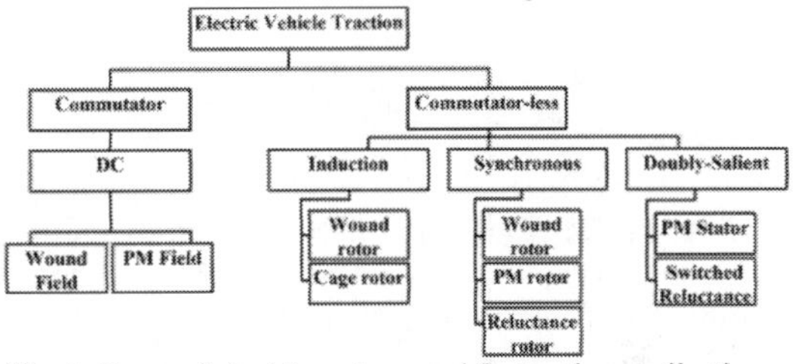

Fig. 1. Types of electric motors used for traction applications

Conventional motors like Induction machines are still widely accepted due to their low costs, high reliability and maintenance-free operation. However, conventional control methods cannot provide the required performance in EVs. With newer drive configurations rapidly developing due to advent of power electronics and control methods [4-12], conventional motors have also been able to provide desired performance, but lack efficiency in the lower speed ranges due to higher losses and have a limited constant power range. Permanent magnet (PM) motors, namely PM-synchronous motor and the Brushless-DC (BLDC) motors' use is becoming more prevalent in the EV market [13], due to their high power density, high efficiency, linear torque-speed nature, efficient thermal operation and availability of power electronics for efficient control, which make their use appropriate for traction applications. PM motors, as their name suggests, employ permanent magnets to generate the required operating torque. Despite recent increases in the price of permanent magnets, they have proved to still be profitable in operation costs. Many

* Corresponding author: vsandeep@cuk.ac.in

https://doi.org/10.1051/e3sconf/20198701022

different types of design topologies for PM motors have been developed for various applications [14-18] to improve their efficiencies, optimize PM sizes and reduce unwanted cogging torque.

The paper aims to design a permanent magnet brushless dc motor for three-wheeler application, which has proved to be one of the more staple modes of transport in East-Asian countries, like India, Thailand, China, Vietnam and others. The design criteria and its principles are discussed including sizing and analytical design methods. Computer aided design and analysis is done using 2D FEM (ANSYS Maxwell) for electromagnetic modelling.

2 Design Methodology

Fig. 2 shows the overview of the design methodology for calculation of average power required for sizing of the electric motor for the EV. Iterative loops are not mentioned here, but have been implemented in the procedure, where ever required.

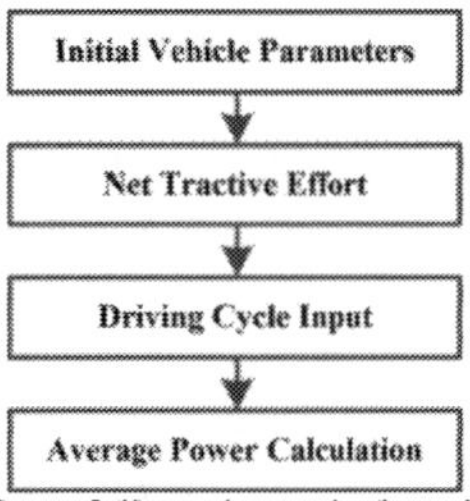

Fig. 2. Methodology followed to calculate the average power needed for the EV

Initial vehicle parameters include the mass of the vehicle under study. The parameters added are, mass of the vehicle (M_v), the coefficients of drag and rolling resistance (C_d , μ_{rr}) etc. The vehicle taken for reference is a popular model of three-wheeler vehicle, the 'Bajaj RE-2S', shown in Fig. 3.

Fig. 3. Bajaj RE-2S three-wheeler as reference model

The parameters for the reference vehicle are presented in Table 1, showing the various metrics used for the force model [2].

Table 1. Values of different parameters of the reference model

Sl.no	*Component/Parameter*	*Values*
1	Vehicle dimensions	1300mm x 2625 mm x 1710 mm
2	Tires (Front-Rear, R)	4.00-8, 4PR
3	Drag coefficient (C_d)	0.45
4	Rolling resistance (μ_{rr})	0.01

These parameters are then input into a force model to calculate the net tractive effort needed to be generated to accelerate, and run the vehicle at required speeds.

The method to calculate the net tractive force on the vehicle is as follows:

a. The net tractive effort on any vehicle comprises of four main forces: Aerodynamic drag force (F_w), rolling resistance force (F_r), acceleration force (F_a), and gradient force (F_g).

b. Aerodynamic force (F_w) is force is produced due to the friction of the vehicle moving through air. It is a function of the body shape and area of the vehicle.

$$F_w = \frac{1}{2}\rho A_f C_d (V + V_w)^2 \qquad (1)$$

c. Rolling resistance (F_r) is mainly caused due to the hysteresis in the tire material. The asymmetric distribution of the ground reaction forces on the tires when the vehicle is moving, causes rolling resistance. Depends on the weight of the vehicle.

$$F_r = \mu_{rr} M_V g \qquad (2)$$

d. Acceleration force (F_a) is the force required to take the vehicle from one speed to another.

$$F_a = M_V a \qquad (3)$$

e. Gradient force (F_g) is the force required to drive the vehicle up a slope. It is caused by the weight component of the vehicle along the downward direction. This force, either adds or opposes the forward motion of the vehicle based on descent or ascent.

$$F_g = M_V g \sin\alpha \qquad (4)$$

The net tractive force (F_{TE}) is the sum of all these forces,

$$F_{TE} = F_w + F_r + F_a + F_g \qquad (5)$$

All the above mentioned forces can be seen in the vehicle dynamic force model in Fig.4.

https://doi.org/10.1051/e3sconf/20198701022

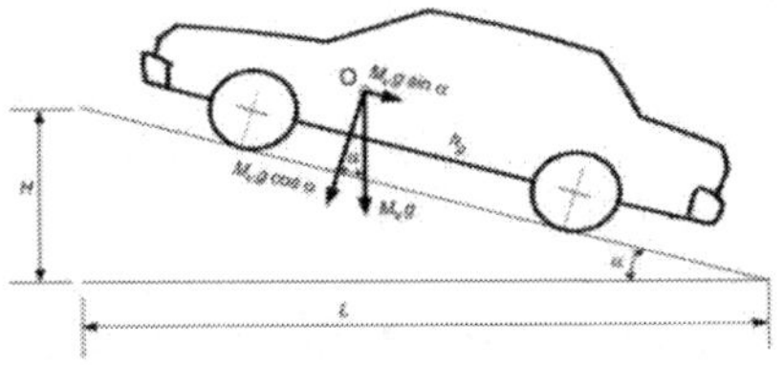

Fig. 4: Vehicle dynamics

Now that we have established all the performance variables, it is now necessary to evaluate the power needed for the vehicle under evaluation. The net force is calculated for a varying driving cycle input to simulate realistic driving behavior of the vehicle under different environments like urban or metropolitan cities. The driving cycle chosen for this study is the Modified Indian Driving Cycle (MIDC), developed by ARAI. This drive is to simulate realistic driving models for Indian roads. Fig. 5 shows the MIDC, as a function of varying velocity versus time.

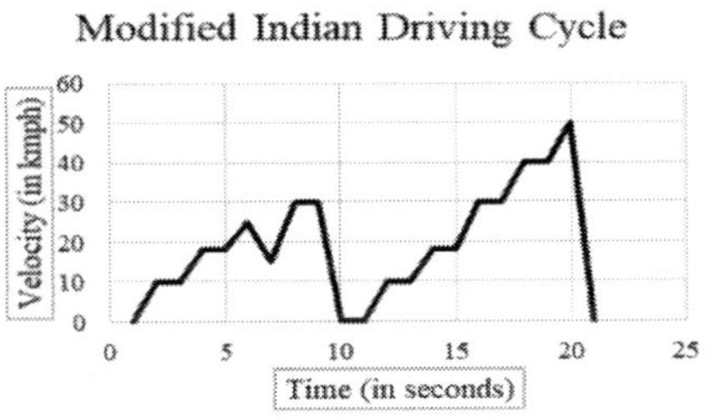

Fig. 5. Modified Indian Driving Cycle

Every point on the graph in Fig.4 has an initial velocity and a final velocity. For each of these points, the net tractive force is multiplied with the velocity difference to get the required power at that point.

The average power can be calculated and estimated as:

$$P = F_{TE} \times V \text{ watts} \quad (6)$$

After calculating the power for each point on the drive cycle, its average value is the required power rating of the motor. The calculated average power rating of the motor, for the vehicle is about 1531 W or about 1.5 kW.

3 Motor Design

The structure of the PMBLDC motor under design is shown in Fig. 6. It shows all the primary dimensions required for sizing and designing the motor. [19-21]
The PMBLDC motor type selected for design is a surface-permanent magnet type, meaning the permanent magnets are attached to the outer surface of the rotor.

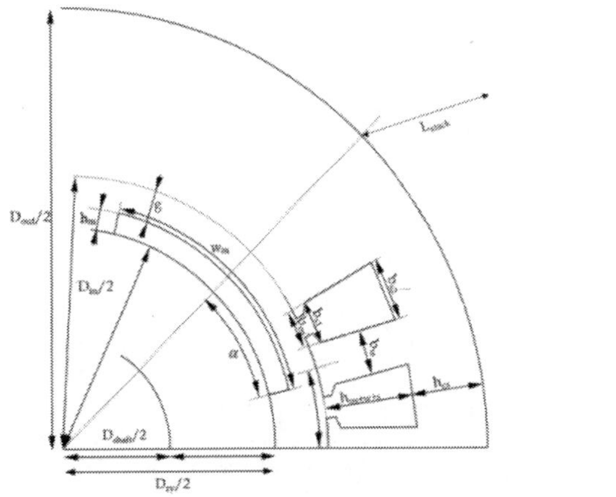

Fig. 6. Quarter motor cross section showing all the dimensions

Fig. 7 shows the procedure for the design of PMBLDC motor of rated power for the three-wheeler. Iterative loops are not mentioned here, but have been implemented in the procedure, where ever required. The followed methodology for design of the electric motor is as follows:

3.1 Prediction of required power

The required power for the motor has been determined using the net tractive effort. The estimated power rating required for the vehicle is about 1.5 kW.

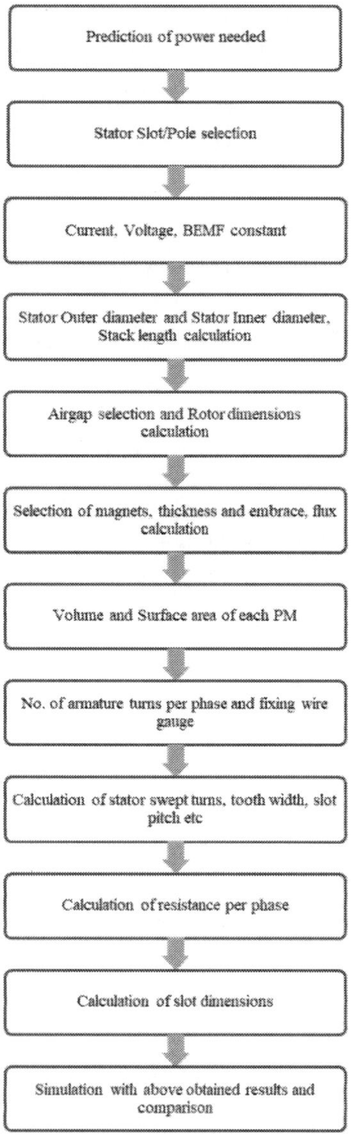

Fig. 7. Detailed procedure for design of the electric motor.

3.2 Stator Slot/Pole Selection

The number of slots on the stator is selected to be 24 (N_s) and the number of rotor poles is selected to be 4 (N_p). Proper slot/pole selection is very important to reduce effects like crawling and/or cogging.

3.3 Ratings of the motor

The motor to be designed has a power rating of 1.5kW, with an input voltage of 120V DC. The required speed range is 3000 to 4000 rpm.

Motor power, $P=2EI_a\eta\cos\phi$ (7)

Now, armature current, $I_a = \dfrac{P}{2E\eta\cos\phi}$ (8)

Substituting all the values, $I_a = 6.57$ A (9)

3.4 Back EMF Constant (K_E)

For a rated speed of 3000 rpm, and an input voltage of 120 V, the back EMF constant, K_E will be equal to 0.3437 V-s/rad.

3.5 Stator Outer and Inner Diameters (D_{in} & D_{out})

The stator outer diameter is chosen to be 120 mm (D_{out}) and the stator inner diameter is calculated to be 69 mm.

3.6 Stack length of the motor (L_{stack})

The following assumptions have been considered to calculate the stack length of the motor.

a. The airgap magnetic flux density is expected to be about 0.9 T.

b. The stator line current density for a 4 pole motor should be approximately A_m=30500 A/m.

c. The winding coefficient k_w is 0.96

d. The no-load EMF to phase voltage ratio ε is 0.83.

The output coefficient (σ_p) now becomes,

$$\sigma_P = \frac{P_{out}.\varepsilon}{D_{in}^2.L_{stack}.n_s} = 0.5\pi^2 k_w A_m B_{mg}\cos\phi \quad (10)$$

Solving, we get σ_p=105334 VAs/m, the length of the stack, L_{stack} is equal to 50 mm.

3.7 Air gap (g) and rotor diameter (D_r)

The air gap is kept at an optimum 0.5 mm, and therefore the inner rotor diameter is equal to 68 mm.

3.8 Magnet thickness (h_m) and rotor yoke diameter (D_{ry})

The thickness of the magnetic pole is chosen to be 2.5 mm, and therefore the rotor yoke diameter (D_{ry}) will be equal to 63 mm.

3.9 Surface area (A_{pole}) and total volume of PM (V_m)

With the above given dimensions, the area of each PM is 2474 mm²/pole and the total volume of PMs is 24740 mm³.

3.10 Armature turns per phase (N_1)

The number of armature turns per phase is equal to

$$N_1 = \frac{A_m p\tau}{m_1 I_a} \quad (11)$$

Substituting all values, the value of N_1 is equal to 118 turns.

3.11 Computation of the flux per pole

The flux in the air gap due to magnets,

$\phi_p = B_m$ x 4.A_{pole} (12)

The flux per pole is equal to

$\phi_{magnet} = B_m$ x A_{pole}

The flux per pole, ϕ_{magnet} is equal to 8.1642e-4 webers for a B_m 3.3 kG.

https://doi.org/10.1051/e3sconf/20198701022

3.12 Slot sizing

After establishing the no. of conductor turns/slot and the no. of parallel paths (a), we can now calculate the dimensions of the slot, the useful slot area, A_{su}, provided we adopt a slot fill factor, K_{fill}. Slot fill factor for round wire ranges from 0.35 to 0.65. The slots designed are rounded trapezoidal in shape.

The useful slot area, A_{su} is given as,

$$A_{su} = \frac{\pi d_{co}^2 a.n_s}{4.K_{fill}} \quad (13)$$

The key dimensions for the stator slot geometry are shown in fig. 8

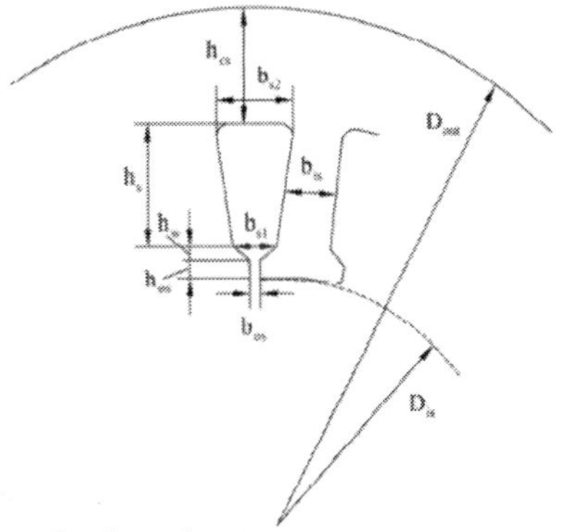

Fig. 8: Stator slot dimensions

Table 2: Slot dimensions

Sl.no	*Dimension*	*Value (in mm)*
1	b_{ts}	5.25 mm
2	b_{os}	2 mm
3	h_{os}	1 mm
4	h_w	2 mm
5	b_{s1}	4.57 mm
6	b_{s2}	4.67 mm
7	h_s	11.77 mm
8	b_{cs}	10.5 mm

4 Analytical Results

After obtaining the required dimensions, analytical results of values such as torque, power, phase current, resistance and back EMF are estimated to be as follows.

Table 3: Analytical results

Sl.no	*Dimension*	*Estimated result*
1	Shaft torque	4.77 Nm
2	Output power	1500 W
3	No load input current	6.57 A
4	Back EMF	120 V
5	Per phase resistance	0.25 ohms
6	No load speed	3400 rpm
7	No load air-gap flux density	0.74998 T
8	Cogging torque	0.657 Nm
9	Full load input current	13.49 A
10	Full load efficiency	92.59%
11	Full load torque	5.23 Nm

Table. 3 shows that the performance of the motor is well within the required ratings. But since analytical methods of calculations, always offer various types of estimation to be considered beforehand to obtain the necessary results.

To get more accurate results, 2D finite element analysis is done on the motor. This offers a high precision method to calculate various parameters of the motor using the finite element method (FE).

5 Final Element Analysis (2D)

Finite element method is employed into this design using ANSYS's Maxwell package, which offers both 2D and 3D FE solution methods for any electromagnetic problem type. The motor can be designed in Maxwell's 2D environment or can be imported from ANSYS's RMxprt package.

After designing the motor in the 2D environment, the software meshes the model, to simplify the problem and obtain solutions. The fineness of the mesh is directly proportional to the number of nodes formed. More the nodes, therefore more accurate the output result.

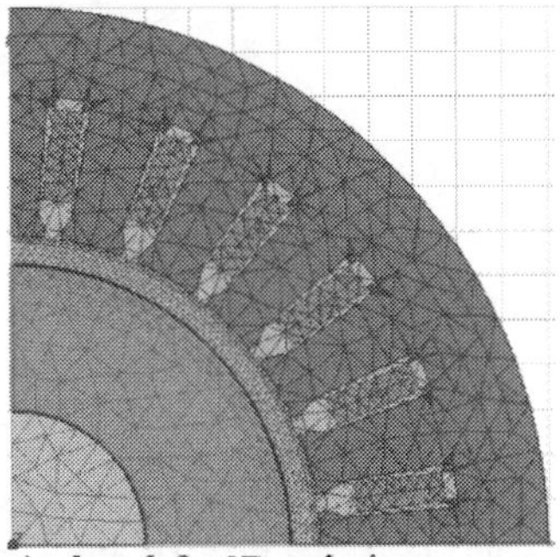

Fig. 9. Optimized mesh for 2D analysis

Fig. 9. shows the optimized mesh for a quarter-fraction model of the motor in ANSYS's Maxwell environment. After solving the 2D FE problem, the following results are obtained for the motor outputs.

5.1 Torque output

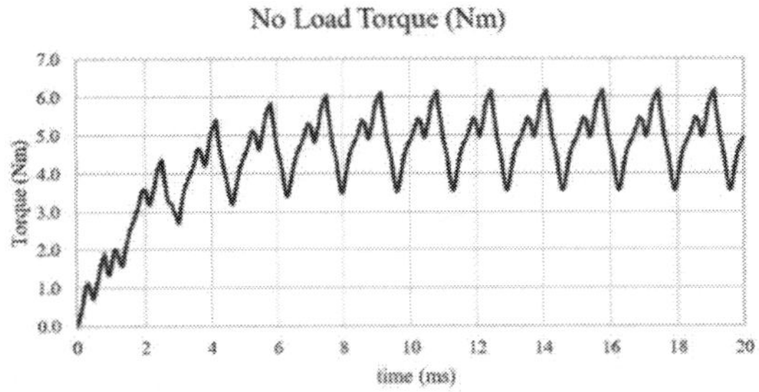

Fig. 10. Output torque

Fig. 10. shows the characteristic torque output for a BLDC motor, the rated torque is averaged at 5 Nm. There are also ripples in the characteristic curve due to high cogging torque in a BLDC motor. These can be reduced using methods to reduce cogging torque.

 https://doi.org/10.1051/e3sconf/20198701022

5.2 Output Speed

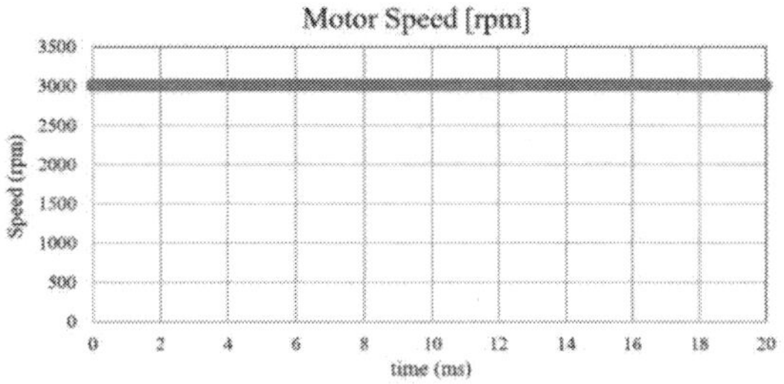

Fig. 11. Output speed

Fig. 11 shows the output speed for the motor, rated at the required speed of 3000 rpm.

5.3 Winding Currents

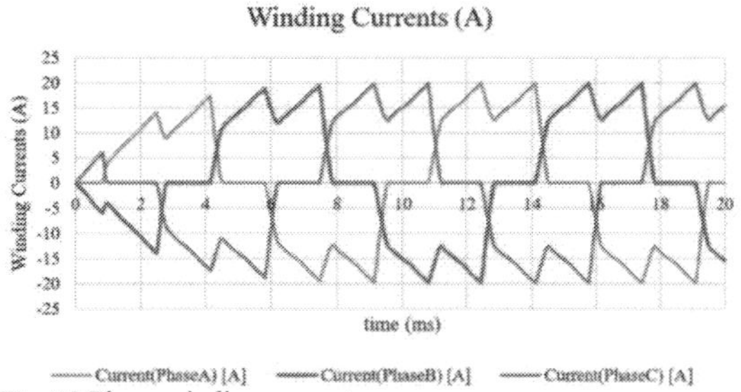

Fig. 12 Phase winding currents

Fig. 12 shows the phase winding currents in the motor.
In addition to this, the FE method also can display accurate distributions of several fields, namely flux density, current density and their vectors. These can be utilized to understand the distribution of flux, saturation, temperature etc. and improve the model by optimizing to reduce the undesired effects.
Fig. 13, 14, and 15 show the plot distribution for flux density, current density, flux lines and current distribution plots.
Fig 13. shows the distribution of flux along the cross section of the motor. The maximum flux density along the cross section does not exceed 2T, which is still lesser than the saturation limits of the ferromagnetic material used. Current density plot (A/m) in Fig.14, and flux distribution in Fig. 15 are both within practical permissible limits.

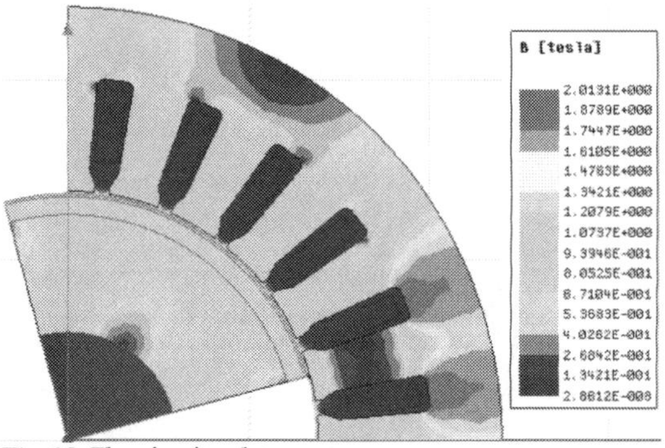

Fig. 13. Flux density plot

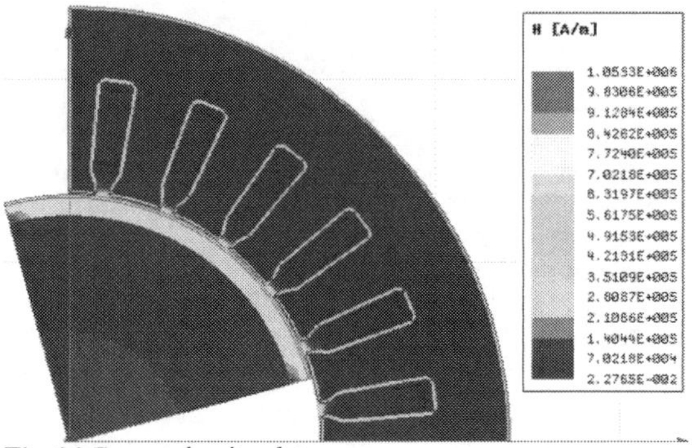

Fig. 14 Current density plot

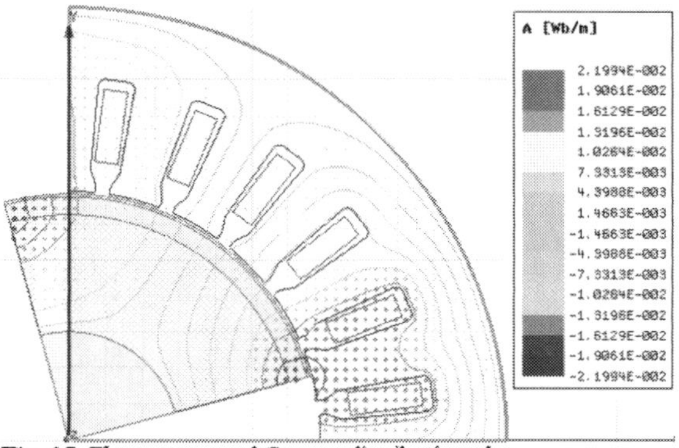

Fig. 15 Flux pattern and Current distribution plot

6 Conclusions

A PMBLDC motor has been proposed and analyzed for the application of three wheeler rickshaw. Complete design of motor is carried along with analytical calculations and finite element analysis is used to analyze the performance of the motor. The results obtained show a good design with reduced losses, higher efficiency, and current density within limits. Further work can be done in optimization of the same motor and be realized in hardware to test actual performance also.

References

[1] Frede Blaabjerg, Dan M. Ionel, Renewable Energy Devices and Systems with Simulations in MATLAB® and ANSYS®, CRC Press, 2017.

[2] Meherdad Eshani, Yimin gao and Ali Emadi, Modern-Electric, Hybrid-Electric and-Fuel-Cell-Vehicles: Fundamentals, theory and design , CRC Press, 2009

[3] A. Emadi, Y.-J. Lee, and K. Rajashekara, “Power electronics and motor drives in electric, hybrid electric, and plug-in hybrid electric vehicles,” IEEE Transactions on Industrial Electronics, vol. 55, no. 6, pp. 2237-2245, June 2008.

[4] X. Nian, F. Peng and H. Zhang, "Regenerative Braking System of Electric Vehicle Driven by Brushless DC Motor," in *IEEE Transactions on Industrial Electronics*, vol. 61, no. 10, pp. 5798-5808, Oct. 2014

[5] M. Krishnamurthy, C. S. Edrington, A. Emadi, P. Asadi, M. Ehsani, and B. Fahimi, "Making the case for applications of switched reluctance motor technology in automotive products," IEEE Transactions on Power Electronics, vol. 21, no. 3, pp. 659-675, May 2006.

[6] J. Cao and B. Cao,"Fuzzy-logic - Based Sliding Mode Controller Design for Position -Sensorless Electric vehicle, " IEEE Trans.Power Electron. vol. 24, no. 10, pp. 2368--- 2378, October 2009

[7] J. Cao, B. Cao, Z. Bai , P. Xu and X. Wu ," Neural network control of Electric Vehicle based on Position –Sensor less Brushless DC Motor," Proceedings of the IEEE Int. Conf. on Automation And logistics, pp. 2029–2034, August 2007.

[8] Z.Q. Zhu, David Howe, "Electric machines and drives for electric hybrid and fuel cell vehicles", *Proceedings of IEEE*, vol. 95, no. 4, pp. 746-765, Apr 2007.

[9] R. Kumar and B. Singh, "Buck-boost converter fed BLDC motor drive for solar PV array based water pumping," *2014 IEEE International Conference on Power Electronics, Drives and Energy Systems (PEDES)*, Mumbai, 2014, pp. 1-6.

[10] B. Singh V. Bist "A BL-CSC Converter Fed BLDC Motor Drive with Power Factor Correction" IEEE Transactions on Industrial Electronics no. 99 2014.

[11] R. Kumar and B. Singh, "Buck-boost converter fed BLDC motor drive for solar PV array based water pumping," *2014 IEEE International Conference on Power Electronics, Drives and Energy Systems (PEDES)*, Mumbai, 2014, pp. 1-6.

[12] B. Singh V. Bist "A BL-CSC Converter Fed BLDC Motor Drive with Power Factor Correction" IEEE Transactions on Industrial Electronics no. 99 2014.

[13] J.F. Gieras and Wing, *Permenent Magnet Motor Technology*, Marcel Dekker Inc, New York, 1997

[14] R. Qu, M. Aydin, T.A. Lipo, "Performance comparison of dualrotor radial-flux and axial-flux permanent-magnet BLDC machines", *in proc. IEEE Int. Elect. Mach. Drives Conf*, pp. 1948-1954, 2003.

[15] K. Sitapati, R. Krishnan, "Performance comparisons of radial and axial field permanent magnet brushless machines", *IEEE Trans. Ind. Appl*, vol. 37, no. 5, pp. 1219-1226, 2001.

[16] N.B. Simisie, H.B. Ertan, "A comparison of torque capabilities of axial flux and radial flux type of brushless dc (BLDC) drives for wide speed range applications", *in proc. IEEE PEDS Honk Kong*, pp. 719-724, 1999.

[17] R Qu, T A Lipo, "Dual-rotor radial-flux torodially wound permanent-magnet machines", *IEEE Trans. Industry Applications*, vol. 39, no. 6, pp. 1665-1673, 2003.

[18] R. Qu, T.A. Lipo, "Design and parameter effect analysis of dual-rotor radial flux toroidally wound permanent magnet machines", *IEEE Trans. Industry Applications*, vol. 40, no. 3, pp. 771-779, 2004.

[19] Duane C. Hanselman, *Brushless Permanent-Magnet Motor Design*, New York: McGraw-Hill Inc., 1994.

[20] T. J. E. Miller, Brushless Permanent-Magnet and Reluctance Motor Drives, Clarendon Press, Oxford, 1989.

[21] J. R. Hendershot Jr. and T. J. E Miller, *Design of Permanent Magnet Motors*, Oxford, U.K: Oxford Univ. Press, 1994

Smart helmet for safe driving

Keesari Shravya[(1)], *Yamini* Mandapati[(2)], *Donuru* Keerthi[(3)], *Kothapu* Harika[(4)], and *Ranjan* K. Senapati[(5)]

[1,2,3,4,5]VNR VignanaJyothi Institute of Engineering and Technology, Bachupally, Hyderabad, Telangana, India

E-mail: shravyakeesari@gmail.com[(1)], yaminimandapati974@gmail.com[(2)], keerthidonuru@gmail.com[(3)], kothapuharika@gmail.com[(4)], ranjan_ks@vnrvjiet.in[(5)]

Abstract. A smart helmet is a type of protective headgear used by the rider which makes bike driving safer than before. The main purpose of this helmet is to provide safety for the rider. This can be implemented by using advanced features like alcohol detection, accident identification, location tracking, use as a hands free device, fall detection. This makes it not only a smart helmet but also a feature of a smart bike. It is compulsory to wear the helmet, without which the ignition switch cannot turn ON. An RF Module can be used as wireless link for communication between transmitter and receiver. If the rider is drunk the ignition gets automatically locked, and sends a message to the registered number with his current location. In case of an accident it will send a message through GSM along with location with the help of GPS module. The distinctive utility of project is fall detection; if the rider falls down from the bike it sends a message.

1 Introduction

In recent times helmets have been made compulsory in Telangana State. Traffic accidents in India have been increased every year. As per Section 129 of Motor Vehicles Act, 1988, every single person riding a two-wheeler is required to wear protective headgear following the standards of BIS (Bureau of Indian Standards). Also drunken driving under the influence (DUI) is a criminal offence according to the Motor Vehicle act 1939, which states that the bike rider will get punishment. Currently bike riders easily escape from the law [1]. These are the three main issues which motivates us for developing this project.

The first step is to identify whether the helmet is worn or not. If helmet is worn then ignition will start otherwise it remains off. For this, Force Sensing Sensor (FSR) sensor is used. The second step is alcohol detection[2]. Alcohol sensor is used as breath analyser which detects the presence of alcohol in rider's breath and if it exceeds permissible limit ignition cannot start. It will send message to the number saying that "Rider is drunk and is trying to ride the bike". MQ-3 sensor is used for this purpose. When these two conditions are satisfied then only ignition starts. The third main issue is accident and late medical help. If the rider has met with an accident, he may not receive medical help instantly, which is one of the main reasons for death. Every second people dies due to delay in medical help, or in the case where the place of accident is unmanned. In fall detection, we place accelerometer in the bike unit. By this mechanism accidents can be detected.

The aim of this project is to make à protection system in a helmet for the safety of bike rider. The smart helmet that is made is fitted with different sensors responsible for detection[3]. There are two main units in this project. Each unit uses a microcontroller. Signal transmission between the helmet unit and bike unit is done using a RF module.

2 Technical Studies

2.1 Force Sensing Resistor (FSR)

Force Sensing Resistor in Figure 1 is placed inside the helmet, where the actual human touch is sensed. It is determined by the helmet unit whether it is worn or not before starting the bike. If this condition is satisfied, then it sends the signal to bike unit to starti. Force Sensing Resistors or FSRs are strong polymer thick film (PTF) devices whose resistance is inversely proportional to force applied to the face of the sensor.

Fig. 1. FSR

https://doi.org/10.1051/e3sconf/20198701023

This sensor is used as human touch control in various applications, such as medical systems, automotive electronics and in robotics and industrial applications. Force sensing resistor is a two-wire sensor with a resistance that changes on applied force.

2.2 MQ-3 Alcohol Sensor

MQ-3 gas sensor shown in Figure 2 is used for identifying the alcohol content from breath. It can be positioned just in front of the mouth. The sensor responds to various molecules in alcohol and determines if the rider is drunk. The sensor also has a potentiometer to adjust the concentration of gases. We calibrate the detector for 0.4mg/L of alcohol concentration in air and use a resistance of 200 KΩ. It has a 4 pins namely GND, VCC, A out and D out. The sensor supports both analog and digital outputs. Here we use digital output of this sensor.

Fig. 2. MQ 3 sensor

2.3 Accelerometer ADXL345

The ADXL345 in Figure 3 is a tinny, tri-axial accelerometer with resolution of 13 bit. The output of accelerometer is digital and uses 16 bit 2's complement data. It can be connected via Serial Peripheral Interface (SPI 3-4 wire) or I2C interface. ADXL345 is used for both measurement of static and dynamic acceleration. In this project we use accelerometer, which measures the static acceleration of gravity. Free-fall sensing unit senses if the bike is falling. And Bike unit take decision that accident is occured or not.

Fig. 3. ADXL345

2.4 RF communication circuit

An RF transmitter module is a small PCB sub-assembly capable of transmitting radio waves and modulating radio wave to carry data (see Figure 4). Transmitter modules are usually implemented alongside a microcontroller which will provide data to the module that can be transmitted. RF transmitters are usually subject to regulatory requirements which dictate the maximum allowable transmitter power output, harmonics and band edge requirements.

An RF receiver module receives the modulated RF signal and demodulates it. There are two modules, such as, super-heterodyne receiver and super-regenerative receiver. Super-regenerative modules are usually low cost and low power designs using a series of amplifiers to extract modulated data from a carrier wave. Super-regenerative modules are generally imprecise as their frequency of operation varies considerably with temperature and power supply voltage. Super-heterodyne receivers have a performance advantage over super-regenerative; i.e. they offer increased accuracy and stability over a large voltage and temperature range. This stability comes from a fixed crystal design which in the past intended to mean a comparatively more expensive product. However, advances in receiver chip design now mean that currently there is little price difference between super-heterodyne and super-regenerative receiver modules.

Fig. 4. RF module

2.5 Vibration sensor

Digital vibration sensor (Figure 5, DFR0027) is a digital Plug and Play sensor block. It has vibration switch, digital input module and dedicated sensor expansion boards with the Arduino in combination. It can sense the weak vibration signals. When the accident occurs, it senses the pressure and sends it to the Arduino which in turn it sends the message to the registered mobile number.

https://doi.org/10.1051/e3sconf/20198701023

Fig. 5. DFR0027 sensor

3 Construction

It is already mentioned that the project is divided into two units namely helmet and bike. In helmet unit, also called the transmitter unit shown in Figure 6 (a), the force sensing resistor is placed on inside upper part of the helmet where actually head will touch with sensor surface. And alcohol sensor is placed on in front of rider's mouth so that it can sense easily. Solar panels are mounted on upper side of helmet which is in direct sunlight. And the battery and regular circuits were fixed inside the helmet. Secondary controller and RF transmitter circuit were also placed inside the helmet. Antenna is located outside the helmet.

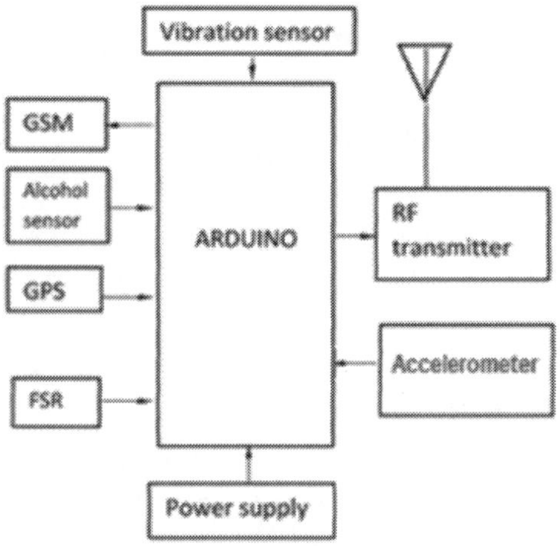

(a)

The receiver unit shown in Figure 6 (b) is placed in the bike. The RF receiver accepts all the data from the helmet (i.e transmitter) unit. Depending on the conditions, if true, the ignition starts and bike moves. The GSM can continuously send the location information of the bike. If any accident occurs, the vibration sensor gets activated and sends the location information to the registered mobile number.

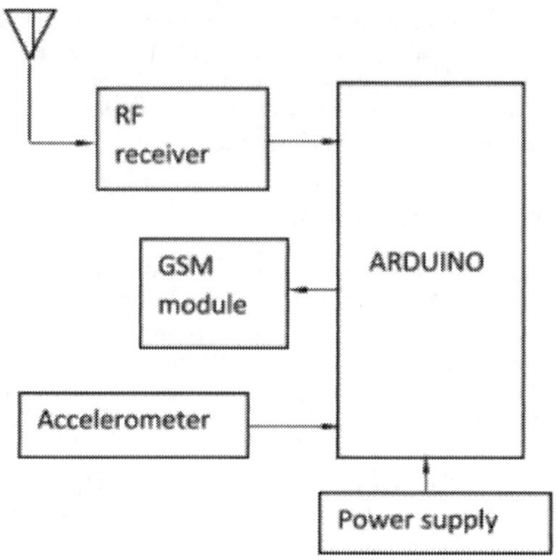

(b)

Fig. 6. (a) Transmitter unit and (b) Receiver unit

4 Working

The first step of project is it initializes all the port and next step is accident detection using accelerometer. If No accident occurs, then it will go to third step. Third step is listening to RF module continuously for Data and interprets data using if conditions. Fourth step is to check weather helmet is wear or not. If Helmet is not wore then display Message "Please wear the helmet" will be displayed. Next step is to check the condition of drunk, if rider is drunk, display message "You are Drunk" and then send the message to stored number with location, and ask for the password. If password is correct then bike will start. If accident is detected in sixth step, then it will stop everything and send a message with location.

5 Design and Realizations

A smart helmet is designed using Arduino with transmitter (Figure 7). If alcohol concentration is present in human breath, then it displays the message on LCD and it sends the SMS to register no. with their current location. (shown in Figure 8). If accident occurs and bike falls, then it displays the message on LCD and sends an SMS to the registered numbered with the current geographical location.

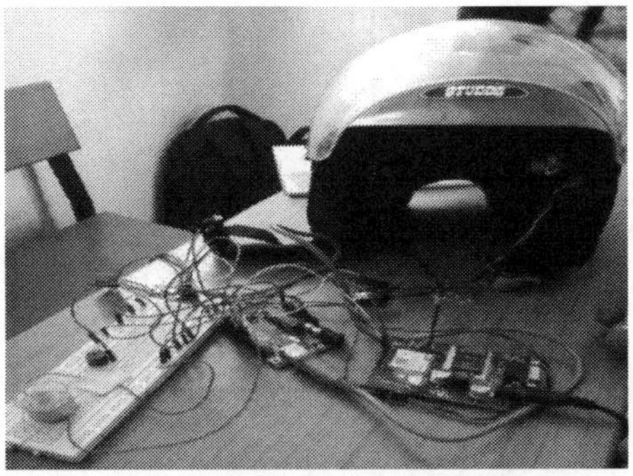

(a)

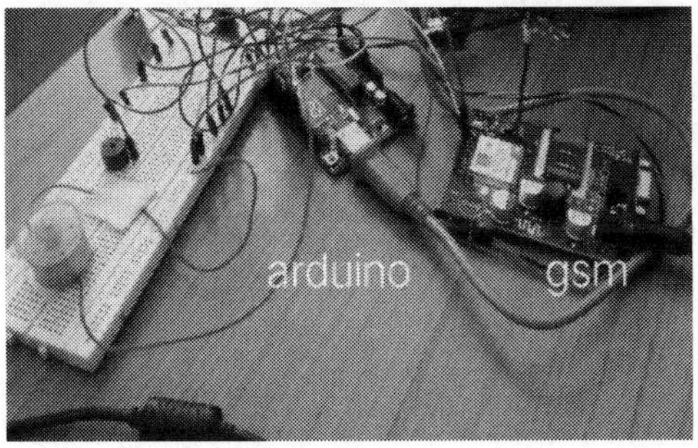

(b)

Fig. 7 (a) Helmet with transmitter and receiver, (b) identification of Arduino and GSM part from (a)

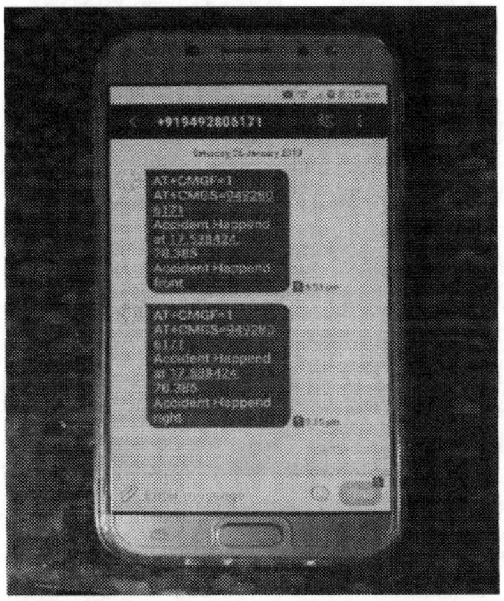

Fig. 8 Display of messages send to the registered mobile number

6 Advantages, applications and future scope

6.1 Advantages

1. Detection of accident in remote area can be easily detected and medical services provided in short time.
2. It will reduce the probability of accidents by simply avoiding drunken drive by using alcohol detector.

6.2 Applications

1. It can be used in real time safety system.
2. We can implement the whole circuit into small VLSI chip that can be embedded into the helmet and bike unit.
3. It can be designed for less power consuming safety system.
4. This safety system technology can further be enhanced in car or other vehicle by replacing the helmet with seat belt.

6.3 Future Scope

1. We can implement various bioelectric sensors on the helmet to measure various activities.
2. We can use small camera for the recording the drivers activity. It can be used for passing message from the one vehicle to another vehicle by using wireless transmitter.

7 Conclusion

The designed Smart helmet ensures the safety of the rider by making it necessary to wear helmet, and also ensures that the rider hasn't consumed alcohol more than the permissible limit. If any of these prime safety rules are violated, the proposed system will prevent the biker from starting the bike. The system also helps in efficient handling of the aftermath of accidents by sending a SMS with the location of the biker to the police station. This ensures that the victims get proper and prompt medical attention, if he/she met with an accident.

References

1. Manjesh N., Sudarshan Raj, Smart Helmet using GSM & GPS Technology for Accident detection and Reporting System, International Journal of Electrical and Electronics Research, **2**, 4 (2014).
2. Abhinav Anand, Alcoholic detection, Department of Electronics and Telecommunications, IJEETC, 2015.
3. Mohd Khairul, Afiq Mohd Rasli, Smart Helmet with sensors for accident prevention, ICEESE, 2013.

A rubust fractional order PID controller for MIMO power systems

Vasampalli Shashidhar[1], and *Dola Gobinda* Padhan[2,*]

[1,2]Gokaraju Rangaraju Institute of Engineering and Technology, Hyderabad, India

Abstract. This paper proposes a novel fractional order PID control scheme for Multi Input and Multi Output (MIMO) power systems. This control scheme utilizes Cuckoo Search algorithm to tune the fractional PID controller to guarantee better closed loop performance in the transmission or distribution networks. Cuckoo Search optimization algorithm is proposed to optimize gain values of the fractional PID controller. The effectiveness of the proposed control strategy has been assessed by simulations in MATLAB/Simulink platform. The robustness of the proposed controller has been validated by varying the plant parameters.

1 Introduction

In the MIMO power systems there are a number of control areas and each control area is obliged to generate power to cover the customers load requirements as well as maintaining the systems frequency and the interchanged power at the scheduled values. Therefore, load frequency controller (LFC) [1], is designed and implemented to automatically balance generated power and load demand in each control area so that the quality of the power delivered is maintained at the requisite level. The analysis and design of load frequency control system of individual generator eventually controlling large interconnections between different control areas plays a vital role in automation of power system. The purpose of LFC is to maintain system frequency very close to a specified nominal value to maintain generation of individual units at the most economical value and to keep the correct value of the line power between different control areas. Many investigations in the area of LFC problem of interconnected

Cuckoo Search is a Swarm Intelligence based Metaheuristic optimization algorithm inspired by the brood parasitic behaviour observed in the cuckoo bird species [2]. Cuckoo Search has been shown to be superior to other algorithms such as GA and PSO in finding a global optimum of many numeric test functions. Cuckoo Search has also been applied to many structural designing optimization problems such as design of wind turbines, vehicular components and welding structures [3]. Cuckoo Search has also been proposed for machine learning techniques like Feed Forward Neural Network Training and Document Clustering

This paper is organized as follows: section 2 presents four area power systems. Section 3 introduces fractional order controller. Section 4 presents Cuckoo Search algorithm. In section 5, simulation results have been discussed in details. Section 6 presents the conclusion.

2 Four Area Power Systems

A four area power system comprises four areas that are interconnected by high voltage transmission lines or tie-lines. The trend of frequency measured in each control area is an indicator of the trend of the mismatch power in the interconnection and not in the control area alone [4]. Parameter uncertainties, model mismatch and unexpected external disturbances are the key challenges for controller design while dealing with LFC issue of power systems. The simple block diagram of four areas interconnected power systems is shown in Fig. 1. Also, the general block diagram representation of control area 'i' of MIMO power systems is shown in Fig. 2. Here, reheat turbine has been considered for Area 1, Area 2 and Area 3. For Area 4, non-reheat turbine is considered for simulation.

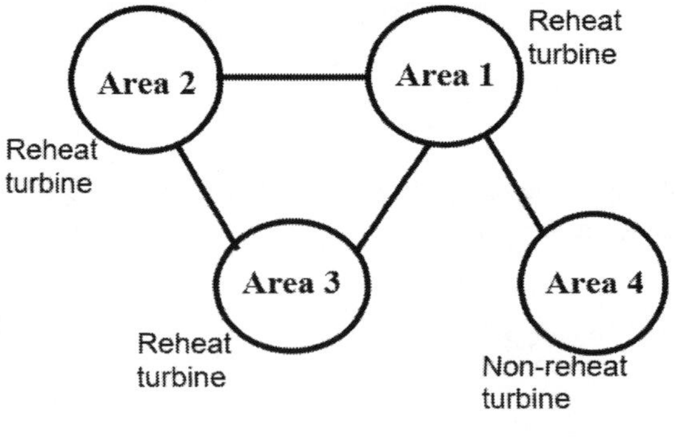

* Corresponding author: dg.padhan@griet.ac.in

https://doi.org/10.1051/e3sconf/20198701024

Fig. 1. Simplified Block diagram of 4 areas inter-connected power systems

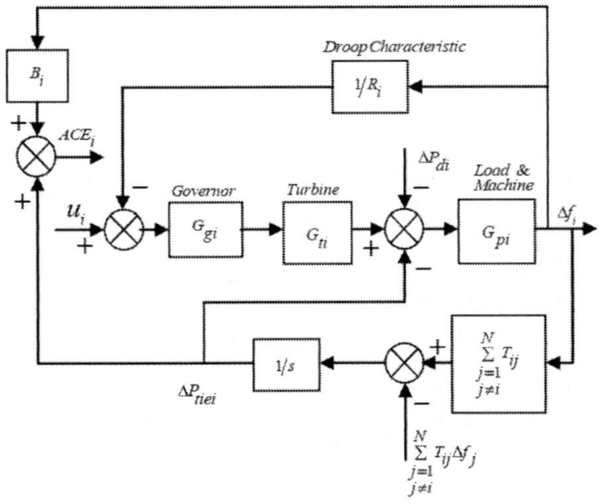

Fig. 2. General Block diagram representation of control area i of MIMO Power systems

For interconnected power systems, the governor dynamics is given by

$$G_g = \frac{1}{T_g s + 1} \quad (1)$$

The dynamics for non-reheat turbine is in the form

$$G_t = \frac{1}{T_t s + 1} \quad (2)$$

The dynamics for reheat turbine is given by

$$G_{tr} = \frac{cT_r s + 1}{(T_r s+1)(T_t s+1)} \quad (3)$$

where is T_r a constant and c is the percentage of the power generated by the reheat process in the total generated power.

The transfer function for load and machine is

$$G_p = \frac{K_p}{T_p s + 1} \quad (4)$$

3 Fractional order $PI^{\lambda}D^{\mu}$ controller

From the literature it is confirmed that, the fractional controller outperforms the integer-order PID Controller.

The control action of the $PI^{\lambda}D^{\mu}$ controller can be expressed in time domain as

$$u(t) = K_p e(t) + K_i D^{-\lambda} e(t) + K_d D^{\mu} e(t) \quad (5)$$

Where $e(t)$ is error signal. By applying Laplace transform to equation (5) with zero initial conditions, the following equation is obtained

$$G_c(s) = K_p + \frac{K_i}{s} + K_d s \quad (6)$$

Several approaches are suggested for tuning of fractional PID controllers over the years [5-6]. However, there is still no well-established, general tuning algorithm.

Nevertheless, optimization techniques may be applied to the tuning problem instead. The parameter set to optimize consists of fractional PID parameters such as (K_p,K_i, K_d, λ and μ). Cuckoo Search algorithm has fewer control parameters to fine tune than other algorithms. Moreover, tweaking of control parameters for a particular problem is not needed [2]. Hence Cuckoo search is more generic and robust for many optimization problems.

The integral time absolute error ($ITAE = \int_0^t t|e(t)|dt$) performance criteria has been used as objective functions for optimization .

4 Cuckoo search algorithm

Cuckoo Search is a nature inspired optimization algorithm based on the cuckoo bird's parasitic breeding behaviour of laying its eggs in the nests of different bird species. The cuckoo bird deposits her eggs in the nest of a host bird whose eggs closely resemble her own eggs and uses the services of host birds to hatch her own eggs. The cuckoo search algorithm simulates the cuckoo bird's intelligent search strategy of finding the best host nest to deposit her egg.

The cuckoos make series of flights searching and evaluating one nest after the other until they find the optimal host nest to spawn their eggs. The series of flights the cuckoos make can be classified into two types namely Lévy/Global Flight and Local Flight. The Lévy Flight is based on the Lévy Probability distribution. Many animals and insects forage using the Lévy flight for finding their prey/food location [3]. During the Local Flight, the cuckoo examines around the host nests which the other cuckoos are using, and finds the fittest nest among them and computes the distance and direction of her Local flight based on that. The flow chart for Cuckoo search algorithm is shown in Fig. 3.

 https://doi.org/10.1051/e3sconf/20198701024

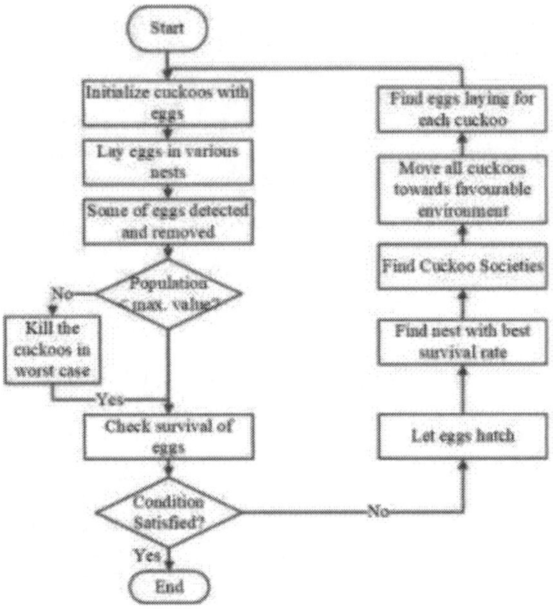

Fig. 3. Flow chart for Cuckoo search algorithm

5 Simulation Results

In the simulation results, area 1, area 2 and area 3 are considered as the area with reheat thermal turbine, and area 4 is represented as the area with non-reheat turbine. The turbine dynamics of area 1, area 2 and area 3 as studied by [4] are considered.

The governer dynamics of area 1, area 2 and area 3 is given by $G_g = 1/(0.2s+1)$

The load and machine dynamics of area 1, area 2 and area 3 is $G_p = 120/(20s+1)$

The turbine dynamics of area 1, area 2 and area 3 is given by $G_{tr} = (0.333*20s+1)/((20s+1)(0.3s+1))$

The governer dynamics of area 4 is given by $G_{gnr} = 1/(0.08s+1)$

The load and machine dynamics of area 4 is $G_{pnr} = 120/(20s+1)$

The turbine dynamics of area 4 is given by $G_{tnr} = 1/(0.3s+1)$

And for all cases droop parameter, R=2.4

The synchronising constants are 0.0707 and the frequency bias constants are 0.425.

Using Cuckoo search algorithm, the parameters of the controller for area 1, area 2 and area 3 are K_p=1.1895, K_i=0.6231, K_d=0.6487, $\lambda = 0.99$ and $\mu = 0.78$. Also the parameters of the controller for area 4 are K_p=1.9822, K_i=0.5242, K_d=0.1756, $\lambda = 1.12$ and $\mu = 0.82$.

To show the performance fractional order PID controller, a step load 0.01 is applied to area1at t=5 s, area 2 at t=100 s, area 3 at t=150 s and followed by a step load 0.01 to area 4 at t= 250 s. Fig. 4 to Fig. 7 illustrates the frequency deviations of the four different areas. Fig. 8 to Fig. 11 shows the tie-line power deviations of the four areas. From the simulation results, it can be seen that the frequency errors, and tie-line power deviations have been driven to zero by proposed controller in the presences of power load changes.

In order to test the robustness of the controller, the variations of the parameters of the non-reheat and reheat turbines in the four areas are assumed to be 50% of their nominal values. However, the controller parameters are not changed with the variations of the system parameters. Fig. 4 to Fig. 7 illustrates the frequency errors of four areas with the variant parameter values. Fig. 8 to Fig. 11 shows the tie-line power errors of the four areas with the variant parameter values. From the simulation results, it can be seen that despite such large parameter variations, the system responses do not show notable difference from the nominal values.Therefore,the simulation results demonstrate the robustness of proposed controller against system parameter variations.

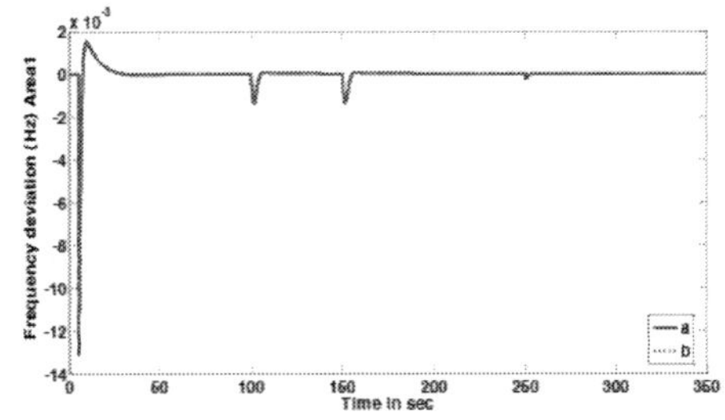

Fig. 4. Frequency deviation for Area 1, solid line-a-nominal response, dotted line-b-perturbed response

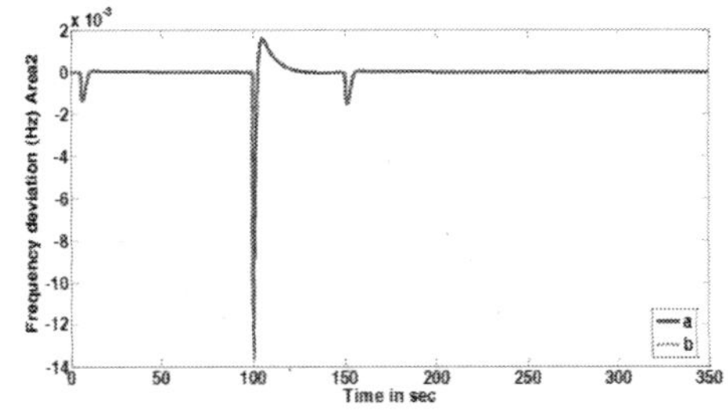

Fig. 5. Frequency deviation for Area 2, solid line-a-nominal response, dotted line-b-perturbed response

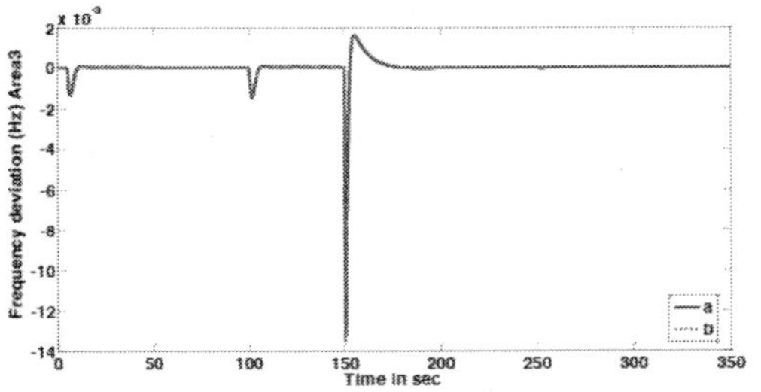

Fig. 6. Frequency deviation for Area 3, solid line-a-nominal response, dotted line-b-perturbed response

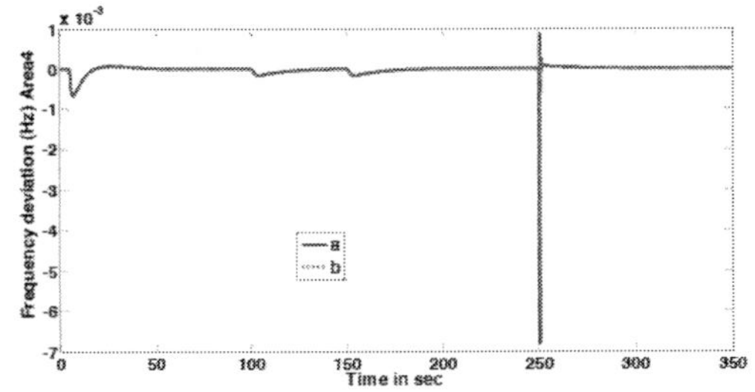

Fig. 7. Frequency deviation for Area 4, solid line-a-nominal response, dotted line-b-perturbed response

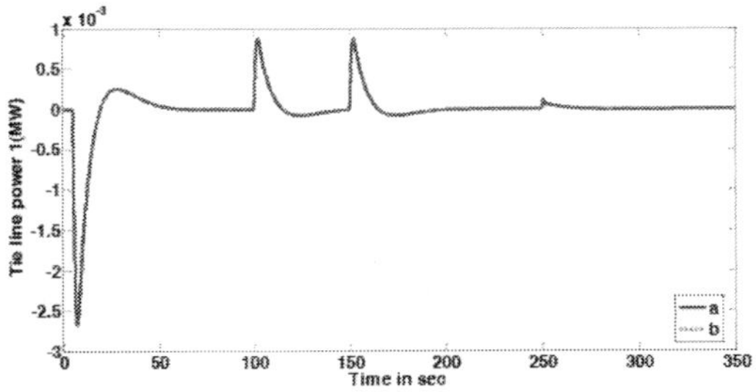

Fig. 8. Tie line power deviation for Area 1 and Area 2, solid line-a-nominal response, dotted line-b-perturbed response

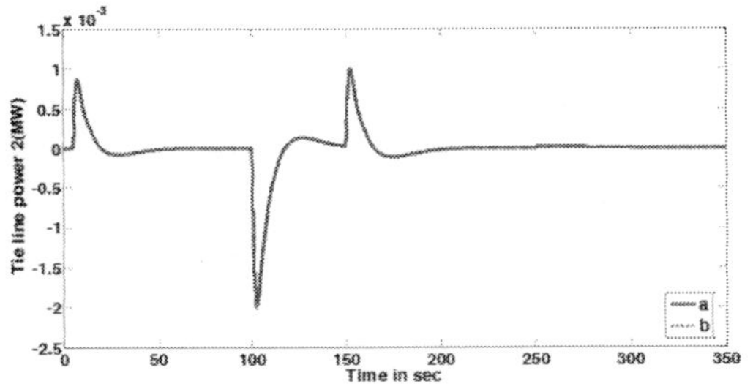

Fig. 9. Tie line power deviation for Area 2 and Area 3, solid line-a-nominal response, dotted line-b-perturbed response

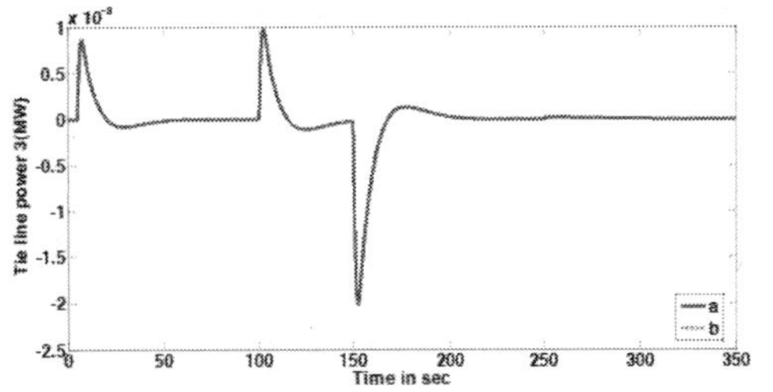

Fig. 10. Tie line power deviation for Area 3and Area 1, solid line-a-nominal response, dotted line-b-perturbed response

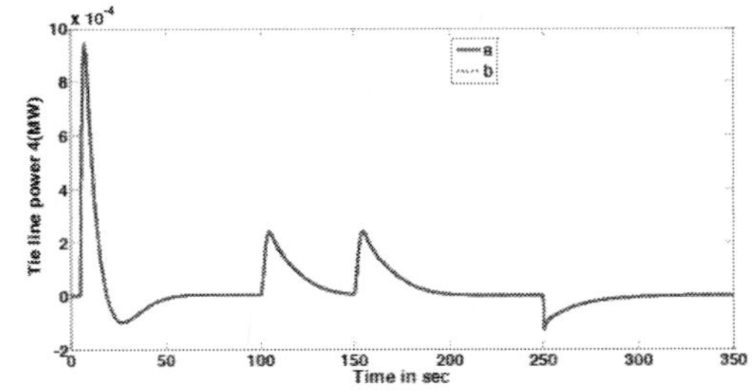

Fig. 11. Tie line power deviation for Area 4 and Area 1, solid line-a-nominal response, dotted line-b-perturbed response

7 Conclusion

In this paper, a fractional $PI^{\lambda}D^{\mu}$ controller is designed using Cuckoo search algorithm for load frequency control of four-area interconnected MIMO power systems. The robustness of the controllers are tested with different plant parameters uncertainty scenarios.

References

1. A. Khodabakhshian , R. Hooshmand, A new PID controller design for automatic generation control of hydro power systems, Electrical Power and Energy Systems **32**, 375–382, (2010)
2. E. Valian, S. Mohanna and S. Tavakoli, 'Improved Cuckoo Search Algorithm for Feed forward Neural Network Training', International Journal of Artificial Intelligence & Applications, **2** (3), 36-43, (2011)
3. A. Yildiz, 'Cuckoo search algorithm for the selection of optimal machining parameters in milling operations', Int J Adv Manuf Technol, **64** (1), 55-61, (2012)
4. D.G. Padhan, S. Majhi, A new control scheme for PID load frequency controller of single-area and multi-area power systems, ISA Transactions, **52**, 242–251, (2013)
5. C.A. Monje, Y.Q. Chen, B.M. Vinagre, D. Xue, V. Feliu, *Fractional order Systems and Controls Fundamentals and Applications*, Springer-Verlag, London, (2010)
6. C.N. Zhao, D.Y. Xue, Y.Q. Chen, A fractional order PID tuning algorithm for a class of fractional order

 https://doi.org/10.1051/e3sconf/20198701024

plants, Proceedings of the International Conference on Mechatronics and Automation. Niagara, Canada, 216-221, (2005)

Non-isolated Modified Quadratic Boost Converter with Midpoint Output for Solar Photovoltaic Applications

Ravivarman *Shanmugasundaram*[1*], Natarajan *Karuppiah*[2], and Raja Gopal *Reddy B* [2]

[1]Professor, Department of Electrical and Electronics Engineering, Vardhaman College of Engineering, Hyderabad, Telangana
[2]Associate Professor, Department of Electrical and Electronics Engineering, Vardhaman College of Engineering, Hyderabad, Telangana

Abstract. This paper presents a boost DC-DC converter topology with non - isolated high gain and output midpoint, to boost the voltage obtained from solar photovoltaic panels. The three-level boost converter is coupled to the output port of the single-switch quadratic boost converter to derive the proposed converter topology. The voltage gain of the proposed converter is greater than that of the classical boost converter. The voltage stress on the switches of the proposed converter is equal to half of the converter output voltage. Static analysis, operating modes, experimental waveforms in continuous current conduction and discontinuous current conduction modes are shown. A 520 W prototype converter was implemented in the laboratory and its results are presented.

1 Introduction

In photovoltaic systems with multiple modules connected in series, partial shadowing can lead to a reduction of voltage and power available from the panels together. In addition, the tracking of maximum power point is complex because the system has multiple peaks, resulting from the local maximum power points. In [1, 2] a comparative investigation of various photovoltaic arrangements with partial shading is presented. Still, the power of a photovoltaic panel is between 100 W and 350 W, and the voltage at maximum power point is 12 V to 42 V. The above values are not sufficient when compared to the voltage required in the dc bus of inverters, which reduces the efficiency of the whole system [3, 4].

To extend the conversion range, the cascaded boost converters was used [5, 6]. However, the resulting structure is large and control is complex. With simplified structure, the quadratic boost converter with a single switch was proposed and analysed in [7-9], and in [9, 10] the drive cell has been associated to aid self-resonant switching, allowing the zero-voltage switching in the converter switch. The voltage gain can be increased by using magnetic coupling, coupled inductors based high gain converters are introduced in [11, 12]. With this technique, the ripples in the inductor current and the reverse recovery problems in the diodes are reduced. However, the voltage stress of the output diode is still high, the leakage inductance and parasitic capacitor of the diode will form a resonant circuit. Additionally, the ripples in the input current is highly discontinuous. By switched - capacitor technique, the topologies proposed in [13, 14] provide increased gain, but this can be achieved by using large number of capacitors.

The conduction and switching losses of quadratic converters are significant, making these converters more suitable for low power applications. Using three-state cells, high-gain converters with reduced input current ripple was proposed in [15, 16]. The solution presents a high gain, where the gain is increased by increasing the turns-ratio of the coupled inductor. A converter topology utilizing three-state cell and voltage multiplier cell was proposed in [17]. This solution may be limited to high power applications as a function of the current flowing through the capacitors. The voltage stresses in the semiconductor devices can be reduced as in [18], where the maximum voltage applied to the switches are equal to the fraction of its converter output voltage. The topology proposed in [19] presents asymmetry in the voltages across the switches when the gain is greater than 4, being characterized as a disadvantage despite the converter can operate with gains higher than four. Other quadratic converter topologies were presented in [20, 21], for driving the Light Emitting Diode, but the step-down voltage version. Besides these applications, a quadratic converter was used as a battery charger in [22]. In this case the system was developed to contemplate a high power-factor. With high voltage gain and low input current ripple, a converter topology based on interleaving technique was proposed in [23]. This article proposes a new topology for high gain DC-DC converter based on quadratic feature, which use low voltage switch and access to the midpoint of the output voltage.

The proposed converter topology is devised by connecting the three-level boost cell [24] shown in Fig. 1 (a) to the output port of the quadratic boost converters [5-7] which is shown in Fig. 1 (b). The resultant proposed converter is shown in Fig. 2.

* Corresponding author: siliconravi@yahoo.com

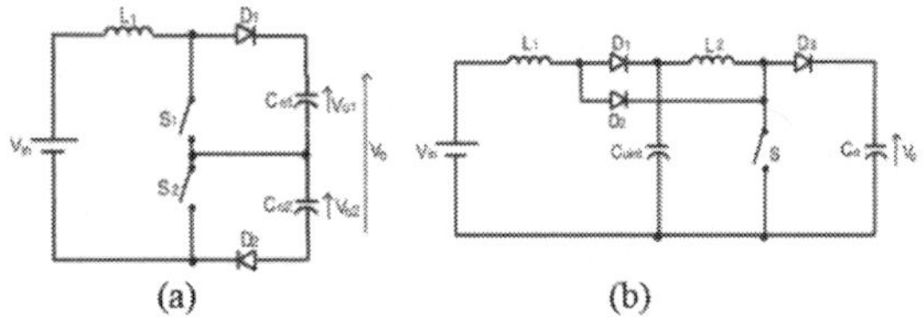

Fig. 1. Boost converter (a) Three level boost converter (b) Quadratic boost converter

Use In the proposed topology, an appropriate modulation strategy was adopted and the output capacitors are equal, $C_{o1} = C_{o2}$, hence the voltage on these capacitors will be half of total output voltage of converter. The switches, S_1 and S_2 are clamped at half of output voltage, thus a significant reduction in voltage stress on switches. In contrast, current flowing through the switches are the sum of the currents of inductors, L_1 and L_2. The switching technique used for operating the proposed converter is a delayed modulation technique, implemented through two 180° lagged pulses with fixed frequency and variable duty ratio. The converter will have two distinct regions of operation, the first relating to operation where duty cycle is lesser than 0.5 and other with a duty cycle greater than 0.5. The control signals of the switches can be seen in Fig. 4 and Fig. 6.

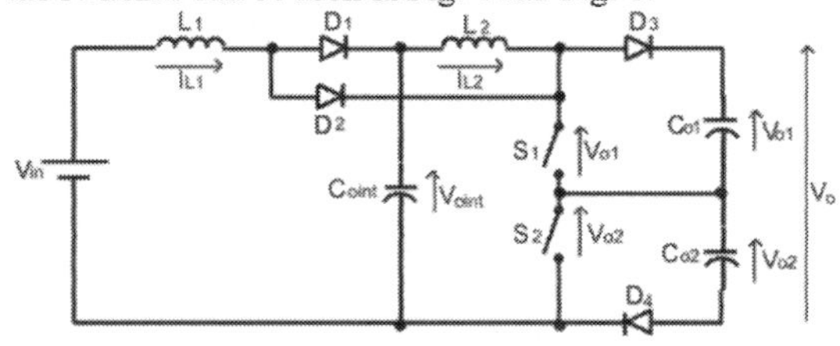

Fig. 2. Proposed boost converter

2 Principle of operation

To understand the proposed converter operation, the analysis is made for Continuous Conduction Mode (CCM) and Discontinuous Conduction Mode (DCM) operations and the converter is investigated in the two regions defined by the duty cycle ratios.

2.1 Operation in CCM The operating modes in CCM for duty ratio less than 0.5 are described below. The four operating modes of proposed converter for D < 0.5 is shown in Fig. 3. The theoretical waveforms related to these modes are shown in Fig. 5.

First step of operation (t_0 - t_1) - (Fig. 3 (a)): Switches, S_1 and S_2 are in off state and the inductor current decreases linearly. While the C_{oint} capacitor and the inductor L_1 are being charged from the input power source, the inductor L_2 and the capacitors C_{o1} and C_{o2} supply power to the load.

Second step of operation (t_1 - t_2) - (Fig. 3 (b)): In this step of operation, S_1 remains in off state and S_2 is switched on. The energies from the input source and inductors are transferred to both the capacitor C_{o1} and the load.

Third step of operation (t_2 - t_3) - (Fig. 3 (a)): This step of operation begins when S_2 is turned off. The topological structure and the operation of this step are similar to the first step of operation.

Fourth step of operation (t_3 - t_4) - (Fig. 3 (c)): During this step of operation, switch S_1 is turned on and the switch S_2 remains in off state. The energy is transferred to the capacitor C_{o2} and load from the input source. From the theoretical waveforms which was shown in Fig. 4, for D < 0.5, the equations for time intervals in terms of duty cycle, D and time period, T are given in (1) and (2).

The voltage gain of the proposed converter in Continuous Conduction Mode for duty cycle less than 0.5 can be obtained by determining the average voltage in inductors L_1 and L_2 during one full cycle of operation. Since this converter is composed of two conversion stages, static gain analysis can be done individually.

$$\Delta t_1 = \Delta t_4 = \frac{T(1-2D)}{2} \quad (1)$$

$$\Delta t_2 = \Delta t_3 = DT \quad (2)$$

At steady state, the average inductor voltage across L_1 is equal to zero, as given by (3), where V_{in} is the input voltage; V_{oint} is the intermediate voltage and V_o is the output voltage. The partial voltage gain of the first step of operation is calculated and given in (4).

In the same way, the partial voltage gain is determined for the second step of operation, and it is given in equations (5) and (6). Multiplying the partial voltage gains (4) and (6), which gives equation (7), the total voltage-gain of the proposed converter in continuous conduction mode for D < 0.5.

$$2\Delta t_1 (V_{in} - V_{o\,\mathrm{int}}) + \Delta t_2 (2V_{in} - V_o) = 0 \quad (3)$$

$$\frac{V_{o\,\mathrm{int}}}{V_{in}} = \frac{1-D}{2D^2 - 2D + 1} \quad (4)$$

$$2\Delta t_1 (V_{o\,\mathrm{int}} - V_o) + \Delta t_2 (2V_{o\,\mathrm{int}} - V_o) = 0 \quad (5)$$

$$\frac{V_o}{Vo\,\mathrm{int}} = \frac{1}{1-D} \quad (6)$$

$$G_{CCM(D<0.5)} = \frac{V_o}{V_{in}} = \frac{1}{2D^2 - 2D + 1} \quad (7)$$

The converter operating in CCM for duty cycle greater than 0.5 is described below. Fig. 4 illustrates the different operation steps and Fig. 6 shows the key waveforms of the proposed converter.

First step of operation (t_0 - t_1) - (Fig. 4 (a)): Switches, S_1 and S_2 are in conduction state and the inductor current increases linearly. At this stage of operation, there is no energy transfer between the source and the load. The load is being powered by capacitors C_{o1} and C_{o2}.

Second step of operation (t_1 - t_2) - (Fig. 4 (b)): The second operating step starts when the switch, S_2 is switched off and diodes D_1 and D_4 are in conducting state. The capacitor C_{oint} is being charged from the supply voltage V_{in} and the energy from inductor, L_1. The inductor L_2 transfers its energy to the capacitor C_{o2}.

https://doi.org/10.1051/e3sconf/20198701025

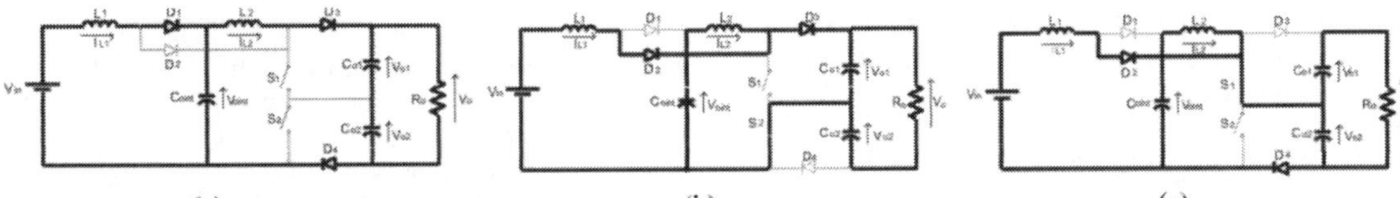

(a) (b) (c)

Fig. 3. Proposed boost converter operating modes in Continuous Conduction Mode (D < 0.5)

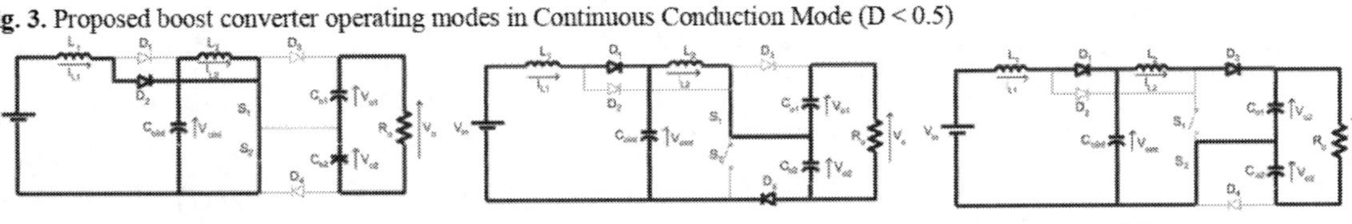

(a) (b) (c)

Fig. 4. Proposed boost converter operating modes in Continuous Conduction Mode (D > 0.5)

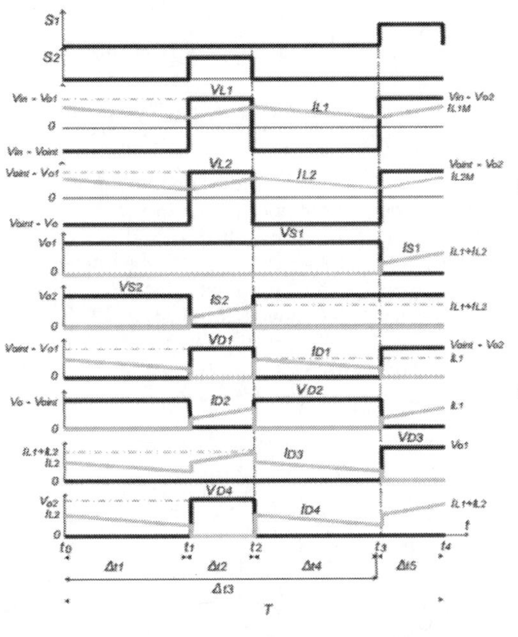

Fig. 5. Theoretical waveforms of the proposed converter in continuous conduction mode (D< 0.5)

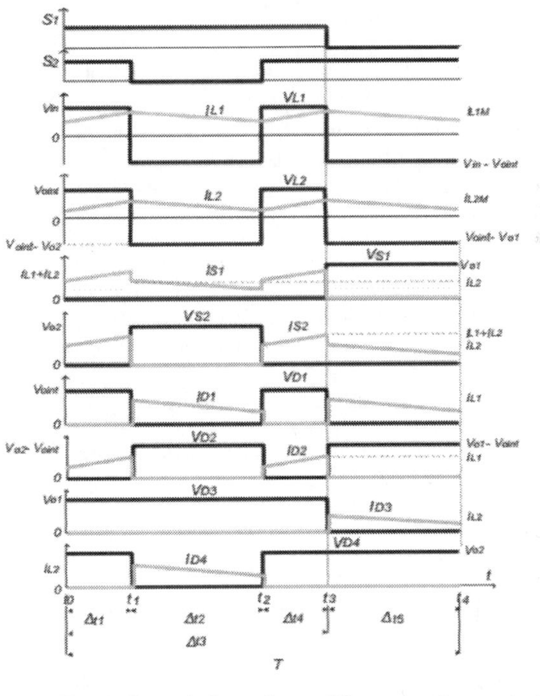

Fig. 6. Theoretical waveforms of the proposed converter in continuous conduction mode (D > 0.5)

Third step of operation (t_2 - t_3) - (Fig. 4 (a)): This step begins when S_2 is turned on again. The topological condition and operation of this step are the same as the first step of operation.

Fourth step of operation (t_3 - t_4) - (Fig. 4 (c)): When the switch, S_1 is turned off, the fourth step of converter operation starts. However, switch, S_2 continues its conduction process. During this operation, the power from input source, energy from L_1, charge across C_{oint} and energy of the inductor L_2 deliver energy to the capacitor C_{o1}. The waveforms shown in Fig. 6 is used to determine the equations of the time intervals of the four operating steps in terms of time period and duty cycle and they are given in equations (8) and (9).

$$\Delta t_1 = \Delta t_4 = \frac{T(2D-1)}{2} \tag{8}$$

$$\Delta t_2 = \Delta t_5 = T(1-D) \tag{9}$$

For D > 0.5, the voltage gain of the proposed converter in continuous conduction mode can be determined by analysing each operating step of converter. At steady state, the average voltage developed in the inductor, L_1 is zero. Thus, the expression can be written as in (10). The partial voltage gain of the proposed converter in first operating step is given by the relationship between V_{in} and V_{oint}. It is determined by (11). In steady state, the average voltage across inductor, L_2 equals zero, as shown in (12). Equation (12) gives equation (13), which is the partial voltage gain of the proposed converter in its second operating step.

The partial voltage gains which are given in equations (11) and (13) are multiplied to determine the total voltage gain of the proposed converter in continuous conduction mode for duty cycle more than 0.5 and it is given in (14).

$$2\left[V_{in}\Delta t_1 + \Delta t_2\left(V_{in} - V_{oint}\right)\right] = 0 \tag{10}$$

$$\frac{V_{oint}}{V_{in}} = \frac{1}{2(1-D)} \tag{11}$$

$$2\Delta t_1 V_{oint} + \Delta t_2\left(2V_{oint} - V_o\right) = 0 \tag{12}$$

$$\frac{V_o}{V_{oint}} = \frac{1}{1-D} \tag{13}$$

$$G_{CCM(D>0.5)} = \frac{V_o}{V_{in}} = \frac{1}{2(1-D)^2} \tag{14}$$

Fig. 7 demonstrates the voltage gain of classic boost converter, basic quadratic boost converter and the proposed converter, all operating in the continuous conduction mode. It is observed from the figure that the voltage gain of the proposed converter is always greater than (or equal at D = 0.5) the gain of the classic boost converter. In this way, one can obtain higher voltages from a lower voltage, without using larger duty cycles, which results in reduction of power losses in the semiconductor devices.

For a given current ripple specifications, it is possible to obtain the value of inductances L_1 and L_2, for both the duty cycle less than 0.5 and greater than 0.5, and it is given in equations (15) to (18).

$$L_{1(D<0.5)} = \frac{V_o\left[4D^3 - 4D^2 + D\right]}{2\Delta I_{L1} f} \tag{15}$$

$$L_{2(D<0.5)} = \frac{V_o\left[D(1-2D)\right]}{2\Delta I_{L2} f} \tag{16}$$

$$L_{1(D>0.5)} = \frac{V_o\left[(1-D)^2(2D-1)\right]}{\Delta I_{L1} f} \tag{17}$$

$$L_{2(D>0.5)} = \frac{V_o\left[(1-D)(2D-1)\right]}{2\Delta I_{L2} f} \tag{18}$$

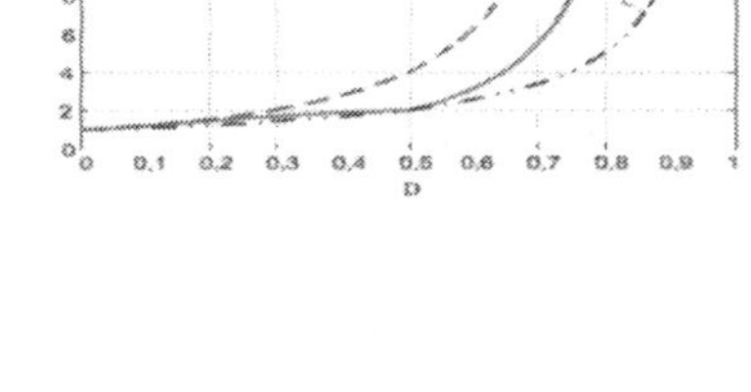

Fig. 7. Voltage gain of classic boost converter, basic quadratic boost converter and proposed converter

The equations of the current ripple in the inductor L_1 can be normalized by dividing (15) and (17) by the input current of the converter. The input current is related to output current, which results in (19) and (20). Thus, we have (21) and (23), where R_o represents load resistance. The current ripple of inductor L2 has been normalized by dividing (16) and (18) by the output current of converter, which results in equations (22) and (24).

$$I_{inmed(D<0.5)} = \frac{I_o}{2D^2 - 2D + 1} \tag{19}$$

$$I_{inmed(D>0.5)} = \frac{I_o}{2(1-D)^2} \tag{20}$$

$$\overline{\Delta I_{L1(D<0.5)}} \frac{L_1 f}{R_o} = \frac{\left(4D^3 - 4D^2 + D\right)\left(2D^2 - 2D + 1\right)}{2} \tag{21}$$

$$\overline{\Delta I_{L2(D<0.5)}} \frac{L_2 f}{R_o} = \frac{D(1-2D)}{2} \tag{22}$$

$$\overline{\Delta I_{L1(D>0.5)}} \frac{L_1 f}{R_o} = 2(1-D)^4(2D-1) \tag{23}$$

$$\overline{\Delta I_{L2(D>0.5)}} \frac{L_2 f}{R_o} = \frac{(1-D)(2D-1)}{2} \tag{24}$$

The normalized relative ripple current of the inductors for the two operating regions are shown in Fig. 8. The voltage ripples in the intermediate capacitor V_{coint} and output filter capacitors V_{Co1} and V_{Co2} are determined from the expressions (25) and (26), respectively. V_{Co1} and V_{Co2} are the voltages in the output filter capacitors, Io is the output current of the converter and f is the switching frequency of the converter.

$$\Delta V_{Coint} = \frac{2I_o\left(V_{oint} - V_{in}\right)}{C_{oint} f V_{in}} \tag{25}$$

$$\Delta V_{Co1} = \Delta V_{Co2} = \frac{I_o\left(2V_{Co1} - V_{oint}\right)}{2C_{o1} f V_{Co1}} \tag{26}$$

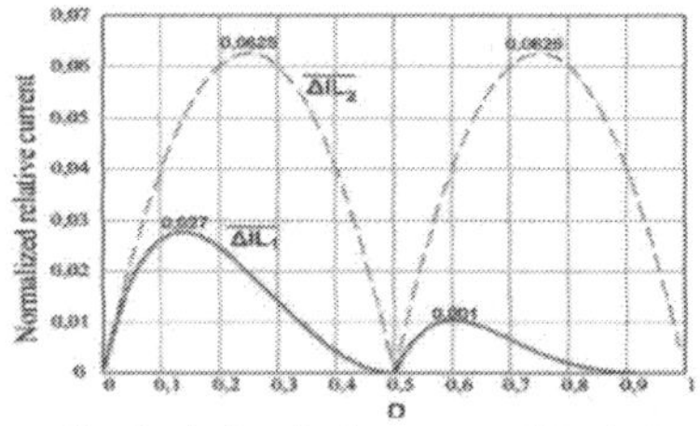

Fig. 8. Normalized relative ripple current of the inductors

2.2 Operation in DCM

In the discontinuous conduction mode, three possible discontinuities occur in the proposed converter operation. The first two possibilities are identified by the discontinuity of current in the inductor L_1 or current in the inductor L_2. While the third possibility is indicated by the discontinuity of currents flowing in both L_1 and L_2. The first two possibilities will be called as first Partial

 https://doi.org/10.1051/e3sconf/20198701025

Discontinuous Conduction Mode (PDCM) and the third possibility will be called as Total Discontinuous Conduction Mode (TDCM). However, considering that the mean current of L_1 is greater than the mean current of L_2 and if the current ripple in both inductors are equal, then the current flowing in inductor L_2 always cancels the current flowing in the inductor L_1. Thus, in this work, the PDCM operation refers to the current discontinuity only in the inductor L_2.

The analysis for the PDCM is presented for the two operating conditions defined by the duty cycle (for $D < 0.5$ and for $D > 0.5$). The six operating steps of the proposed converter in the PDCM for the duty cycle less than 0.5 are explained below. The circuit diagrams of the converter for different operation steps and the theoretical waveforms of the proposed converter are shown in Fig. 9 and Fig. 10.

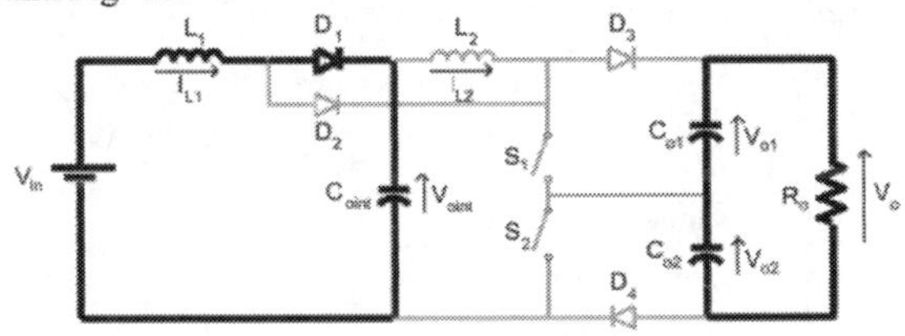

Fig. 9. Proposed boost converter operating modes in Partial Discontinuous Conduction Mode ($D < 0.5$)

First operation step (t_0 - t_1) - (Fig. 3 (a)): The switches, S_1 and S_2 remain in off state and the inductor current decreases linearly. While the source V_{in} and L_1 send energy to C_{oint}, the inductor L_2 and the capacitors C_{o1} and C_{o2} feed energy to the load.

Second operation step (t_1 - t_2) - (Fig. 9): This operation step starts when the current of L_2 reaches zero, blocking D_3 and D_4. In this step, the C_{oint} capacitor continues to be charged with the energy of V_{in} and L_1, however the load is fed by the capacitors C_{o1} and C_{o2}.

Third operation step (t_2 - t_3) - (Fig. 3 (b)): When the switch S_2 is turned on and switch S_1 remains in off state, this third step of operation starts. The current flowing through the inductors L_1 and L_2 increases linearly, and energy is transferred to C_{o1} and the load.

Fourth operation step (t_3 - t_4) - (Fig. 3 (a)): During this operation step, switches S_1 and S_2 are in off state. The topological condition and operation of this step are the same as first mode.

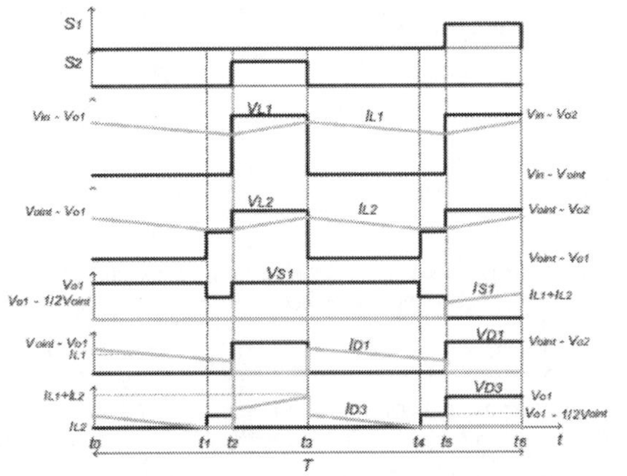

Fig. 10. Theoretical waveforms of the proposed converter in partial discontinuous conduction mode ($D < 0.5$)

Fifth operation step (t_2 - t_3) - (Fig. 9): This mode of operation starts, when the L_2 current reaches zero. The topological condition and operation of this step are the same as the second step of operation.

Sixth operation step (t_3 - t_4) - (Fig. 3 (c)): In this operation step, the switch S_1 is turned on, and switch S_2 remains in off state. During this operating mode, current flowing through the inductors L_1 and L_2 increase again. The capacitor C_{o2} and the load receives the energy from the inductors.

The six converter operating steps in PDCM for duty cycle greater than 0.5 are described below. The converter operating modes and the theoretical waveforms are shown in Fig. 11 and Fig. 12.

First step of operation (t_0 - t_1) - (Fig. 5 (a)): In this mode of operation, the switched S_1 and S_2 are in conducting state and the current flowing through the inductors increases linearly. D_1, D_3 and D_4 are in off state. The load is being powered by the capacitors *Co1* and *Co2*.

Second operation stage (t_1 - t_2) - (Fig. 5 (b)): In this operating step S_1 remains conducting and S_2 is turned off. The currents in the inductors decrease linearly. The source V_{in} and inductor L_1 transfers the energy to C_{oint}, however L_2 transfers the energy to C_{o2} and load.

Third operation step (t_2 - t_3) - (Fig. 11 (a)): This stage begins when the L_2 current reaches zero, reverse biasing D_3 and D_4. The supply voltage V_{in} and the inductor L_1 charges C_{oint} continues to charge with the energy obtained from V_{in} and L_1, and the load is fed by the capacitors C_{o1} and C_{o2}.

Fourth operation phase (t_3 - t_4) - (Fig. 5 (a)): this operation mode starts when S_2 is turned on again. In this step, the inductors store energy and capacitors C_{o1} and C_{o2} feed the load.

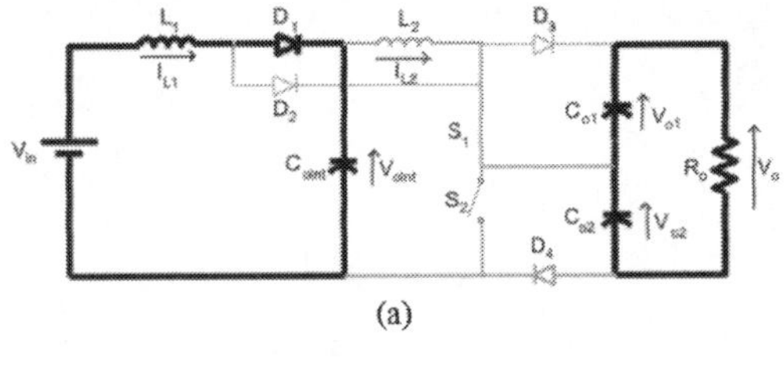

(a)

 https://doi.org/10.1051/e3sconf/20198701025

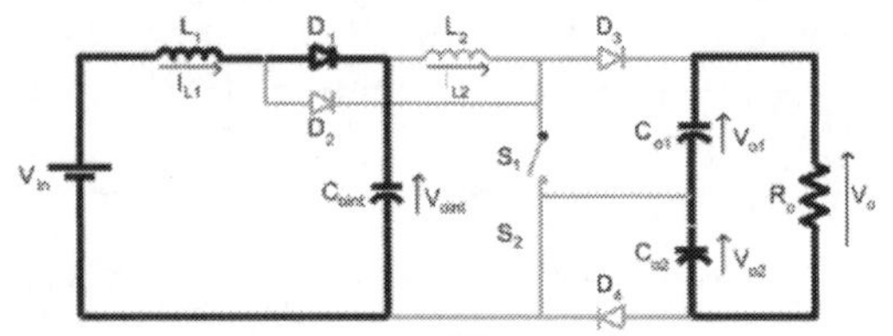

(b)

Fig. 11. Proposed boost converter operating modes in Partial Discontinuous Conduction Mode (D > 0.5)

Fifth stage operation (t_4 - t_5) - (Fig. 5 (c)): When the switch S_1 is turned off, the converter shifts to this operating mode. In this mode, the source V_{in} and L_1 charge the capacitor C_{oint}, while the inductor L_2 delivers the energy to C_{o1} and the load.

Sixth stage operation (t_5 - t_6) - (Fig. 11 (b)): This mode occurs when the current in L_2 reaches zero, and the diode D_3 is reverse biased. The topological state and operation of this step are similar to the third step.

Based on the converter operating mode waveforms shown in Fig. 10 and Fig. 12, the mean values of the voltages in the inductors L_1 and L_2 are calculated, we obtain (27) and (28), which gives the equations for the total voltage gain of the proposed converter in partial discontinuous conduction mode. The parameter γ_2 is a current parameter which represents the external factors that influence the static behaviour of the converter.

$$G_{PDCM(D<0.5)} = \left[1+\frac{D^2}{\gamma_2(1-2D)}\right]\left[\frac{1-D}{2D^2-2D+1}\right] \quad (27)$$

$$G_{PDCM(D>0.5)} = \left[2+\frac{(2D-1)^2}{\gamma_2}\right]\left[\frac{1}{2(1-D)}\right] \quad (28)$$

3. Experimental results

A laboratory prototype of the proposed converter with a power rating of 520 W was built in order to demonstrate its operation and performance. The parameters used to design the proposed converter is given in Table 1. The power rating and rated voltage values are equal to an array of four solar photovoltaic panels, each having 130W and 17 V at maximum power. The switching frequency was chosen with the purpose of comparing the constructive aspects of this converter with the work carried out in [19]. The design parameters of the inductors and the capacitors presented in the proposed converter were calculated by considering the duty cycle, D = 0.79. The maximum current ripples of the inductors were limited to 20% of the mean value of the current in each inductor. The voltage ripples on the capacitors were limited to 1% of the mean voltage on the respective capacitor.

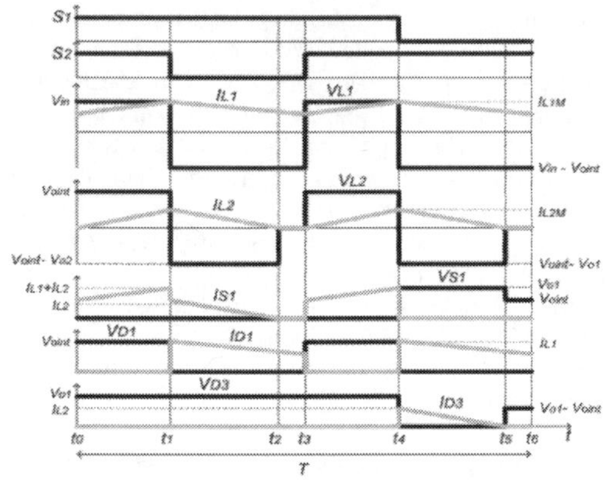

Fig. 12. Theoretical waveforms of the proposed converter in partial discontinuous conduction mode (D > 0.5)

Table 1. Selection of Proposed Converter Parameters.

Design Considerations	Selected Values
Rated power, Po	520 W
Input voltage, Vin	34 V
Output voltage, Vo	380 V
Switching frequency, f	50 kHz

Fig. 13 shows the experimental waveforms of the proposed converter. The input voltage is 34 V and the mean current flowing through the inductor is 15.38 A with ripple of 9 %, which is allowable as shown in Fig. 13 (a). The intermediate voltage across the capacitor is almost 80 V and the mean current flowing through the inductor L_2 is 6.47 A, with ripple of 12 % as shown in Fig. 13 (b). The voltage stress across the switches S_1 and S_2 and maximum reverse blocking voltage across the diodes D_3 and D_4 are 190 V, which is equal to half of the total output voltage of the converter as shown in Fig. 13 (c) and 13 (d). The proposed converter has midpoint in the output port, so that the inverter can be directly connected to it without using any extra components. The voltage across each output capacitor is equal to 190 V as shown in Fig. 13 (e). The output voltage and current of the proposed converter for given specification is shown in Fig. 13 (f), and they are 380 V and 1.37 A, respectively. Fig. 14 shows the efficiency curve of the converter as a function of output power. The efficiency reaches its maximum value at the output power of 350W. Although the efficiency value is low when compared to other single-stage and high-voltage converters, similar results were found in other high gain and low input voltage converters.

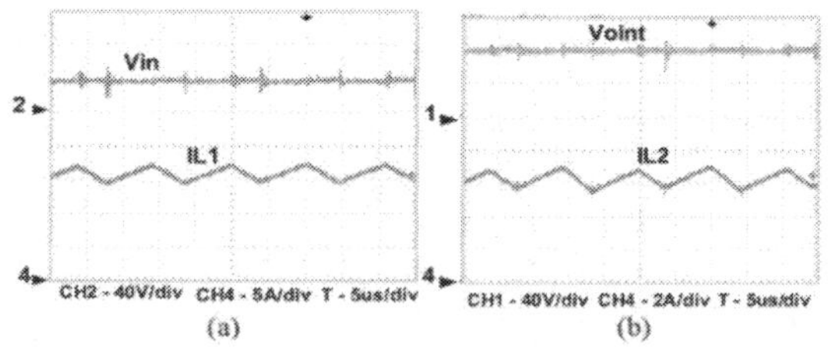

(a) (b)

 https://doi.org/10.1051/e3sconf/20198701025

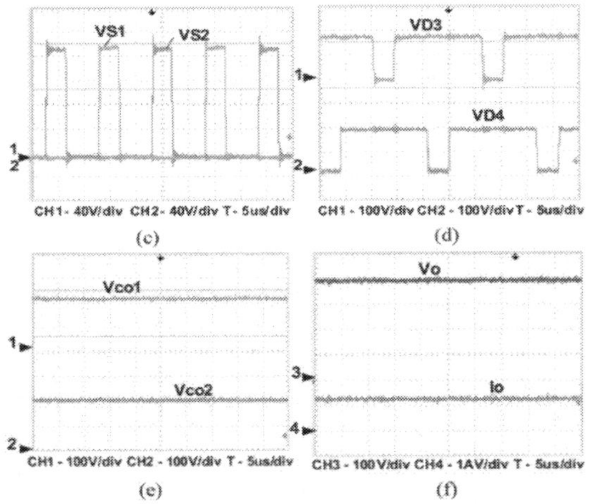

Fig. 13. Experimental waveforms of the proposed converter

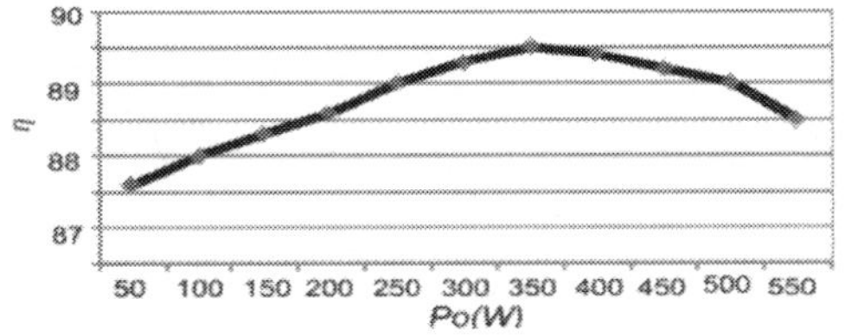

Fig. 14. Efficiency – output power characteristics of the proposed boost converter

References

1. R. D. de Oliveira Reiter, L. Michels, J. R. Pinheiro, R. A. Reiter, S. V. G. Oliveira, A. Péres, Comparative analysis of series and parallel photovoltaic arrays under partial shading conditions, *2012 10th IEEE International Conference on Industry Applications, Fortaleza,* pp. 1-5, (2012)
2. C. T. K. Kho, J. Ahmed, S. Kashem, Y. L. Then, A comprehensive review on PV configurations to maximize power under partial shading, *TENCON 2017 - 2017 IEEE Region 10 Conference, Penang*, pp. 763-768, (2017)
3. S. M. Chen, T. J. Liang, L. S. Yang, J. F. Chen, A Safety Enhanced, High Step-Up DC–DC Converter for AC Photovoltaic Module Application, IEEE Trans. on Power Electronics, **(27)** 4, pp. 1809-1817, (2012)
4. F. L. Tofoli, D. d. C. Pereira, W. Josias de Paula, D. d. S. Oliveira Júnior, Survey on non-isolated high-voltage step-up dc–dc topologies based on the boost converter, IET Power Electronics, **(8)** 10, pp. 2044-2057, (2015)
5. M. Lotfi Nejad, B. Poorali, E. Adib, A. A. Motie Birjandi, New cascade boost converter with reduced losses, IET Power Electronics, **(9)** 6, pp. 1213-1219, (2016)
6. J. Lu, Y. Wang, X. Li, Isolated high step-up DC–DC converter with integrated cascade structure, IET Power Electronics, **(11)** 7, pp. 1143-1152, (2018)
7. D. Maksimovic, S. Cuk, Switching converters with wide DC conversion range, IEEE Transactions on Power Electronics, **(6)** 1, pp. 151-157, (1991)
8. Tsai-Fu Wu, Te-Hung Yu, Unified approach to developing single-stage power converters, IEEE Transactions on Aerospace and Electronic Systems, **(34)** 1, pp. 211-223, (1998)
9. L. H. S. C. Barreto, E. A. A. Coelho, V. J. Farias, L. C. de Freitas, J. B. Vieira, An optimal lossless commutation quadratic PWM Boost converter, *APEC. Seventeenth Annual IEEE Applied Power Electronics Conference and Exposition, Dallas, TX*, **(2)**, pp. 624-629, (2002)
10. S. Sharifi, M. Jabbari, H. Farzanehfard, A New Family of Single-Switch ZVS Resonant Converters, IEEE Transactions on Industrial Electronics, **(64)** 6, pp. 4539-4548, (2017)
11. E. S. Hass, F. C. Corrêa, C. B. Nascimento, A novel non-isolated high DC-DC voltage based on boost converter employing coupled inductor, *2017 Brazilian Power Electronics Conference (COBEP), Juiz de Fora*, pp. 1-6, (2017)
12. H. Moradi Sizkoohi, J. Milimonfared, M. Taheri, S. Salehi, High step-up soft-switched dual-boost coupled-inductor-based converter integrating multipurpose coupled inductors with capacitor-diode stages, IET Power Electronics, **(8)** 9, pp. 1786-1797, (2015)
13. O. Abutbul, A. Gherlitz, Y. Berkovich, A. Ioinovici, Step-up switching-mode converter with high voltage gain using a switched-capacitor circuit, IEEE Transactions on Circuits and Systems I: Fundamental Theory and Applications, **(50)** 8, pp. 1098-1102, (2003)
14. E. H. Ismail, M. A. Al-Saffar, A. J. Sabzali, High Conversion Ratio DC–DC Converters With Reduced Switch Stress, IEEE Transactions on Circuits and Systems I: Regular Papers, **(55)** 7, pp. 2139-2151, (2008)
15. G. V. T. Bascope, I. Barbi, Generation of a family of non-isolated DC-DC PWM converters using new three-state switching cells, *2000 IEEE 31st Annual Power Electronics Specialists Conference. Conference Proceedings, Galway*, **(2)** pp. 858-863, (2000)
16. R. P. Torrico-Bascopé, L. F. Costa, G. V. Torrico-Bascopé, Generation of new nonisolated high voltage gain DC-DC converters, *2011 IEEE 33rd International Telecommunications Energy Conference (INTELEC), Amsterdam*, pp. 1-8, (2011).
17. R. P. Torrico-Bascopé, F. L. Tofoli, C. G. Castelo Branco, C. M. T. Cruz, Three-state switching cell (3SSC)-based non-isolated dc–dc boost-type converter with balanced output voltage and wide voltage conversion range, IET Power Electronics, **(11)** 7, pp. 1217-1223, (2018)
18. X. Hu, G. Dai, L. Wang, C. Gong, A Three-State Switching Boost Converter Mixed With Magnetic Coupling and Voltage Multiplier Techniques for High Gain Conversion, IEEE Transactions on Power Electronics, **(31)** 4, pp. 2991-3001, (2016)

 https://doi.org/10.1051/e3sconf/20198701025

19. Y. R. de Novaes, A. Rufer, I. Barbi, A New Quadratic, Three-Level, DC/DC Converter Suitable for Fuel Cell Applications, *2007 Power Conversion Conference - Nagoya, Nagoya*, pp. 601-607, (2007).
20. T. Yan, J. Xu, X. Liu, G. Zhou, J. Gao, Flicker-free transformerless LED driving circuit based on quadratic buck PFC converter, Electronics Letters, (**50**) 25, pp. 1972-1974, (2014)
21. J. M. Alonso, J. Viña, D. Gacio, L. Campa, G. Martínez, R. Osorio, Analysis and design of the quadratic buck-boost converter as a high-power-factor driver for power-LED lamps, *IECON 2010 - 36th Annual Conference on IEEE Industrial Electronics Society*, pp. 2541-2546, (2010)
22. V. F. Pires, D. Foito, A. Cordeiro, A DC–DC Converter With Quadratic Gain and Bidirectional Capability for Batteries/Supercapacitors, IEEE Transactions on Industry Applications, (**54**) 1, pp. 274-285, (2018)
23. R. Gules, L. L. Pfitscher, L. C. Franco, An interleaved boost DC-DC converter with large conversion ratio, *2003 IEEE International Symposium on Industrial Electronics*, (**1**), pp. 411-416, (2003)
24. D. Maksimovic, R. Erickson, Universal-input, high-power-factor, boost doubler rectifiers, *Applied Power Electronics Conference and Exposition, 1995. APEC '95. Conference Proceedings 1995., Tenth Annual, Dallas, TX,* (**1**), pp. 459-465, (1995).

Capacity Fade Modeling of Li-Ion Battery using Evolutionary Algorithm

Tamilselvi S[1*], *Karuppiah* N[2], and *Rajagopal Reddy* B[3]

[1]Associate Professor, Department of Electrical and Electronics Engineering, SSN College of Engineering, Chennai, Tamilnadu, India, tamilselvis@ssn.edu.in
[2]Professor, Department of Electrical and Electronics Engineering, Vardhaman College of Engineering, Shamshabad, Hyderabad, India, natarajankaruppiah@gmail.com
[3]Associate Professor, Department of Electrical and Electronics Engineering, Vardhaman College of Engineering, Shamshabad, Hyderabad, India, reddy.rajagopal@gmail.com

Abstract. Renewable sources are seasonal and cannot be considered as available energy source as their generation varies with time. The insufficient forecasting techniques lead to thought of storage of energy. Even though many techniques of energy storage are available, batteries play a vital role as the time taken to start delivering the stored energy is very less. The life period of the battery depends upon the charging and discharging characteristics which in turn depend on the internal parameters such as life period, charge rate, discharge rate of the battery. The energy stored in the battery can be calculated by finding these parameters. In this paper these parameters are estimated for a Sony lithium ion battery by evolutionary algorithm CMA-ES under different Charging and discharging rates. As the batteries are charged and discharged there is capacity loss in the battery. This loss is modelled by modified Arrhenius equation on practical conditions. Capacity loss of the sample battery is modelled for five different cycles starting from 50th cycle to 100th cycle in an interval of 10 cycles. The results are validated with those of manufacturer catalogue. The optimized battery capacity loss are found to coincide with the measured values.

1 Introduction

The demand for electrical energy has increased at a rate of 6% per annum over the past 10 years and the approximate per capita energy consumption is 1075 KWh [1] in India. During peak hours, the demand is 159 GW against availability of 156 GW which is a shortage of 3.2% [2]. The rapid growth in the field of renewable energy resources has increased the overall installed capacity of the country. But these renewable energy sources are seasonal and cannot be considered as available energy source. Insufficient forecasting techniques also lead to thought of using energy storage device (ESD) at available period. From the power plant usage data it is seen that most of the power plants are underutilized during the off-peak hours. If some energy storage device can be used to store the energy during the off-peak hours, then an energy gap of 500 MW during peak power demand can be met. Therefore, an Energy Storage System (ESS) is required to solve the issue of energy gap during peak hours. In order to handle the peak power demand, one possible solution can be to ramp-up the power generation capacity, but this option would require significant infrastructure cost of the power plants.
Even though many techniques of energy storage are available, batteries play a vital role and the time taken to deliver the stored energy is very less. But the efficiency of the available batteries in the market is very less. Even though these existing batteries are charged during off peak hours, the extracted energy transfer from batteries to grid during peak demand hours may not be sufficient. The reason for this issue is that the available power in the batteries is dependent on the characteristics of the battery. Every battery has its own characteristics and the battery characteristics depend on the internal parameters such as resistance, capacitance and the open circuit voltage. These internal parameters vary with respect to life period, charge rate, discharge rate of the battery. The frequent charging and discharging of batteries, to solve the power shortage issue, result in the degradation of the internal circuit parameters of these batteries [3, 4]. The parameters such as state of charge, depth of discharge, charge rate and discharge rate get affected. Due to these problems the capacity of the batteries gets decreased. This degradation in capacity of the battery is proportional to charge or discharge rate and temperature. Thus the life period of the battery depends upon the charging and discharging characteristics of the battery. Hence it is necessary to study and analyze the battery characteristics for increasing efficiency.
The different types of batteries used are lead acid, lithium-ion, alkaline battery etc. Lithium ion batteries are the most efficient of all [5].There are four types of battery models

* Corresponding author: tamilselvi.manjuraj@gmail.com

available. They are Electro-chemical models, Experimental models, Mathematical models and Electric circuit models. Of all these four, electric circuit models are best suited to represent the characteristics of EV battery. Wardburg impedance model is taken as the base model. It consists of an open circuit voltage in series with resistance and parallel RC circuit. To model a battery the internal circuit parameters are to be estimated. But the estimation of internal circuit parameters in a battery is tedious due to their non-linear behaviour. Soft computing techniques are used nowadays to extract the parameters of the battery.

Praveen Kumar, Pavol Bauser presented a paper "Parameter extraction of Battery Models Using Multi objective Optimization Genetic Algorithms"-14th International Power Electronics and motion Control Conference -2010[6]. In this work a model of the battery and a methodology for determining the parameters of a battery model using the catalogue data of the batteries. An Battery Model (BM) and Capacity Model (CM) was done by Kannan Thirugnanam, Himanshu Saini and Praveen Kumar Indian Institute of Technology Guwahati/Electronics and Electrical Engineering, Guwahati-781039, India, in their paper "Mathematical Modelling of Li-Ion Battery for Charge/Discharge Rate and Capacity Fading Characteristics using Genetic Algorithm Approach." And published in IEEE journal 2012 [7]. Using Battery modelling and Capacity Modelling, the amount of capacity fading is analyzed while the power is stored in the battery or transferred to the grid. The BM is developed for Li-ion battery based on the Electrical Equivalent Circuit (EEC) and the CM is proposed based on processed energy for different charging/discharging rate. The proposed model is simulated using MATLAB Simulink environment and the parameters are optimized using binary Genetic Algorithm (GA) approach to compare the results with manufacturer's catalogue data. In 2014, in the paper "Mathematical Modelling of Li-Ion Battery Using Genetic Algorithm Approach for V2G Applications", Kannan Thirugnanam, *Student Member, IEEE*, Ezhil Reena Joy T. P., Student Member, IEEE, Mukesh Singh, Student Member, IEEE, and Praveen Kumar, Member, IEEE- IEEE transactions on Energy conversions-2014 [8] presents an electric circuit-based battery and a capacity fade model suitable for electric vehicles (EVs) invehicle-to-grid applications. The circuit parameters of the battery model (BM) are extracted using genetic algorithm-based optimization method. A control algorithm has been developed for the battery, which calculates the processed energy, charge or discharge rate, and state of charge limits of the battery in order to satisfy the future requirements of EVs. A complete capacity fade analysis has also been carried out to quantify the capacity loss with respect to processed energy and cycling. Min Chen, Student Member IEEE and Gabriel A. Rincon-Mora, Senior Member, IEEE published a paper "Accurate Electrical Battery Model capable of predicting Runtime and I-V Performance" IEEE transactions on Energy conversions[9].A simplified model neglecting the effects of self discharge, cycle number, temperature in low power LI-ion batteries was presented. Liao Chenglin, Li Huiju, Wang Lifang Institute of Electrical Engineering Chinese Academy of Sciences Beijing, China presented a paper in 2009 "A Dynamic Equivalent Circuit Model of LiFePO4 Cathode Material for Lithium Ion Batteries on Hybrid Electric Vehicles "-Journal of Power Sources, Elsevier[10].In this paper, the dynamic characteristics of the batteries are investigated over different current profiles. A dynamic equivalent circuit model was also proposed to describe the dynamic characteristics of the batteries. In the year 2004 the "Modelling of lithium ion cells - A simple equivalent-circuit model approach" Bor Yann Liaw, Ganesan Nagasubramanian, Rudolph G. Jungst, Daniel H. Doughty [11] presented an equivalent-circuit-based battery model, capable of simulating charge and discharge behaviour of lithium-ion batteries (LiB). The model, although simple in concept, can simulate complex discharge behaviour with high fidelity. "A dynamic lithium-ion battery model considering the effects of temperature and capacity fading" O. Erdinc, B. Vural and M. Uzunoglu, *Member, IEEE* (2009) [12]. In this paper, a dynamic model of lithium-ion battery has been developed with MATLAB/Simulink® in order to investigate the output characteristics of lithium-ion batteries. Dynamic simulations are carried out, including the observation of the changes in battery terminal output voltage under different charging/discharging, temperature and cycling conditions, and the simulation results are compared with the results obtained from several recent studies. In 2006 J.J. Esperilla, J. Fe´lez, G. Romero, A. Carretero ETS Ingenieros Industries, Universidad Polite´cnica de Madrid (UPM), Jose Gutierrez Abascal, 2, 28006 Madrid, Spain in their paper [13]."A model for simulating a lead-acid battery using bond graphs" presented a model of a lead-acid battery developed with bond graphs. The bond graph structure is used to reproduce the behaviour of reversible electrochemical cells in charging conditions or in discharging conditions. The work presented here has been applied to the particular case of standard 12V lead-acid battery, which is widely used in automotive industries, traction, and electrical or hybrid vehicles.

In this paper the internal parameters are extracted for Sony make, 5 V, 2 Ahr Lithium ion battery and a modelling has been done for its capacity fading after 50 cycles of operation. Covariance Matrix Adaptation Evolutionary Strategy (CMA-ES) has been applied to optimize the internal circuit parameters of a battery, under different charging and discharging rates. The algorithm results are compared with the catalogue values given by the manufacturer for validation. The obtained battery characteristics and capacity fading results coincide with the measured (manufacturer catalogue) characteristics in this paper.

The paper is organized as follows. Section 2 describes the proposed battery model. Evolutionary algorithm is explained in Section 3. The details of battery parameter extraction and capacity fading are discussed in results and discussion of Section 4. The model validation has been done in the Section 4 by comparing the proposed model result with manufactures' data. Section 5 concludes the research work.

https://doi.org/10.1051/e3sconf/20198701026

2 Battery model

An Electric Equivalent Circuit for Li-ion battery [6, 7] is shown in Fig. 1. The charging and discharging characteristics are exponentially increasing and decreasing with respect to rate of charging or discharging. Hence a polynomial equation with exponential function has been used to represent the characteristics of battery.

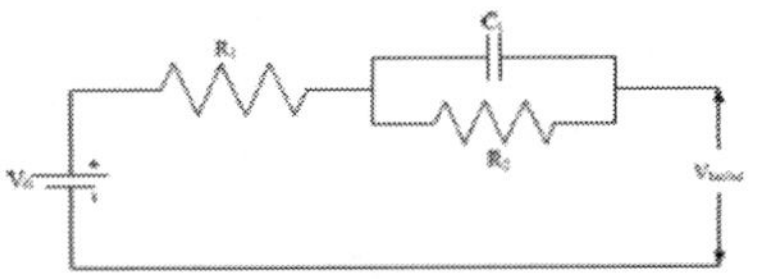

Fig. 1 Electric Equivalent Circuit for Li ion battery

2.1 Internal battery parameters

The general polynomial equations for the calculation of internal parameters of the battery are given below [3]. A set of 31 coefficients are used in the above set of equations to represent the parameters in terms of polynomial equations. The battery parameters can be derived by replacing x and y with Cr and SOC_{cr}, for charging process, whereas for discharging process with Dr and $(1 - DOD_{cr})$.

$$R_1 = (a_1 + a_2 x + a_3 x^2)e^{-a_4 y} + (a_5 + a_6 x + a_7 x^2) \quad (1)$$

$$R_2 = (a_8 + a_9 x + a_{10} x^2)e^{-a_{11} y} + (a_{12} + a_{13} x + a_{14} x^2) \quad (2)$$

$$C = -(a_{15} + a_{16} x + a_{17} x^2)e^{-a_{18} y} + (a_{19} + a_{20} x + a_{21} x^2) \quad (3)$$

$$V_0 = (a_{22} + a_{23} x + a_{24} x^2)e^{-a_{25} y} + (a_{26} + a_{27} y + a_{28} y^2 + a_{29} y^3) - a_{30} x + a_{31} x^2) \quad (4)$$

where R_1, R_2, C are internal parameters and are represented in terms of polynomial equations; V_0 is the voltage between the terminals of battery in open circuit condition.

2.2 Battery charging/discharging voltage

The battery terminal voltage for charging/discharging scenario with respect to time under constant current is given in Eq. 5 and Eq. 6 [3]. The accurate behaviour of any type of battery can be represented by Eq. 5 and Eq. 6, if the parameters are accurately extracted. The nonlinear behaviour can be captured by these equations

$$V_{C_i}^C = ((\frac{Q_r}{C} + I_c R_2)\exp(-\frac{t_c}{R_2 C})) + V_0 - (I_c(R_1 + R_2)) \quad (5)$$

$$V_{d_j}^C = ((\frac{Q_r}{C} + I_d R_2)\exp(-\frac{t_d}{R_2 C})) + V_0 - (I_d(R_1 + R_2)) \quad (6)$$

where Q_r is the remaining capacity of the battery; t_c, I_c, t_d and I_d are charge time, charge current, discharge time and discharge current, respectively.

2.3 Charge/Discharge rate and SOC calculations

The charge or discharge rate algorithm is used to determine the amount of energy stored or extracted from the battery. The C_r and SOC_{cr} of the battery vary depending on the present condition of the battery. The battery status is checked and the current charge rate (C^{crt}) of the battery is calculated by the control algorithm developed inside the battery. User defined C_r limit (C^{lmt}) and initial battery SOC (SOC_{ini}) are also taken into account .The C_r^{crt} and Dr^{crt} of the battery can be calculated from Eq. (7) & (8).

$$C_r = C_r^{crt} = \frac{I_c}{Q_r} \quad (7)$$

$$D_r = D_r^{crt} = \frac{I_d}{Q_r} \quad (8)$$

The above are calculated based on present status of the battery, which is the ratio of current and remaining capacity of the battery. The minimum of charge rate based on the C^{lmt} and C^{crt} charge are found by the algorithm to regulate the charge current of the battery. This type of control algorithm is used for discharging scenario. The SOC_{cr} and DOD_{cr} can be calculated from Eq. 9 to Eq. 10.

$$SOC_{cr} = SOC_{ini} + (\frac{I_c \Delta t_c}{Q_r 3600}) \quad (9)$$

$$DOD_{cr} = DOD_{ini} + (\frac{I_d \Delta t_c}{Q_r 3600}) \quad (10)$$

Where, SOC_{ini} is the initial SOC of the battery; SOC_{max} and DOD_{max} are the maximum user defined SOC and DOD limits respectively.

3 Capacity Fade Model

The energy storage capacity of battery decreases due to frequent charging or discharging. In this paper, the capacity loss is calculated at an interval of 10 cycles from 50th cycle based on variable charge or discharge rate. A cycle is defined as an interval during which the battery charges till SOC_{max} and discharges up to DOD_{max}. When the battery experiences different charge and discharge cycles, it fails to meet its performance thus reducing its life expectancy. Therefore, it is required to study the capacity fade to quantify the capacity loss with respect to number of cycles.

3.1 Battery power and processed energy

The battery power for charging (P_c) and discharging (P_d) scenario are given in Eq. (11) and Eq. (12). The amount of stored energy (E_{stor}) during charging process depends on increase in V_{ci}^{C} and SOC_{cr}, which is given in Eq. (13).

$$P_c = V_{c_i}^{C} I_c \tag{11}$$

$$P_d = V_{d_j}^{C} I_d \tag{12}$$

$$E_{stor} = V_{c_i}^{C} Q_r \Delta SOC_{cr} \tag{13}$$

where ΔSOC_{cr} is the change in current SOC_{cr}. The processed energy (PE_c) for charging scenario is given in Eq. (14).

$$PE_c = \sum E_{stor} \tag{14}$$

The available energy (E_{avail}) in the battery during discharging process decreases V_{di}^{C} with increase in DOD_{cr}, which can be calculated using Eq. (15).

$$E_{avail} = V_{d_j}^{C} Q_r \Delta DOD_{cr} \tag{15}$$

where, ΔDOD_{cr} is the change in current DOD_{cr}. The processed energy (PE_d) for discharging scenario is given in Eq. (16).

$$PE_d = \sum E_{avail} \tag{16}$$

The total processed energy (E_{total}) of the battery in a cycle is calculated using Eq. (17).

$$E_{total} = \sum PE_c + PE_d \tag{17}$$

Eq. (11) to Eq. (17) represents the real-time performance of the battery during charging and discharging process. Simulations are done based on these equations for the developed battery model and have been validated with the manufacturers' catalogue.

3.2 ABCD Parameters of capacity fade model

Capacity fade of the battery can be quantified using Arrhenius equation for Constant charge or discharge rate [16]. The mathematical model used to quantify the capacity fade is based on Arrhenius equation and is given in Eq. (18) [14, 15]. Here, A is pre-exponential factor, Q_l represents the capacity loss (Ah), E_a denotes the activation energy (J), R is the gas constant T is temperature (K), z is the adjustable factor.

$$Q_l = A\exp(-\frac{E_a}{RT})t^2 \tag{18}$$

Eq (18) is used for calculating capacity fade under constant C_r. However, in real-time grid support, the charge or discharge rate of the battery would vary with respect to peak power demand and node voltage variations. This causes capacity loss of the battery. In real-time, the batteries charge or discharge rate varies with respect to variable node voltage conditions, SOC_{cr} or DOD_{cr} limits, and temperature and also on the system constraints and user preferences. Due to this, the capacity fades changes with respect to the varying C_r, D_rSOC, DOD limits and temperature. To account this loss, a mathematical model is developed in this work to predict the capacity fade at different charge and discharge rate. The total capacity fade can be quantified and written in terms of nominal capacity (Q), C_r, D_r, PE_c, PE_d, pre-exponential (A, B) and adjustable (C, D) factors, gas constant (R) and temperature (T) which is given in Eq. (19). In Eq. (19), Q_{lk}^{C} denotes calculated capacity loss and Q_r the remaining capacity of the battery can be calculated from Eq. (20).

$$Q_{l_k}^{c} = A\exp(-\frac{CPE_c Q_r C_r SOC_{cr}}{RT}) + B\exp(-\frac{DPE_d Q_r D_r DOD_{cr}}{RT}) \tag{19}$$

$$Q_r = Q - Q_{l_k}^{c} \tag{20}$$

where, Q_r is the remaining capacity of the battery which is updated value of the capacity loss at every cycle. The pre-exponential and adjustable factors are calculated by CMAES. The capacity loss characteristics have been optimized by the fitness function ($F'(x)$). $F'(x)$ is maximized, when the difference between the measured (catalogue value) ($QlkM$) and calculated ($QlkC$) capacity loss is minimized. The mathematical representation of $F'(x)$ is given in Eq. (21). Simulations are done to verify the developed capacity fade model and have been validated with manufacturers' catalogue.

$$F'(x) = \frac{1}{1 + \sum_{k=1}^{n} (Q_{l_k}^{M} - Q_{l_k}^{C})} \tag{21}$$

The battery manufacturers' charge/discharge rate characteristics have been extracted using the mathematical equations described in this Section. The control algorithm decides the amount of energy stored into the battery or extracted from the battery based on the cur- rent energy status and user defined C_r/D_r limit of the battery. The input of the control algorithm are charge/discharge current, SOC/DOD limits and polynomial coefficients for charging/discharging scenario (a_1 − a_{31}). Based on these inputs, the BM calculates SOC_{cr}/DOD_{cr}, charge/discharge voltage (VC/VC), current (I_c/I_d), power (P_c/P_d) and energy (PE_c/PE_d). Capacity loss is calculated using energy of the battery, pre-exponential factor, gas constant, temperature and adjustable factor.

3.3 Fitness function

Battery parameters are extracted with population based search using bio inspired evolutionary algorithms based optimization technique. The aim of using evolutionary algorithms is due to the fact that it needs only manufacturers C_r and D_r Characteristics for giving polynomial coefficient in relatively less iteration. Also, evolutionary algorithm is more flexible in extracting the battery parameters with any initial values, while other analytical techniques are not capable of obtaining feasible solutions. It is easy to understand and can be optimized using fitness function. Different types of manufacturers' data of Li-ion batteries are considered for extraction purpose. The main objective of the algorithm is to optimize the battery parameters polynomial coefficients ($a_1 - a_{31}$) to evaluate the equations given in Eq. (1) to Eq. (6) and optimize the adjustable factors and exponential factors (A,B,C,D) to evaluate the equations given in Eq (19). The requirement after generation of random solution set is the measure of quality of solution set. This can be achieved by establishing a fitness function F(x), which is rated by each solution according to its fitness. The difference between the measured ($\boldsymbol{V^M_{c_i}}$or$\boldsymbol{V^M_{d_j}}$) and calculated voltage for charging and discharging cases ($\boldsymbol{V^c_{c_i}}$or$\boldsymbol{V^c_{d_j}}$) are determined with Eq.(5) and Eq.(6) respectively. Based on the calculated fitness value, the best list gets updated and the lowest fitness is discarded. The mathematical formulation of fitness function f(x) for charging and discharging scenario is in Eq. (11) – (12).

Minimize f(x)

$$f(x) = \begin{cases} \sum_{i=0}^{n} \left|\left(V^M_{c_i} - V^C_{c_i}\right)\right| \\ \sum_{j=0}^{m} \left|\left(V^M_{d_j} - V^C_{d_j}\right)\right| \end{cases} \quad (22) - (23)$$

4 Covariance Matrix Adaptation Evolution Strategy

A recent approach to adapting the search direction is Covariance Matrix Adaptation Evolution Strategy (CMA-ES). It is important property is invariance against the linear transformations in the continuous search space. CMA-ES outperformed its other similar learning algorithms in CEC2005 benchmark functions [16, 17] and BBOB-2009 benchmark functions [18]. Hence, an attempt is made to optimize the Battery Modelling design using CMA-ES in this research work. CMA learns all pair wise dependencies between the variables and increases the probability to repeat the successful steps. One evolution path, enhances CMA procedure in place of single successful search steps and facilitates possibly much faster increase of favourable directions. CMA identifies the function landscape which is convex-quadratic one with the concept of Hessian matrix (*H*). The CMA-ES estimates the inverse Hessian matrix (H^{-1}) in the form of a covariance matrix of the search distribution within an iterative procedure. The aim of CMA is to approximate the matrix, '*H*' and to closely suit the search direction to the contour lines of the objective function to be optimized [19]. The CMA influences the scale of the distribution. Nevertheless, additional step-size, control is necessary [20]. So, the step size update is also introduced to enhance the scaling adaptation and in particular to facilitate the increase of distribution spread, which is very difficult with CMA only. Step size adaptation aims to make consecutive movements of the distribution mean orthogonal in expectation and prevents the premature convergence.

5 Results and Discussion

The Li-ion batteries manufactured by SONY US18650, 5V, 2Ahr considered for parameter extraction problem. For Li-ion battery, SOC verses battery terminal voltage for charging, and SOC verses battery terminal voltage for discharging at different charge/discharge rates for various cycles are taken from manufacturer catalogue. To obtain the parameters (a_1-a_{31}), measure the charging voltage, and discharging voltage at different charge/discharge rate. The parameters for charging and discharging characteristics are calculated from GA and CMA-ES. The mathematical formulation of fitness function for obtaining one set of the parameters for charging and discharging scenario are given in Equations (11, 12). The results of established mathematical model for battery at different charge/discharge rate are discussed in this section. Battery charge/discharge rate characteristics were obtained for the cycles 50 to 100 in the intervals of 10 cycles and compared with the measured (manufacturer catalogue) data.

The plots of charge curves obtained for charge/discharge for cycle No 50 are calculated using Equation (5, 6). The experimental data's are measured from the battery manufacturer's catalogue. It is evident from the plots that the simulated curves are in close agreement with the measured data for CMA-ES. The polynomial coefficients obtained by conducting simulations for charging and discharging characteristics after 50,60,70,80 and 100 are listed in the Table 1. The results for the capacity fading i.e pre adjustable factors(A,B) ,Adjustable Factors(C,D) and the Capacity of the battery fading have been listed in Table 2.The plot for the cycle Vs remaining capacity of the battery is drawn in Fig 5. From the figure it can be seen that the capacity of the battery reduces as the number of cycle increases. The remaining capacity of the battery after each cycle can be found from this graph.

5.1 Sony US18650 battery charging characteristics after 50 cycles

The variation of charging/discharging voltage with respect to increasing SOC_{cr} and DOD_{cr} are shown in Figures 1 and 2 respectively by CMA-ES method. The simulations are observed for SOC_{cr} /DODcr from 0-100%.The difference between the calculated and catalogue values varies from 0 to 22.5 mV. It is clear from the plot that the battery can be discharged only upto 90% of total capacity. Beyond this the voltage of the battery rapidly decreases. So the maximum DOD for the battery at 0.5 Dr is 90 %. The fitness Value for charge and discharge rate is calculated using CMA-ES as 8.6482V and 8.1738V respectively.

5.2 Sony US18650 battery charging characteristics after 60 cycles

The variation of charging/discharging voltage with respect to increasing SOC_{cr} and DOD_{cr} are in Figures 3 and 4 respectively by CMA-ES method. The simulations are observed for SOC_{cr} / DODcr from 0-100%. The difference between the calculated and catalogue values varies from 0 to 22.5 mV. It is clear from the plot that the battery can be discharged only upto 90% of total capacity. Beyond this, the voltage of the battery rapidly decreases. So the maximum DOD for the battery is 90 %. The fitness Value for charge and discharge rate is calculated using CMA-ES as 393.2175V and 8.1738V respectively. The measured energy that can be discharged from the battery are 5.770 Wh and 5.7881Wh

6 CONCLUSION

In this work an accurate circuit based battery modeling have been done for a real type of battery available in market. The novelty of the model is that the optimization procedure is done by Real GA while binary GA has been used earlier. The battery polynomial coefficients are extracted using CMAES based optimization methods for Sony Us 18650 battery after 50 cycles in an interval of 10 cycles upto 100 cycles .The results has been compared with the measured values given by the manufacturer and the results calculated by the equations are accurate. Five cycles charge rate and discharge rate characteristics of Sony US 18650 battery have been obtained and compared with manufacturer's data for validation. The capacity fading of the battery after 50 cycles has been modeled at an interval of 10 cycles up to 100 cycles and graph has been drawn which accurately represents the catalog values. The proposed battery models are simple and it accurately represents the measured (catalogue value) charge and discharge curves of the manufacturers data sheets. The simulation and measured (catalogue value) results are in good agreement. It has been seen that the methodology presented in this work extracts accurate results and this can be extended to obtain the Capacity fading of different batteries after every cycle of charging and discharging. Future works can be carried for modeling of other types of batteries with the same procedures.

References

1. https://factly.in/indias-per-capita-electricity-consumption-lowest-among-brics-nations/"Energy Calculation per Capita," http://www.indexmundi.com/g/ r.aspx/ v=81000
2. "Electricity Generation in India and Power Sector at a Glance ALL INDIA" http://en.wikipedia.org/wiki/Electricity sector in Indiaandhttp:// www. powermin.nic.in/ Indian electricity scenario/introduction.htm.
3. C. M. Shepherd, "Design of primary and secondary cells - part 2. An equation describing battery discharge," *J Electroche. Soc.*, vol. 112, pp. 657–664, Jul. 1965.
4. O. Tremblay, L.-A. Dessaint, and A.-I. Dekkiche, "A generic battery model for the dynamic simulation of hybrid electric vehicles," *IEEE Int. Vehicle Power and Propulsion Conf. (VPPC'07)*, pp. 284–289, Sept. 2007.
5. E. Sortomme and M. El-Sharkawi, "Optimal combined bidding of Vehicle-to-Grid ancillary services," *IEEE Trans. on Smart Grid*, vol. 3, no. 1, pp. 70–79, Mar. 2012.
6. "Parameter extraction of Battery Models Using Multi objective Optimization Genetic Algorithms"- Praveen kumar, Pavol Bauser -14 th International Power Electronics and motion Control Conference -2010.
7. "Mathematical Modeling of Li-Ion Battery for Charge/Discharge Rate and Capacity Fading Characteristics using Genetic Algorithm Approach." And published in IEEE journal 2012
8. Mathematical Modeling of Li-Ion Battery Using Genetic Algorithm Approach for V2G Applications" , Kannan Thirugnanam, *Student Member, IEEE*, Ezhil Reena Joy T. P., *Student Member, IEEE*, Mukesh Singh, *Student Member, IEEE*, and Praveen Kumar, *Member, IEEE*- IEEE transactions on Energy conversions-2014 K. Deb, "Optimization for engineering design: Algorithms and examples," *Prentice Hall, India*, 1998
9. Min Chen,Student Member IEEE and Gabriel A.Rincon-Mora,Senior Member ,IEEE published a paper "Accurate Electrical Battery Model capable of predicting Runtime and I-V Performance"IEEE trans on Energy conversions
10. Liao Chenglin, Li Huiju, Wang Lifang Institute of Electrical Engineering Chinese Academy of Sciences Beijing, China presented a paper in 2009 "A Dynamic Equivalent Circuit Model of LiFePO4 Cathode Material for Lithium Ion Batteries on Hybrid Electric Vehicles "-Journal of Power Sources, Elsveir
11. Bor Yann Liaw, Ganesan Nagasubramanian, Rudolph G. Jungst, Daniel H. Doughty, "Modeling of lithium ion cells—A simple equivalent-circuit model approach
12. "A dynamic lithium-ion battery model considering the effects of temperature and capacity fading "O. Erdinc, B. Vural and M. Uzunoglu, *Member, IEEE (2009)*
13. J.J. Esperilla, J. Fe´lez *, G. Romero, A. Carretero ETS Ingenieros Industries, Universidad Polite´cnica de Madrid (UPM), Jose Gutierrez Abascal, 2, 28006 Madrid, Spain
14. P. Ramadass, B. Haran, R. White, and B. N. Popov, "Mathematical modeling of the capacity fade ofli-ion cells," *J Power Sources*, vol. 123, pp. 230–240, 2003.
15. C. M. Shepherd, "Design of primary and secondary cells - part 2. an equation describing battery discharge,"*J Electroche. Soc.*, vol. 112, pp. 657–664, Jul. 1965.
16. Suganthan, P, Hansen, N, Liang, JJ, Deb, K, Chen, YP, Auger, A, & Tiwari, S 2005, 'Problem, definitions and evaluation Criteria for the CEC 2005', in Special Session on Real-Parameter Optimization, Technical report, Nanyang Technological University, Singapore.
17. Hansen, N, Auger, A, Ros, R, Finck, S, & Pošík, P 2010, 'Comparing results of 31 algorithms from the black-box optimization benchmarking BBOB-2009', Proceedings of the twelth annual conference companion on Genetic and evolutionary computation, GECCO, pp. 1689-1696.
18. Hansen, N, & Ostermeier, A 2001, 'Completely de randomized self-adaptation in evolution strategies', Evolutionary Computation, 9(2) 159-195.
19. Hansen, N, Auger, A, Ros, R, Finck, S, & Pošík, P 2010, 'Comparing results of 31 algorithms from the black-box \optimization benchmarking BBOB-2009', Proceedings of the twelth annual conference companion on Genetic and evolutionary computation, GECCO, pp. 1689-1696
20. Hansen, N 2006, 'The CMA evolution strategy: a comparing review', in towards a new evolutionary computation, Springer Berlin, pp. 75-102.
21. Liao Chenglin, Li Huiju, Wang Lifang Institute of Electrical Engineering Chinese Academy of Sciences Beijing, China presented a paper in 2009 "A Dynamic Equivalent Circuit Model of LiFePO4 Cathode Material for Lithium Ion Batteries on Hybrid Electric Vehicles "-Journal of Power Sources, Elsevier.
22. "Modeling of lithium ion cells—A simple equivalent-circuit model approach Bor Yann Liaw, Ganesan Nagasubramanian, Rudolph G. Jungst, Daniel H. Doughty.

 https://doi.org/10.1051/e3sconf/20198701026

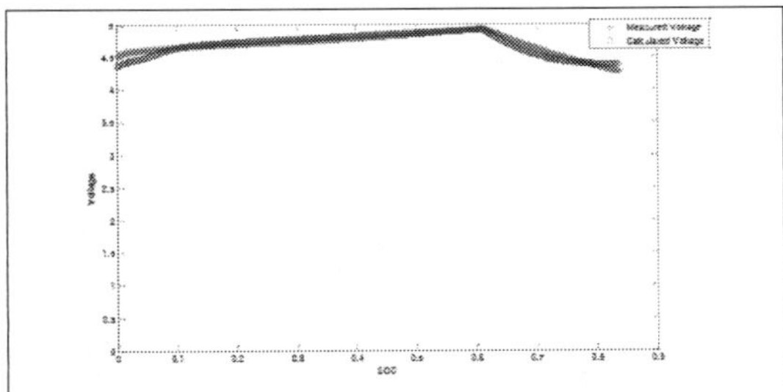

Figure 2 Battery Charging Characteristics After 50 Cycles

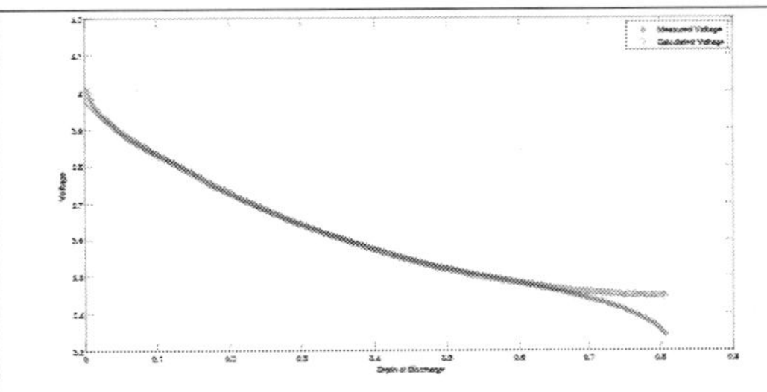

Figure 3 Battery Discharging Characteristics After 50 Cycles.

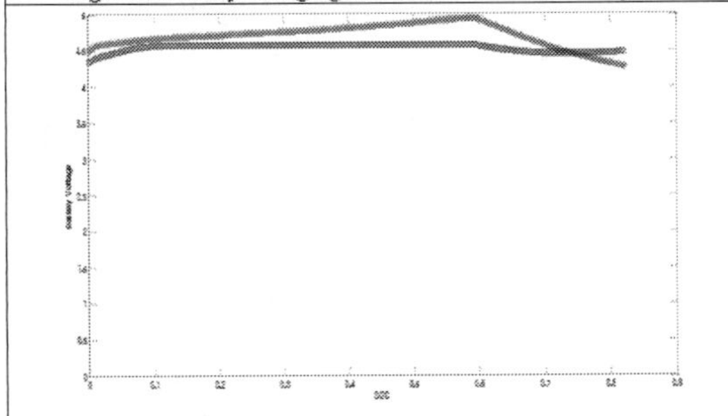

Figure 4 Battery Charging Characteristics After 60 Cycles

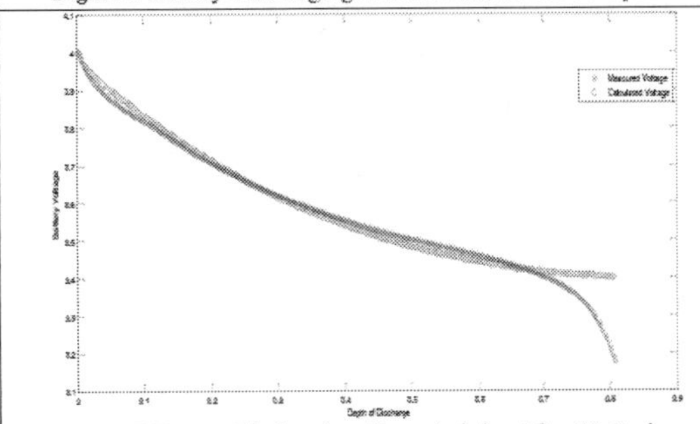

Figure 5 Battery Discharging Characteristics After 60 Cycles.

Table 1. Capacity Polynomial Coefficients for charging/discharging scenario

Coeff	Charging Scenario Cycle No					Discharging Scenario Cycle No						
	50	**60**	**70**	**80**	90	**100**	**50**	**60**	**70**	**80**	90	**100**
a_1	0.000	0.133	0.000	0.000	0.000	0.000	0.000	0.000	0.000	0.000	0.000	0.021
a_2	0.000	0.000	0.370	0.050	0.000	0.000	0.000	0.000	0.000	0.000	0.000	0.000
a_3	0.339	0.000	0.000	0.000	0.000	0.534	0.000	0.000	0.027	0.000	0.000	0.000
a_4	44.305	50.000	46.010	3.165	50.000	48.448	26.573	42.816	26.731	50.000	27.818	50.000
a_5	0.000	0.000	0.000	0.038	0.000	0.000	0.000	0.000	0.000	0.000	0.000	0.000
a_6	0.000	0.000	0.000	0.000	0.000	0.000	0.000	0.000	0.000	0.000	0.000	0.000
a_7	0.000	0.000	0.000	0.000	0.000	0.000	0.000	0.000	0.000	0.000	0.000	0.000
a_8	0.000	0.226	0.008	0.307	0.163	0.067	0.000	0.000	0.000	0.000	0.000	0.173
a_9	0.303	0.207	0.007	0.164	0.000	0.000	0.000	0.000	0.000	0.000	0.000	0.149
a_{10}	0.517	0.000	0.000	0.418	0.000	0.201	0.000	0.000	0.000	0.000	0.000	0.000
a_{11}	29.006	38.224	26.828	46.608	50.000	16.703	43.329	44.710	50.000	0.000	49.640	49.061
a_{12}	0.000	0.004	0.000	0.000	0.000	0.042	0.000	0.000	0.059	0.799	0.000	0.000
a_{13}	0.000	0.000	0.035	0.000	0.000	0.000	0.000	0.859	0.000	0.000	0.000	0.000
a_{14}	0.000	0.000	0.000	0.137	0.000	0.286	0.969	0.000	0.733	0.000	0.858	0.000
a_{15}	520.021	1000	0.000	95.470	394.12	332.68	354.80	1000	812.11	350.27	449.01	0.000
a_{16}	163.403	453.90	948.98	329.11	0.000	0.000	604.828	1000.000	246.990	42.707	381.947	340.079
a_{17}	281.682	1000.000	183.508	863.322	0.000	256.458	0.000	152.764	0.000	538.446	173.985	358.668
a_{18}	11.338	50.000	50.000	12.776	41.635	46.509	15.217	50.000	14.753	24.148	16.643	50.000
a_{19}	1000.000	853.942	1000.000	757.967	353.886	613.630	1000.000	1000.000	993.682	1000.000	1000.000	776.075
a_{20}	897.165	1000.000	0.000	873.389	651.093	0.000	1000.000	1000.000	995.364	1000.000	1000.000	766.410
a_{21}	630.715	945.054	1000.000	1000.000	1000.000	547.346	1000.000	1000.000	1000.000	1000.000	1000.000	771.229
a_{22}	0.168	0.076	0.000	1.000	0.081	0.182	0.000	0.042	0.041	0.000	0.000	0.540
a_{23}	0.000	0.725	0.000	0.947	0.776	0.631	0.000	0.000	0.000	0.000	0.000	0.000
a_{24}	0.138	0.000	0.000	0.443	0.000	0.000	0.000	0.000	0.000	0.000	0.000	0.679
a_{25}	0.000	0.000	50.000	0.000	0.000	0.000	50.000	50.000	39.645	35.153	29.595	1.436
a_{26}	3.834	4.417	4.443	2.217	3.802	3.616	5.000	5.000	5.000	5.000	5.000	2.822
a_{27}	0.451	0.000	0.000	0.561	0.420	0.000	0.848	0.711	0.623	0.590	0.660	0.055
a_{28}	0.000	0.000	0.000	0.751	0.014	0.710	0.000	0.000	0.000	0.000	0.000	0.114
a_{29}	0.000	0.000	0.000	0.208	0.014	0.000	0.000	0.000	0.000	0.000	0.000	0.256
a_{30}	0.000	1.000	0.371	0.321	0.943	0.000	1.000	1.000	1.000	1.000	1.000	0.688
a_{31}	1.000	0.536	1.000	0.349	0.363	0.000	0.000	0.000	0.000	0.000	0.000	0.594

Table 2. Capacity Fading

Cycle	50	60	70	80	90	100
A	0.17	0.114	0.114	0.174	0.264	0.186
B	0.152	0.235	0.3	0.259	0.224	0.329
C	5.994	5.089	3.277	5.28	7.637	4.454
D	7.797	4.469	2.723	7.784	8.244	7.357

* Corresponding author: tamilselvi.manjuraj@gmail.com

https://doi.org/10.1051/e3sconf/20198701026

Q_{rm} (meas)Ah	1.685	1.656	1.589	1.576	1.523	1.495
Q_{rm}(calc) Ah	1.685	1.656	1.589	1.576	1.523	1.495

Peak Detection Control Strategy based STATCOM for Suppression of Power Quality Problems

J. Bangarraju[1], *V.* Rajagopal[2], *V.* Sharada[2] and *K.*Lakshmaiah[3]

[1]B V Raju Institute of Technology, Narsapur, Medak, Telangana, India
[2]Stanley College of Engineering and Technology for Women, Hyderabad, Telangana, India
[3]I&CAD Department, Warangal, Telangana, India

Abstract. This paper explains about control of VSC using uni-polar switching based Static Compensator (STATCOM) using Peak Detection Control Strategy for elimination of PQ Problems. The proposed STATCOM are operated in PFC (Power Factor Correction) mode and ZVR (Zero Voltage Regulation) mode. Under these operation modes of STATCOM is able to suppression of harmonics, PFC at source, reactive power compensation, load balancing and compensation of neutral current. A non-isolated star/three individual single-phase transformers are connected at PCC (Point of Common Coupling) to mitigate neutral current. The Peak Detection Control Strategy is simple to implement for estimating switching pulses for VSC. This peak detection control strategy based STATCOM is modelled, simulated results are validated using MATLAB R2013b.

1 Introduction

Recent years Power Distribution Network (PDN) suffering from high neutral current, reactive power burden, waveform distortion and load balancing are main PQ problems. These PQ problems are primarily occurs due to non-linear loads, lagging power factor balanced or unbalanced loads etc. These PQ (Power Quality) problems are mitigated using advancement of active compensators [1]-[2]. The different topologies modeling and design analysis of active compensator are discussed [3]-[4].To maintain balance supply currents in 3-phase 4-wire system, mitigation of neutral current and its control are necessary for maintaining balanced source currents. Different transformer connections such as T-Connection, Scott Connection, Star-delta, Star-Hexagon, Zig-Zag transformer etc based STATCOM are used for neutral current compensation [5]-[7].This proposed star/three single phase transformer configuration will reduce rating of VSC. The performance of STATCOM very much depends on control strategy and proposed transformer also reduce size of STATCOM. The d-q and p-q method based control strategies are used for harmonics estimation without estimating separate components which reduces calculation burden [8]. A application of non-linear robust control strategy and mathematical for active filters [9]. A review on different control algorithms to APFs (Active Power Filters) and different performance strategies are compared [10]. The control algorithms are I-Φ p-q theory, Instantaneous Reactive Power Theory, Integral generalized control staretgy, Synchronous Reference Frame Theory, and adaptive control startegies [11].In general, the performance analysis and design of STATCOM are observed based on IEC and IEEE standards [12]-[13].

The preset day improvement of shunt APFs are considered as STATCOM to mitigate current related PQ problems. The mathematical analysis of STATCOM using '-' sequence component. The analysis of zero and '-' sequence component of voltage estimation elimination of current sensors using balanced theory [14]-[15].

The peak detection control strategy is implemented using MATLAB R2013b for suppression of harmonics, PFC at source, reactive power control, load balancing and mitigate neutral current.

2 Design and Analysis of STATCOM

The basic schematic diagram of six-leg STATCOM is connected to 3-Φ AC supply with an impedance of (Z_s) is depicted in Fig.1. In this system configuration, the main feature of six leg of VSC is to doubles switching of waveform which reduce filter circuit requirement. For design and implementation of peak detection control algorithm, sensed AC mains phase voltage variables are (v_{sa}, v_{sb},v_{sc}), AC mains currents (i_{sa}, i_{sb}, i_{sc}), distribution load currents (i_{La}, i_{Lb}, i_{Lc}) and DC bus voltage (v_{dc}). A six-leg VSC is connected to one interfacing inductor (L_f) at each leg of VSC. Interfacing inductor voltage drop and DC bus voltage are dependent on each other. The interfacing inductor voltage drop must be less than 8% of rated value for successful operation.

A series connected R-C passive filter (R_f and C_f) is installed at PCC (Point of Common Coupling), which is used to mitigate high frequency noise switching at PCC. The compensating currents (i_{ca}, i_{cb}, i_{cc}) are injected by proposed STATCOM to suppression of PQ problems. The IGBT switches rating is selected in VSC is based on compensation of STATCOM.

3 Peak Detection Control Strategy

A block diagram of Peak detection control strategy is shown in Fig.1. It is implemented on a six-leg VSC based STATCOM for PQ improvement. The proposed peak detection control algorithm is robust and it can easily reduce harmonic distortion at the source. The basic equations of peak detection control strategy for generation of different control signals are given.

A. Estimation of In-Phase Unit Vectors and Amplitude of PCC Voltage

The in-phase unit vector components and magnitude of PCC voltages (V_{tp}) of distribution system and are determined as

$$v_{tp} = \sqrt{\left\{ \tfrac{2}{3}\left(v_{sa}^2 + v_{sb}^2 + v_{sc}^2\right)\right\}} \tag{1}$$

The three phase voltages (V_{sa}, V_{sb}, V_{sc}) are at the PCC. The in-phase unit vector components of PCC voltages (u_{ia}, u_{ib} and u_{ic}) are determined as

$$u_{ia} = \frac{v_{sa}}{v_{tp}};\ u_{ib} = \frac{v_{sb}}{v_{tp}};\ u_{ic} = \frac{v_{sc}}{v_{tp}}; \tag{2}$$

The quadrature phase unit vectors (u_{qa}, u_{qb} and u_{qc}) are determined using in-phase unit vectors (u_{ia}, $u_{ib}$and u_{ic}) as

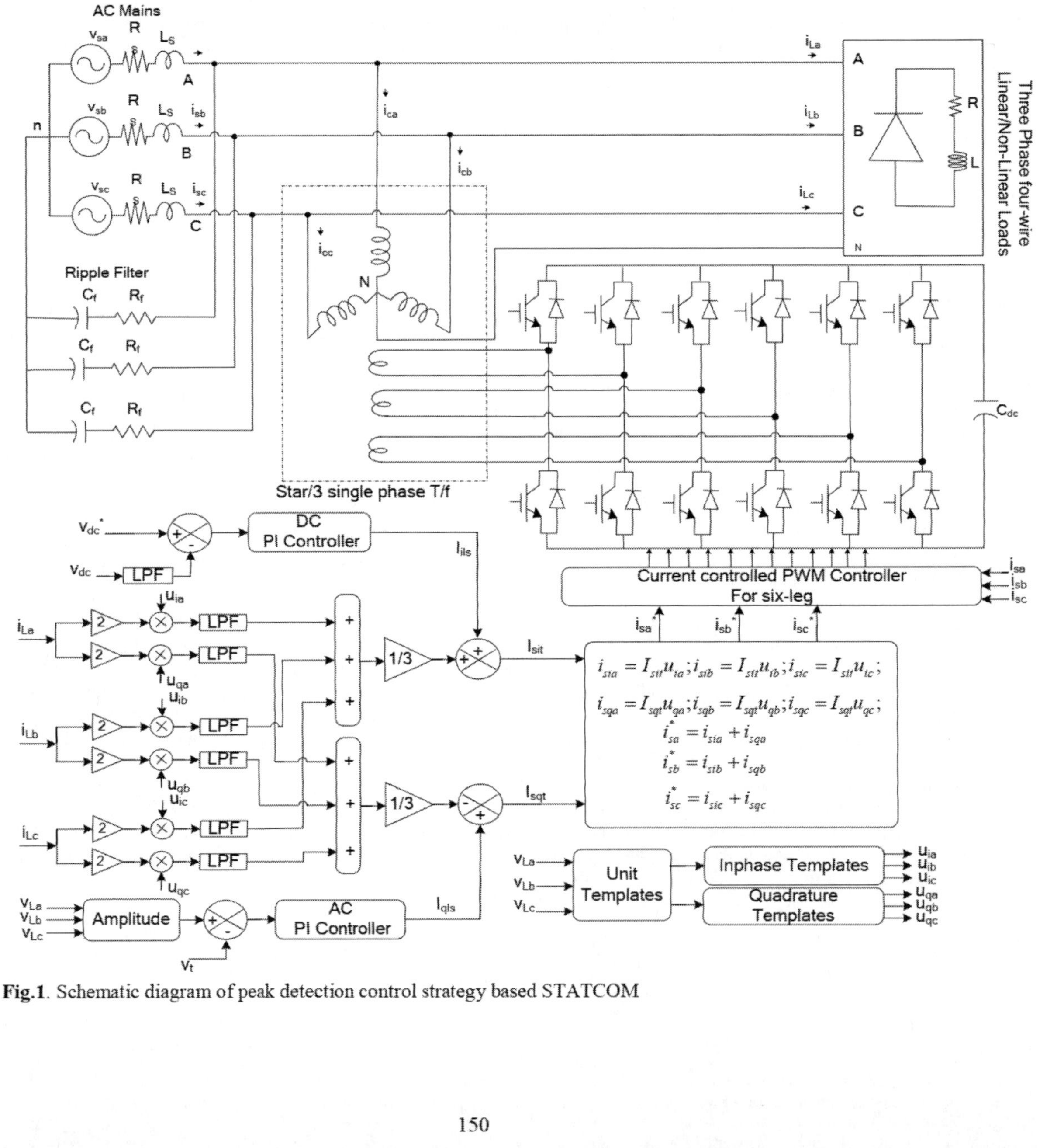

Fig.1. Schematic diagram of peak detection control strategy based STATCOM

$$u_{qa} = \left(-u_{ib} + u_{ic}\right)/\sqrt{3} \tag{3}$$

$$u_{qb} = \left(3u_{ia} + u_{ib} - u_{ic}\right)/2\sqrt{3} \tag{4}$$

$$u_{qc} = \left(-3u_{ia} + u_{ib} - u_{ic}\right)/2\sqrt{3} \tag{5}$$

B. Active and reactive power component of load current Estimation

In General, the distribution load current (i_L) is distorted and lagging because of application of highly non-linear and reactive loads. These distribution load current can be express as

$$i_{La}(t) = I_{dc} + I_{Lia} + I_{Lqa} + I_{hd} \tag{6}$$

Where I_{dc} is the DC component magnitude of load current and this is very small because of load current waveform is half wave symmetry. Where I_{Lia} and I_{Lqa} are the 'a' phase fundamental active and reactive power component of load current. The I_{hd} is corresponding to load current harmonics of i_{La}.
Equation (6) can be represented as

$$i_{La}(t) = I_{dc} + I_{Lia}\sin\omega t + I_{Lqa}\cos\omega t$$
$$+\sum_{n=1}^{\infty} I_{2k}\sin(2\omega \mathrm{k}t + \Phi_{2k}) \tag{7}$$
$$+\sum_{m=1}^{\infty} I_{2m+1}\sin\{(2m+1)\omega \mathrm{t} + \Phi_{2m+1})\}$$

To extract peak magnitude of phase 'a' fundamental active power component of load current (I_{Lia}). The 'a' phase load current (i_{La}) is multiplied with in-phase unit vector u_{ia} (sinωt) [16].

$$i_{Lia} = i_{La}(t).u_{ia} = i_{La}(t).\sin\omega t \tag{8}$$
$$= I_{dc}\sin\omega t + \left[\left(I_{Lia}/2\right)\{1-\cos 2\omega t\}\right] + \left[\left(I_{Lqa}/2\right)\{\sin\omega t\}\right]$$
$$+\sum_{k=1}^{\infty}\tfrac{I_{2k}}{2}\left[\cos\{(2k-1)\omega t + \Phi_{2k}\} - \cos\{(2k+1)\omega t + \Phi_{2k}\}\right]$$
$$+\sum_{m=1}^{\infty}\tfrac{I_{2m+1}}{2}\left[\cos\{(2m\omega t) + \Phi_{2m+1}\} - \cos\{(2m+2)\omega t + \Phi_{2m+1}\}\right] \tag{9}$$

From the equation (8) it is clear that i_{Lia} consists of DC component and oscillatory components. The i_{Lia} is proportional to $I_{Lia}/2$. A LPF with cut-off frequency (f_c) not more 15 Hz is used to extract magnitude of fundamental active power component load current and also gain factor of '2' is used to determine (I_{Lia}'). Similar procedure is used to extract 'b and c' phase fundamental active power component load currents (I_{Lib}' and I_{Lic}').The average magnitude of fundamental active power component of load currents (I_{LiA}) is estimated as

$$I_{LiA} = \left(I'_{Lia} + I'_{Lib} + I'_{Lic}\right)/3 \tag{10}$$

To extract peak magnitude of phase 'a' fundamental reactive power component of load current (I_{Lqa}). The 'a' phase load current (i_{La}) is multiplied with quadrature-phase unit vector u_{qa} (cosωt)

$$i_{Lqa} = i_{La}(t).u_{qa} = i_{La}(t).\cos\omega t \tag{11}$$
$$= I_{dc}\cos\omega t + \left[\left(I_{Lqa}/2\right)\{1+\cos 2\omega t\}\right] + \left[\left(I_{Lia}/2\right)\{\sin\omega t\}\right]$$
$$+\sum_{k=1}^{\infty}\tfrac{I_{2k}}{2}\left[\sin\{(2k+1)\omega t + \Phi_{2k}\} + \sin\{(2k-1)\omega t + \Phi_{2k}\}\right]$$
$$+\sum_{m=1}^{\infty}\tfrac{I_{2m+1}}{2}\left[\sin\{(2m+2)wt + \Phi_{2m+1}\} + \sin\{(2m)\omega t + \Phi_{2m+1}\}\right] \tag{12}$$

From the equation (11) it is clear that i_{Lqa} consists of DC component and oscillatory components. The i_{Lqa} is proportional to $I_{Lqa}/2$. A LPF with cut-off frequency (f_c) not more 15 Hz is used to extract magnitude of fundamental reactive power component load current and also gain factor of '2' is used to determine (I_{Lqa}'). Similar procedure is used to extract 'b and c' phase fundamental reactive power component load currents (I_{Lqb}' and I_{Lqc}').The average magnitude of fundamental reactive power component of load currents (I_{LqA}) is estimated as

$$I_{LqA} = \left(I'_{Lqa} + I'_{Lqb} + I'_{Lqc}\right)/3 \tag{13}$$

C. Active and Reactive power component of reference source current estimation

A small active power flows in to DC bus capacitor of VSC for control of DC bus voltage of STATCOM. The current required for DC bus VSC is extracted using DC bus PI controller to eliminate steady state error.

$$I_{ils}(k) = I_{ils}(k-1) + \mathrm{K}_{ip}\{V_{ed}(k) - V_{ed}(k-1)\} + K_{ii}V_{ed}(k) \tag{14}$$

Where V_{ed}=-V_{dc}+V_{dc}^* is the voltage error in DC bus. V_{dc} and V_{dc}^* are sensed filtered voltage and reference voltage of DC bus of capacitor based STATCOM. The DC bus PI controllers gain constants (K_{ip} and K_{ii}).
The active power component of source current (I_{sit}) is sum of average magnitude fundamental active power component of load current (I_{LiA}) and required current for DC bus of STATCOM (I_{ils}).

$$I_{sit} = I_{LiA} + I_{ils} \tag{15}$$

The proposed STATCOM can be operated in PFC or ZVR modes. The PCC voltage may increase or decrease from rated voltage based on loading. In ZVR mode, the PCC voltage is maintained to rated voltage by supplying additional reactive power. An AC bus PI controller is used to determine reactive power component to be supplied in ZVR mode of STATCOM. The extra component of reactive power of source current required for ZVR is given by output of AC bus PI controller.

$$I_{qls}(k) = I_{qls}(k-1) + \mathrm{K}_{qp}\{V_{tpe}(k) - V_{tpe}(k-1)\} + K_{qi}V_{tpe}(k) \tag{16}$$

Where V_{tpe}=-V_{tp}+V_{tp}^* is the voltage error magnitude in AC bus. V_{tp} and $V_{tp}^*$are sensed magnitude of PCC voltage and reference PCC voltage. The AC bus PI controllers gain constants (K_{qp} and K_{qi}).The average magnitude fundamental reactive power component of load current (I_{LqA}) is subtracted from output of AC bus PI controller (I_{qls}) to estimate total reactive power component of source current(I_{sqt})

$$I_{sqt} = -I_{LqA} + I_{qls} \tag{17}$$

D. Reference supply currents estimation and estimation of switching pulses

Using active power component of source current (I_{sit}) and in- phase unit vector components (u_{ia}, u_{ib} and u_{ic}) are used to estimate reference active power components of supply currents (i_{sia} ,i_{sib} ,i_{sic})

$$i_{sia} = I_{sit}u_{ia};\, i_{sib} = I_{sit}u_{ib};\, i_{sic} = I_{sit}u_{ic}; \tag{18}$$

Similarly, Using reactive power component of source current (I_{sqt}) and quadrature- phase unit vector components (u_{qa}, u_{qb} and u_{qc}) are used to estimate reference reactive power components of supply currents (i_{sqa} ,i_{sqb} ,i_{sqc})

$$i_{sqa} = I_{sqt}u_{qa};\, i_{sqb} = I_{sqt}u_{qb};\, i_{sqc} = I_{sqt}u_{qc}; \tag{19}$$

The total reference supply currents (i_{sa}^*, i_{sb}^*, i_{sc}^*) are obtained sum of reference active powers components of supply currents (i_{sia},i_{sib},i_{sic}) and reactive powers components of supply currents (i_{sqa} ,i_{sqb} ,i_{sqc})

$$i_{sa}^* = i_{sia} + i_{sqa} \tag{20}$$

$$i_{sb}^* = i_{sib} + i_{sqb} \tag{21}$$

$$i_{sc}^* = i_{sic} + i_{sqc} \tag{22}$$

4 Results and Discussion

The performance of peak detection control strategy based STATCOM simulated using MATLAB R2013b under SIMULINK. The proposed STATCOM is operated uni-polar switching to validate dynamic and steady state performance under PFC and ZVR modes in MATLAB.

A. Response of peak detection control strategy based STATCOM under linear load in PFC mode

The response of peak detection control strategy based STATCOM is studied in PFC mode at linear loads is shown in Fig.2. The response of PDN parameters are presented in Fig.2 in terms of source voltage (v_{sabc}), balanced source current(i_{sabc}), load current (i_{Labc}), STATCOM injecting current(i_{cabc}), load neutral current of PDN(i_{Ln}),source neutral current(i_{sn}), dc bus voltage magnitude (V_{dc}) and amplitude of terminal voltage(v_{tp}) under uni-polar switching operations of STATCOM. Upto t=0.8sec, it was clearly observed from Fig.2 that the source voltage (v_{sabc}) and source current (i_{sabc}) are highly balanced, free from harmonics, the angle between source voltage and source current is zero, source neutral current is maintained to zero. At t=0.8sec, one phase is suddenly removed but response parameters of PDN shows highly balanced, free from harmonics, the angle between source voltage and source current is zero, source neutral current is maintained to zero. Again at t=0.9 sec, disconnected one phase reapplied, it was observed same. The amplitude of DC bus voltage (v_{dc}) and amplitude of terminal voltage (v_{tp}) are maintained at 400V and 320V in Fig.2 .

B. Response of peak detection control strategy based STATCOM under linear load in ZVR mode

The response of peak detection control strategy based STATCOM is discussed in ZVR mode at linear loads is depicted in Fig.3. The response of PDN parameters are discussed in Fig.3 in terms of source voltage (v_{sabc}), balanced source current(i_{sabc}), load current (i_{Labc}), STATCOM injecting current(i_{cabc}), load neutral current of PDN(i_{Ln}),source neutral current(i_{sn}), dc bus voltage magnitude (V_{dc}) and amplitude of terminal voltage(v_{tp}) under uni-polar switching operations of STATCOM. Up to t=0.8sec, it was clearly observed from Fig.3 that the source voltage (v_{sabc}) and source current (i_{sabc}) are highly balanced, free from harmonics, reactive power control, source neutral current is maintained to zero. At t=0.8sec, one phase is suddenly removed but response parameters of PDN shows highly balanced, free from harmonics, reactive power control, source neutral current is maintained to zero. Again at t=0.9 sec, disconnected one phase reapplied, it was observed same. The amplitude of DC bus voltage (v_{dc}) and amplitude of terminal voltage (v_{tp}) are maintained at 400V and 339V in Fig.3.

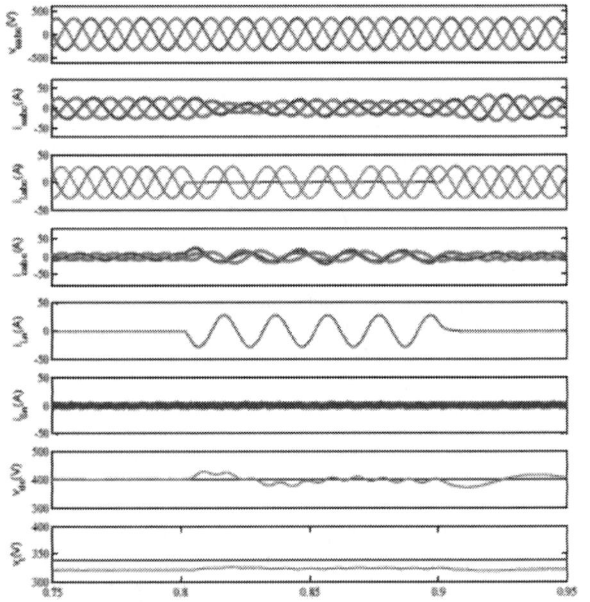

Fig.2 Response of peak detection control strategy based STATCOM under linear loads in PFC mode

C. Response of peak detection control strategy based STATCOM under non-linear load in PFC mode

The response of peak detection control strategy based STATCOM is studied in PFC mode at non-linear loads is shown in Fig.4. The response of PDN parameters are presented in Fig.4 in terms of source voltage (v_{sabc}), balanced source current(i_{sabc}), load current (i_{Labc}), STATCOM injecting current(i_{cabc}), load neutral current of PDN(i_{Ln}),source neutral current(i_{sn}), dc bus voltage magnitude (V_{dc}) and amplitude of terminal voltage(v_{tp})

https://doi.org/10.1051/e3sconf/20198701027

under uni-polar switching operations of STATCOM. Up to t=0.8sec, it was clearly observed from Fig.4that the source voltage (v_{sabc}) and source current (i_{sabc}) are

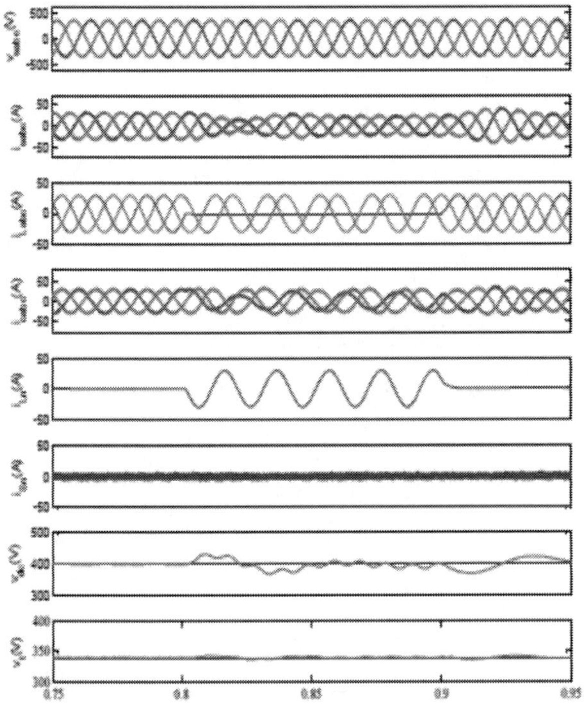

Fig.3 Response of peak detection control strategy based STATCOM under linear loads in ZVR mode

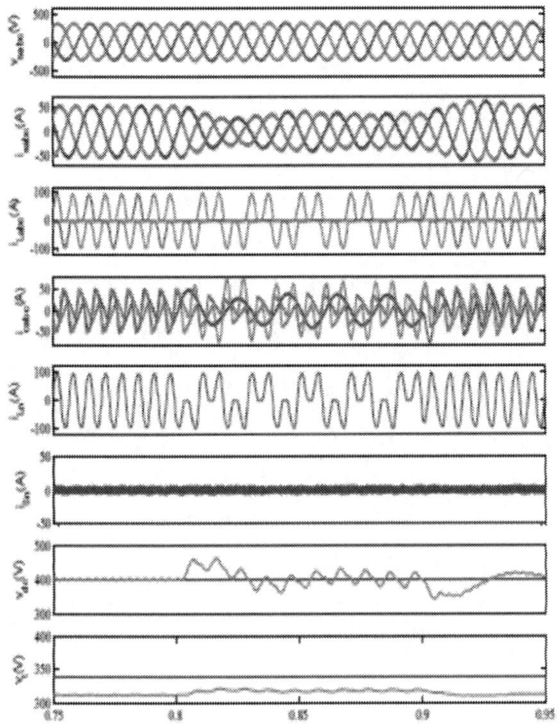

Fig.4 Response of peak detection control strategy based STATCOM under linear loads in PFC mode

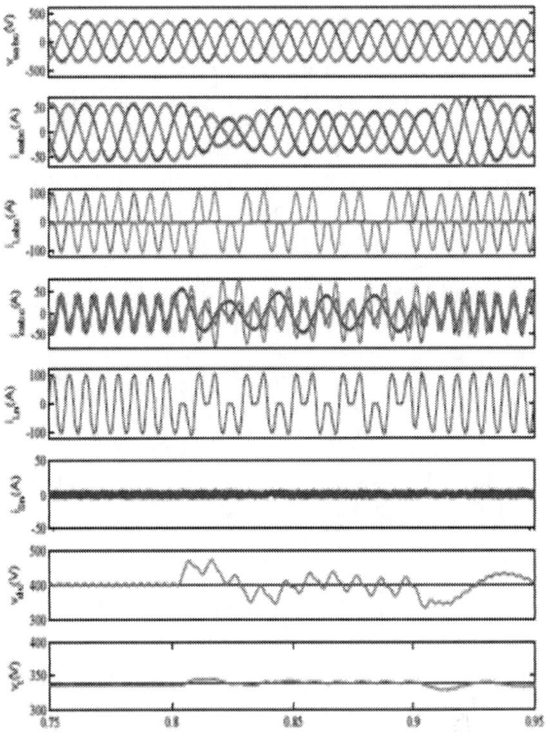

Fig.5 Response of peak detection control strategy based STATCOM under linear loads in ZVRs mode

highly balanced, free from harmonics, the angle between source voltage and source current is zero, source neutral current is maintained to zero. At t=0.8sec, one phase is suddenly removed but response parameters of PDN shows highly balanced, free from harmonics, the angle between source voltage and source current is zero, source neutral current is maintained to zero. Again at t=0.9 sec, disconnected one phase reapplied, it was observed same response as previous in Fig.4. The amplitude of DC bus voltage (v_{dc}) and amplitude of terminal voltage (v_{tp}) are maintained at 400V and 320V in Fig.4.

D. Response of peak detection control strategy based STATCOM under non-linear load in ZVR mode

The response of peak detection control strategy based STATCOM is discussed in ZVR mode at non-linear loads is depicted in Fig.5. The response of PDN parameters are discussed in Fig.5 in terms of source voltage (v_{sabc}), balanced source current(i_{sabc}), load current (i_{Labc}), STATCOM injecting current(i_{cabc}), load neutral current of PDN(i_{Ln}),source neutral current(i_{sn}), dc bus voltage magnitude (V_{dc}) and amplitude of terminal voltage(v_{tp}) under uni-polar switching operations of STATCOM. Up to t=0.8sec, it was clearly observed from Fig.5 that the source voltage (v_{sabc}) and source current (i_{sabc}) are highly balanced, free from harmonics, reactive power control, source neutral current is maintained to zero. At t=0.8sec, one phase is suddenly removed but response parameters of PDN shows highly balanced, free from harmonics, reactive power control, source neutral current is

 https://doi.org/10.1051/e3sconf/20198701027

maintained to zero. Again at t=0.9 sec, disconnected one phase reapplied, it was observed same. The amplitude of DC bus voltage (v_{dc}) and amplitude of terminal voltage (v_{tp}) are maintained at 400V and 339V in Fig.5. Harmonic spectra of proposed STATCOM is shown in Fig.6 and during this condition fundamental source voltage (v_{sa}) is 349.6V and its THD is 2.28%, fundamental source current(i_{sa}) is 50.63A and THD is 2.64% whereas fundamental load current(i_{La}) is 48.98A and THD is 91.85%

5 Conclusion

The proposed peak detection control strategy based STATCOM has been implemented using MATLAB R2013b in power distribution system. The peak detection control strategy used for extracting the active and reactive components at fundamental of load currents. These currents, DC bus voltage and PCC voltage are used estimation of reference source currents. The performance of STATCOM gives satisfactory results in MATLAB for suppression of harmonics, PFC at source, reactive power control, load balancing and mitigate neutral current as per PQ standards. The dc bus voltage and PCC voltage are maintained zero overshoot even for small and large disturbances in the load.

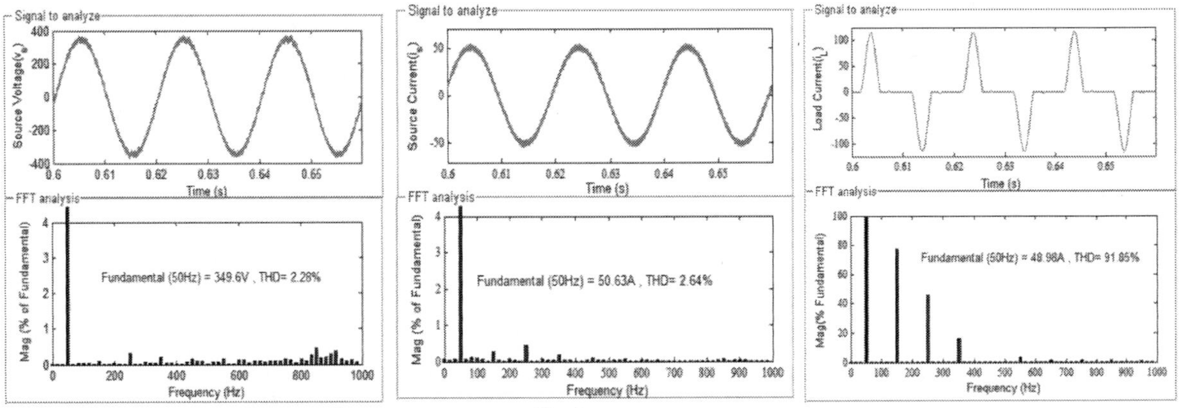

Fig. 6 Harmonic spectra of source voltage, source current and load current

References

[1] Ali Emadi, Abdolhosein Nasiri and Stoyan B. Bekiarov, Uninterruptible
Power Supplies and Active Filters, CRC Press, New York, 2005.

[2] R. C. Dugan, M. F. Mc Granaghan and H. W. Beaty, "ElectricPower Systems Quality", 2ed Edition, McGraw Hill, New York,2006.

[3] Pericle Zanchetta, Mark Sumner, Maria Marinelli, Francesco Cupertino,
"Experimental modeling and control design of shunt active power filters," Journal of Control Engineering Practice, vol. 17, pp. 1126-1135, 2009.

[4] Yu-Jen Liu, G.W. Chang, Rong-Chin Hong and Chi Yun Chao, "Applications of real-time simulation techniques for shunt active power filter design," in Proc. of IEEE Power and Energy Society General Meeting, 2009, pp.1-5.

[5] H.-L. Jou, K.-D.Wu, J.-C.Wu, andW.-J.Chiang, "A three-phase four-wire power filter comprising a three-phase three-wire active filter and a zig-zag transformer," IEEE Trans. Power Electron., vol. 23, no. 1, pp. 252–259, Jan. 2008.
B. Singh, P. Jayaprakash, T.R. Somayajulu, D.P. Kothari, A. Chandra and K. Al-Haddad, "Integrated three-leg VSC with a zig-zag transformer based three-phase four-wire DSTATCOM for power quality improvement," in Proc. of 34th Annual Conference of IEEE Industrial Electronics, 2008, pp.796-801.

[6] B. Singh, V. Garg, and G. Bhuvaneswari, "A novel T-connected autotransformer-based 18-pulse AC–DC converter for harmonic mitigation in adjustable-speed induction-motor drives," IEEE Trans. Ind. Electron., vol. 54, no. 5, pp. 2500–2511, Oct. 2007.

[7] S. Leng, Il-Yop Chung, C.S. Edrington and D.A. Cartes, "A new harmonic distortion measurement algorithm for power quality evaluation and compensation,"IEEE Power and Energy Society General Meeting, 2011, pp.1-8.
LorenzoMarconi, Fabio Ronchi and Andrea Tilli,"Robust nonlinear control of shunt active filters for harmonic current compensation,"Journal of Automatica, vol. 43, pp. 252-263, 2007

[8] K. Vardar, E. Akpınar and T. Surgevil, "Evaluation of reference current
extraction methods for DSP implementation in active power filters," Journal of Electric Power Systems Research, vol. 79, pp. 1342-1352,
2009.

[9] R. Singh, A.K. Singsh and P. Kumar, "Comparison of three evolutionary
algorithms for harmonic mitigation using SAPF," in Proc. of 6th IEEE
International Conference on Industrial and Information Systems, 2011, pp.392-397.

[10] Electromagnetic Compatibility (EMC)—Part 3: Limits-Section 2:Limits for Harmonic Current Emissions (Equipment Input Current >16 A per Phase) IEC1000-3-2 Document1st ed. , (1995)

[11] IEEE Recommended Practices and requirement for Harmonic Control
on Electric Power System, IEEE Std.519, 1992.
An Luo, Lu Fang, Xianyong Xu, Shuangjian Peng, Chuanping Wu and Houhui Fang, "New control strategy for DSTATCOM without current sensors and its engineering application,"Journal of Electrical Power and Energy Systems, vol. 33, pp. 322-331, 2011.
A. Ghosh and G. Ledwich, Power Quality Enhancement using Custom Power Devices, Springer International Edition, Delhi, 2009.

[12] Cardenas, V., Moran, L., Bahamondes, A., Dixon, J.: 'Comparative
analysis of real time reference generation techniques for four-wire shunt active power filters'. Proc. IEEE 34th Annual Power Electronics Specialist Conf., 2003, vol. 2, pp. 791–796

Path tracking mobile robot using steppers

Ankit Deo[1,*], *Ayush Gupta*[2], *Himanshu Khemani*[1] and *Rashmi Ranjan* Das[3]

[1]School of Electrical Engineering Vellore Institute of Technology, Vellore
[2]School of Mechanical Engineering Vellore Institute of Technology, Vellore
[3]Associate professor Vellore Institute of Technology

Abstract. In control of mobile robots, precision plays a key role in path tracking. In this paper we have intended to use hybrid stepper motors for precise control of the two wheeled robot. A control algorithm was developed to control the robot along different trajectories. We have found that stepper motors are more accurate for path tracking than normal DC motors with wheel encoders and one can obtain the implicit coordinates of the robot in runtime more precisely. Getting the precise coordinates of the robot at runtime can be used in various SLAM and VSLAM techniques for more accurate 3D mapping of the environment.

1 Introduction

Robotics has become an indispensable part in our lives from strategic military operations to mundane applications of household and cleaning. Among the various applications of robotics path tracking have been one of the most emerging fields in terms of research. Legged mechanism can be used for most precise mapping of the trajectory. However controlling and operating legged mechanism robots are much more complicated than wheeled robots. Apart from the control systems legged mechanism robots are relatively slower. Differential drive odometry techniques with DC motors and encoders are used for path tracking algorithms. Wheel encoder output gives velocity in which the two wheels move. The bot can be moved along a certain curve by giving the two wheels different velocities. Such a bot will have problems moving along a straight line since both the wheels are independent. Moving both wheels with same velocity is a challenge. So our research is based on a better approach of using stepper motors rather than DC for precise rotation of the wheels and in turn the orientation of the bot. To keep track of orientation of the bot at a particular point a magnetometer is also used. The error caused in the actual orientation of the bot given by magnetometer and the orientation achieved due to movement of stepper motor is also taken in consideration. In traditional differential drive odometry readings from the encoders are sent as a feedback to the system to determine the motion of the wheels and the position of the bot. In our stepper model of the robot stepper motors are used for the precise motion of the bot. The wheels can be independently controlled to change the orientation of the bot at any given instant of time.

2 Types of stepper motors

Based on construction and working there are three types of stepper motor –

2.1. Variable reluctance stepper motor

Soft iron multi tooth rotor and a wound stator is used in the variable reluctance stepper motor[3]. The rotor rotates at a specific angle which minimizes the reluctance between opposite windings in the stator of the stepper motor. This type of stepper motor offers supreme level angular resolution but has low torque.

2.2 Permanent magnet stepper motor

Permanent magnets are added to the motor's structure instead of teeth. The rotor is magnetized with alternating north and south poles aligned along the straight lines parallel to the rotor shaft.[4] Permanent magnet stepper motor has better torque characteristics due to increased magnetic flux density provided by the magnetized rotor poles[4]. It has low cost and low resolution as compared to other types.it is an incremental device which responds to discrete commands by rotating the output shaft in small equal steps for each of the discrete command.

2.3 Hybrid stepper motors

Features of both Variable reluctance motor and permanent motor are included in a hybrid stepper motor for better efficiency .[7]It has very good angular resolution , speed and torque.Hybrid motor contains a multi tooth rotor which has axially magnetized concentric magnet surrounding its shaft. The magnetic flux is guided to preferred locations in the air gap by the path provided by the teeth on the rotor. This increases the detent and holding torque properties of the motor when compared with other two types of stepper motor.

* Corresponding author: ankitdeo7@gmail.com

https://doi.org/10.1051/e3sconf/20198701028

2.3 Stepper vs DC

Normal Dc motors have continuous output whereas stepper motors have step outputs. [2]For precise intrinsic coordinate mapping of our two wheeled mobile robot we found that stepper motors are indeed better than normal dc motors equipped with wheel encoders which give rotations as a feedback in terms of ticks. Which acts as a closed loop control system and renders instability to the overall system.

3 Solidworks modelling

Traditional differential drive wheeled robots are maneuvered with the help of DC motors. Standard chassis are readily available for these kinds of bots. We have designed our own model of chassis as chassis incorporating stepper motors are not designed previously. [1]Our design is a simple three wheel configuration in which the rear wheels are controlled using stepper motors and the front wheel is a castor wheel. This configuration helps for smooth point roatation and lateral movement of the bot.
For our research we made solidworks model of each part i.e. wheels, motors, chassis, battery, etc. separately and then joined them in as assembly model in 1:1 scale. The weight and dimensions for each part is manually entered so that the centre of gravity of the assembly is automatically calculated by solidworks . The position of centre of gravity is critical because the magnetometer is placed in orientation with the centre of gravity to measure the orientation of the robot. The assembly model of the robot was saved in xml file format to import into MATLAB to run simulation.

Fig. 1. Solidworks model with the exact specifications exported to simulink.

4 Solidworks model imported in MATLAB and SIMULINK.

Sim mechanics a library in MATLAB Simulink environment is used to import any kind of solidworks model into Simulink environment. Our bot with three revolute [5]joints two at the rear attached with the stepper and one at the front attached with the castor wheel is imported in the matlab environment fig. Parameters of both the wheels were varied to observe the behavior of the entire model in Simulink environment

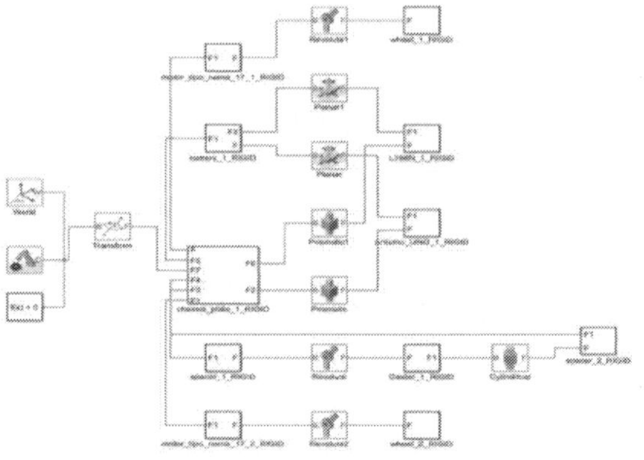

Fig. 2. Simulink model

which is shown in figure 2. Following results were obtained after performing the simulation. For our research we used a stepper motor with 1.8 degrees of rotation in a step. For path planning algorithms we need to know the coordinates of the bot at every given frame. The linear displacement of the bot can be calculated by measuring the angular displacements of the stepper motor shaft. At every frame the bot will have some change in its orientation. This change is measured from a magnetometer. The readings obtained from the

magnetometer acts as an input for our rotation matrices. Subsequent rotation matrices are multiplied in MATLAB Simulink model to get orientations at any given frame of reference.[8] The rotation matrix along with the given displacement factor and scaling factor acts as the input for our homogeneous transformation matrix. We have multiplied the homogeneous matrices in series just like rotation matrices in series in MATLAB Simulink environment. The implicit coordinates of the bot can be calculated at any given point through this.

5 Trajectory Mapping

Every trajectory can be expressed in terms of mathematical function. In our research we have considered continuous and differentiable trajectory functions. Trajectories expressed in terms of y=f(x) are sampled along the x axis. Continuous curved trajectories can be sampled via simple commands in MATLAB.Our simulation shows an example of a sine wave sampled along the x axis. At every sampled point x we have found the corresponding value of f(x) in our case sin(x).for very two consecutive points Xn and Xn-1, we get the corresponding f(Xn) and f(Xn-1) values. Using coordinates f(Xn) , f(Xn-1), Xn , Xn-1 and the distance formula, we can obtain the linear distance the bot needs to move between two subsequent points in order to follow the desired trajectory. Distance d is calculated using the equation 1.

$$d=\sqrt{(x_1-x_2)^2-(y_1-y_2)^2} \qquad (1)$$

https://doi.org/10.1051/e3sconf/20198701028

Using these four coordinates we can determine the change in orientation the bot needs to go through, to reach the adjacent point. In our research we have calculated the change in orientation the bot needs to go through between every two adjacent points, in order to reach the next adjacent point.

$$\text{steps per revolution of stepper} = 360°/1.8° \quad (2)$$

At every step the bot is programmed to change its orientation, Instead of changing it's orientation only at the sampled points. This helps to minimise the error in trajectory mapping since the curve traced after subsequent changes in orientation is relatively smoother.[6] For changes in orientation angles lesser than 1.8 degrees we have used half stepping and 1/8th stepping in stepper motors.
½ stepping and 1/8th stepping in stepper motors: Implementation of half stepping and 1/8th stepping helps in precise motion of our bot, since stepping action is not limited to 1.8 degrees. The error is plotted in table 1. For movements that require change orientation between points lesser than 1.8 degrees we have implemented ½ stepping and 1/8th stepping. Illustrated using an example where we considered relatively smoother trajectory in comparison to the sine wave. F(x)= x^2.This is shown in figure 3.

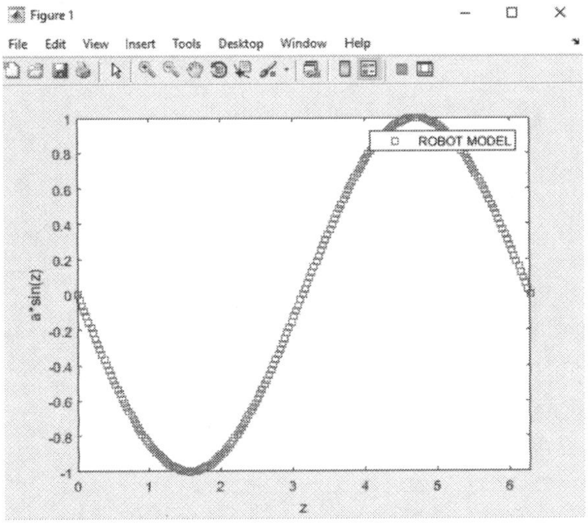

Fig. 3. Path generation (plotted in MATLAB).

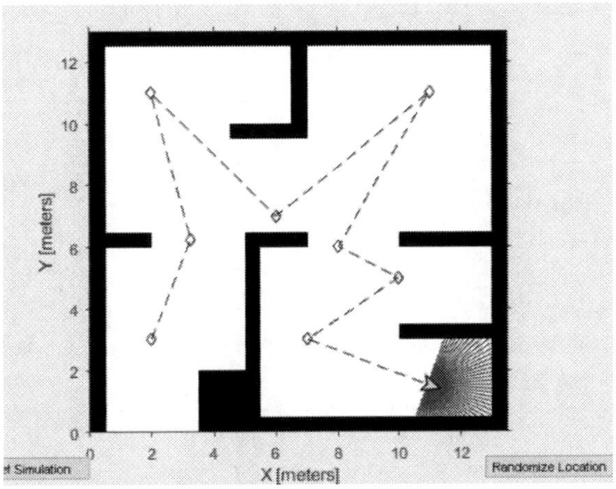

Fig. 4. Path generation using LIDAR simulation. (in MATLAB).

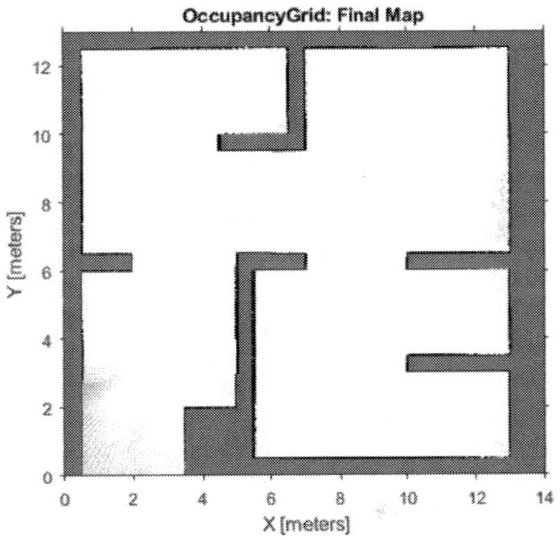

Fig. 5. Binary occupancy grid generated by LIDAR simulation .

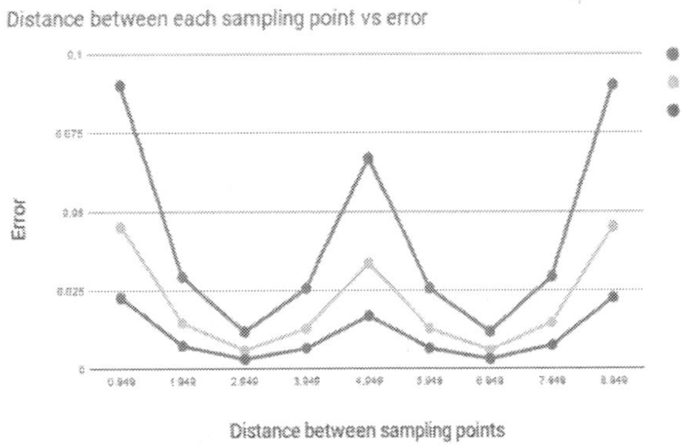

Fig. 6. Error plot for different stepping of stepper motors ..

Table. 1. Data for path tracking

z(in m)	theta (rad)	theta (deg)	Theta/ 1.8	no of rotations	no of steps for stepper motor	no of steps for stepper motor(1/2 stepping)	no of steps for stepper motor (1/4 stepping)	error	for full stepping (cm)	for full stepping(1/2 stepping)(cm)	for full stepping(1/4 stepping)(cm)
0.949	0.7441	42.655	23.69745	4.46979	893.958	1787.916	3575.832	0.958	0.0902436	0.0451218	0.02256
1.949	0.4555	26.111	14.50636	9.17979	1835.95	3671.916	7343.832	0.310	0.0292776	0.0146386	0.00731
2.949	-0.168	-9.659	-5.366242	13.8897	2777.95	5555.916	11111.832	0.124	0.0117188	0.0058592	0.00292
3.949	-0.643	-36.91	-20.50636	18.5997	3719.95	7439.916	14879.832	0.271	0.0255658	0.0127829	0.00639
4.949	-0.775	-44.43	-24.68789	23.3097	4661.95	9323.916	18647.832	0.710	0.0669576	0.0334786	0.01673
5.949	-0.643	-36.91	-20.50636	28.0197	5603.95	11207.916	22415.832	0.271	0.0256036	0.0128017	0.00640
6.949	-0.168	-9.659	-5.366242	32.7297	6545.95	13091.916	26183.832	0.124	0.0117188	0.0058592	0.00292
7.949	0.4555	26.117	14.50636	37.4397	7487.95	14975.916	29951.832	0.310	0.0292776	0.0146386	0.00731

Conclusion

In this research we were able to successfully traverse a sine wave trajectory with 10 points of sampling and 100 points of sampling. Errors observed were much lesser which is plotted in figure 6. than what we observed for traditional differential drive robots with normal DC motor. For angular resolutions lesser than 1.8 degrees concept of half stepping, micro stepping and one eighth stepping can be used. Using stepper motors made the system more energy efficient and stable as it is an open loop control system. Using the coordinates calculated after multiplying rotation matrices we used them for VSLAM. A LIDAR simulation in MATLAB shown in figure 4. We obtained a binary occupancy grid showing free path and obstacles as shown in figure 5.

References

1. Ilya K and N.Harris M . Developments in nonholonomic control problems. IEEE Control Systems 1995; 15(6):20 – 36.
2. Lydia EK, Petr S, Jean CL and Mark HO. Probabilistic roadmaps for path planning in high-dimensional configuration spaces. IEEE Trans.
3. Robotics and Automation 1996; 12(4):566–580.
4. Jean PL, Paul EJ, Michel T, Richard and Murray M. A motion planner for nonholonomic mobile robots. IEEE Trans. Robotics and
5. Automation 1994; 10:577–593.
6. Ankit Deo, Raaghavi Ravisankar. Differential Drive odometry for two wheeled mobile robot. IJRSET.
7. Trajectory tracking and control of differential drive robot for predefined regular geometrical path Global Colloquium in Recent Advancement and Effectual Researches in Engineering, Science and Technology (RAEREST 2016)
8. Kuniaki K, Liang M, Jianru X, Chengwei Z and Nanning Z. A path generation for automated vehicle based on Bezier curve and via-points. Robotics and Autonomous System. 2015; 74:243–252. [6] Steven ML. Rapidly-exploring random trees: A new tool for path planning. Tech. Rep. Computer Science Dept. Iowa State University 1998.
9. James JK and Steven ML. RRT-Connect: An Efficient Approach to Single-Query Path Planning, ICRA 2000; 1-7. [8] Ákos N, Gábor C and Domokos K. Path Planning and Control of Differential and Car-like Robots in Narrow Environments. SAMI 2015;103-108.
10. Bong SP, Sung JY, Jin BP, and Yoon H C. A Simple Adaptive Control Approach for Trajectory Tracking of Electrically Driven Nonholonomic Mobile Robots. IEEE Trans. Control Syst. Technol .2010; 18(5):1199-1206.
11. Elie M, Maarouf S and Hamadou S. A higher level path tracking controller for a four-wheel differentially steered mobile robot. Robotics and Autonomous Syst. 2006; 54:23–33.

Variable Switch Regenerative Braking Technique for PM BLDC Motor Driven Electric Two-wheeler

Manish Kumar Dubey[1,*] and *Phaneendra Babu* Bobba[2]

[1]Research Centre Imarat, Kalam Missile Complex, DRDO, Hyderabad
[2]EEE Department, GRIET, Hyderabad, India

Abstract. In this paper PM BLDC motor driven electric two-wheeler is proposed. The operation of PMBLDC motor in all four quadrants along with various regenerative braking control strategies are simulated MATLAB/ SIMULINK. Three braking methods are proposed for electric vehicle application. Factors like recovered energy, braking time, maximum braking current are compared. Depending on the simulation results switching between different braking methods is proposed for efficient and reliable braking operation. Moreover, by using variable switch braking technique one can extract more braking energy.

1 Introduction

Environmental pollution, rapid depletion and increase in the cost of fossil fuels are the significant factors which are stimulating the use of Electric Vehicles for the transportation purpose. These vehicles would also enable meeting the demands for electrical power due to the increasing use of electronic features to improve vehicle performance, fuel economy, emissions, passenger comfort, and safety. In electric vehicles the challenges are long range, efficient power electronics circuits and their control and electrical machines.

In an internal combustion engine powered vehicle, a multi-gear transmission is required to shape the speed torque profile of the engine to obtain an approximate constant power profile. In modem ICE vehicles are usually equipped with complex automatic transmission, which consists of multi-gear box. When vehicles are operated with more start and stop driving pattern, such as driving in cities, the average fuel consumption efficiency of the torque converter is very low (less than 60%). Therefore, necessity of multi-gear transmission is unfavourable.
The desirable characteristics of an electric motor drive for traction application in an electric and hybrid electric vehicle are high starting torque at low speed region for fast acceleration, hill climbing and obstacle negotiation, and low torque at high speed for normal driving. Permanent Magnet Brushless DC motor (PM BLDC) is very efficient and compact motor, with good reliability and simple control. These motors find application mainly in two-wheeler and three-wheeler electric vehicles.

The bottleneck of the development of EV technologies is the energy sources. At present and in the foreseeable future, the viable EV energy sources are batteries, fuel cells, supercapacitors and ultrahigh-speed flywheels. Batteries are relatively the most mature source for EV application. They offer either high specific energy (HSE) or high specific power (HSP), but not both. An HSE energy source is favourable for long driving range, whereas as HSP energy source is desirable for high acceleration rate and high hill climbing capability. The electric power source for most of the two-wheelers and three-wheelers is batteries. As batteries power storage capacity is limited, these electric vehicles suffer from problem of short range. The braking performance is the basic factor of vehicle safety. In urban driving and hilly-terrain area driving, significant amount of energy is consumed in braking. The electrification of the vehicle drive train makes it feasible to recover some of the energy lost in braking by employing regenerative braking. This recovered energy can be used to increase the range of the vehicle. Depending on number of switches active during regeneration there are three types of regenerative braking methods. These are single switch braking (SSB), double switch braking (DSB) and triple switch braking (TSB).

2 Regenerative Braking, methods and comparison

One of the most important features of EVs, HEVs, and fuel cell vehicles (FCVs) is their ability to recover significant amounts of braking energy. The electric motors in EVs, HEVs, and FCVs can be controlled to operate as generators to convert the kinetic or potential energy of vehicle mass into electric energy that can be stored in the energy storage and then reused. The braking performance of a vehicle is an important factor in vehicle safety. A successfully designed braking system for a vehicle must always meet the distinct demand of

* Corresponding author: manish.current83@gmail.com

quickly reducing vehicle speed and maintaining vehicle direction controllable by the steering wheel. The former requires the braking system to be able to supply sufficient braking torque on all wheels. Generally, the braking torque required is much larger than the torque that an electric motor can produce, especially in heavy braking. In EVs, HEVs, and FCVs, mechanical friction braking systems must coexist with electrical regenerative braking. Therefore, this is a hybrid braking system. As in the hybrid propulsion system, there are many configurations and control strategies. However, the final goal of the design and control of such systems is to ensure the vehicle's braking performance and its ability to recover as much braking energy as possible [1].

The importance of regenerative braking can be seen by knowing the fact that in typical urban areas energy required for braking operation is almost 34% of the total traction energy, while in large cities this may reach up to 80% of the total traction energy which is very high. Hence, recovering energy from the braking process can increase the range of the EV by significant distance [2].

2.1 Regenerative Braking in Brushless DC Motor

For an EV with brushless dc motor when the acceleration command is issued, the electric machine is operated in the motor mode (i.e. starting, acceleration, or power assisting). The output torque of the machine is controlled by adjusting the direction and amplitude of the armature current by a DC to AC inverter. In contrast, if the machine is operated in the regenerative mode (i.e. deceleration, or continuous charging), namely the deceleration command is sensed, the dynamic energy from the vehicle is transferred to the battery or capacitance via the alternator and an AC to DC converter [3].

Complete block diagram of PMBLDC motor based electric vehicle along with its control in motoring and braking region is shown in Fig:1. In this block diagram hybrid braking has been used. At lower speed braking torque developed by motor is not enough to stall or brake the vehicle, hence mechanical braking is preferred at lower speeds. Using this duty cycle PWM signals are generated for braking by braking decoder and PWM generator. Hence, controlled braking can be achieved by using control strategy shown in above Fig 1. If required braking torque is more than the maximum torque produced at that speed (this can be obtained by using speed- torque characteristic of motor as shown in Fig:1.) then mechanical braking must be applied. During braking mode depending on the braking torque required by driver braking current reference is generated. This reference current is compared with actual motor torque and by help of current controller duty cycle is generated.

2.1.1 Regenerative braking with single-switch

In this method only one power switch is operated within each commutation. The fig showing the operation with this type of switching is shown below. When braking command is given the lower switch is turned on such that the current circulates through the two switches (shown with red segments in Fig 2).In this case energy is being stored in the corresponding phase inductor. When the power switch is turned off the reversal of power takes place due to the stored energy in the inductor and hence battery can be charged during this duration (shown with dotted blue segments in Fig 2) [4]

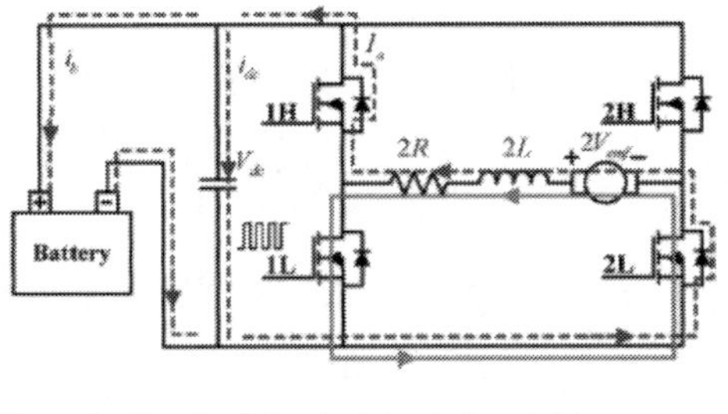

Figure 2. Circuit of the single-switch switching

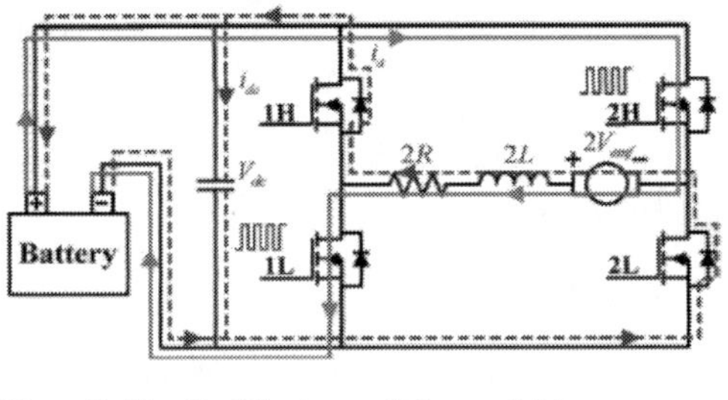

Figure 3. Circuit of the two-switches switching

2.1.2 Regenerative braking with two-switches

In this method two power switches are operated within each commutation. When the speed of the vehicle is relatively high, the issue of how to recover the maximum kinetic energy of the vehicles is of most importance. In contrast, for low speed operation, braking torque is the major concern. Regenerative braking with two-switches is good for low speed operation. The operation is explained with Figure 3 When braking command is given in this case the two power switches operates in such a way that, the polarity of induced voltage and the battery is additive and a large current flows through the windings of the motor, also the charging of phase inductor takes place (current path is shown in Fig 3 with red color). When switches are turned off the power reversal takes place. (shown in Figure 3 with blue dotted segment) [4-5]

2.1.3 Regenerative braking with three-switches

In this method three power switches are operated within each commutation. The principle of the three-switches regenerative braking is like that of the conventional dynamic braking. The current path when lower switches are turned on and off is shown in figure 4 [5].

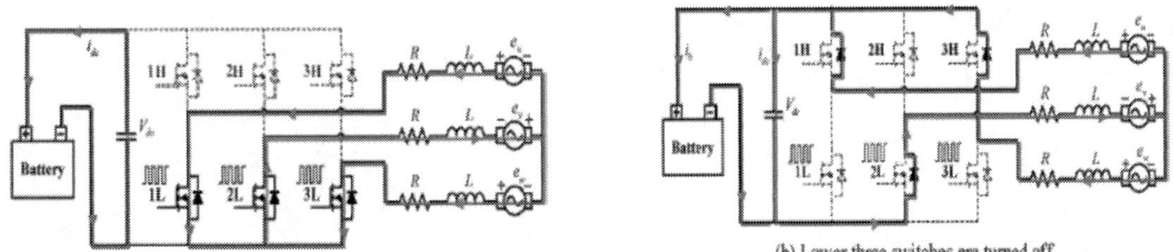

(a) Lower three switches are turned on.

(b) Lower three switches are turned off.

Figure 4. Current paths of the three-switches switching in 30 electric degrees [5]

2.3 Simulation Results

Table 1. Single Switch braking with varying duty cycle and fixed speed

S No.	Nomenclature			Duty Cycle				
		1	0.8	0.6	0.49	0.4	0.3	0.2
1	Regenerative energy (J)	2.85	27.4	27.4	27	27	27	27
2	Regenerative Braking time (Sec)	0.0433	0.085	0.085	0.167	0.167	0.167	0.167
3	Battery Current (A)	60	37	37	14.62	14.5	14.6	14.5
4	Maximum Current (A)	41	37	37	14.62	14.5	14.6	14.5

Table 2. Single Switch braking with varying speed and fixed Dutycycle

S No.	Nomenclature	Speed (rpm)					
		300	275	250	225	200	175
1	Available Energy (J)	65	54.61	45.14	36.56	28.89	22.11
2	Regenerative energy (J)	27.8	22.71	18.025	13.47	9.066	4.51
3	Regenerative Braking time (Sec)	0.085	0.077	0.0686	0.0576	0.0464	0.028
4	Battery Current (A)	37	32.07	26.84	21.55	16.4	11.55
5	Maximum Phase Current (A)	37	23.76	22.96	21.56	12	10.77

From table 1 and 2 , it is observed that braking energy is maximum when the speed is maximum and the braking current is maximum when duty cycle is maximum. Similarly, two and three switch braking methods and the performance in each case is simulated and the results are shown in section 2.4 .

2.4 Comparaison of regenerative braking methods

Performance of three braking methods when duty ratio is varied and speed kept constant are compared in the following graphs

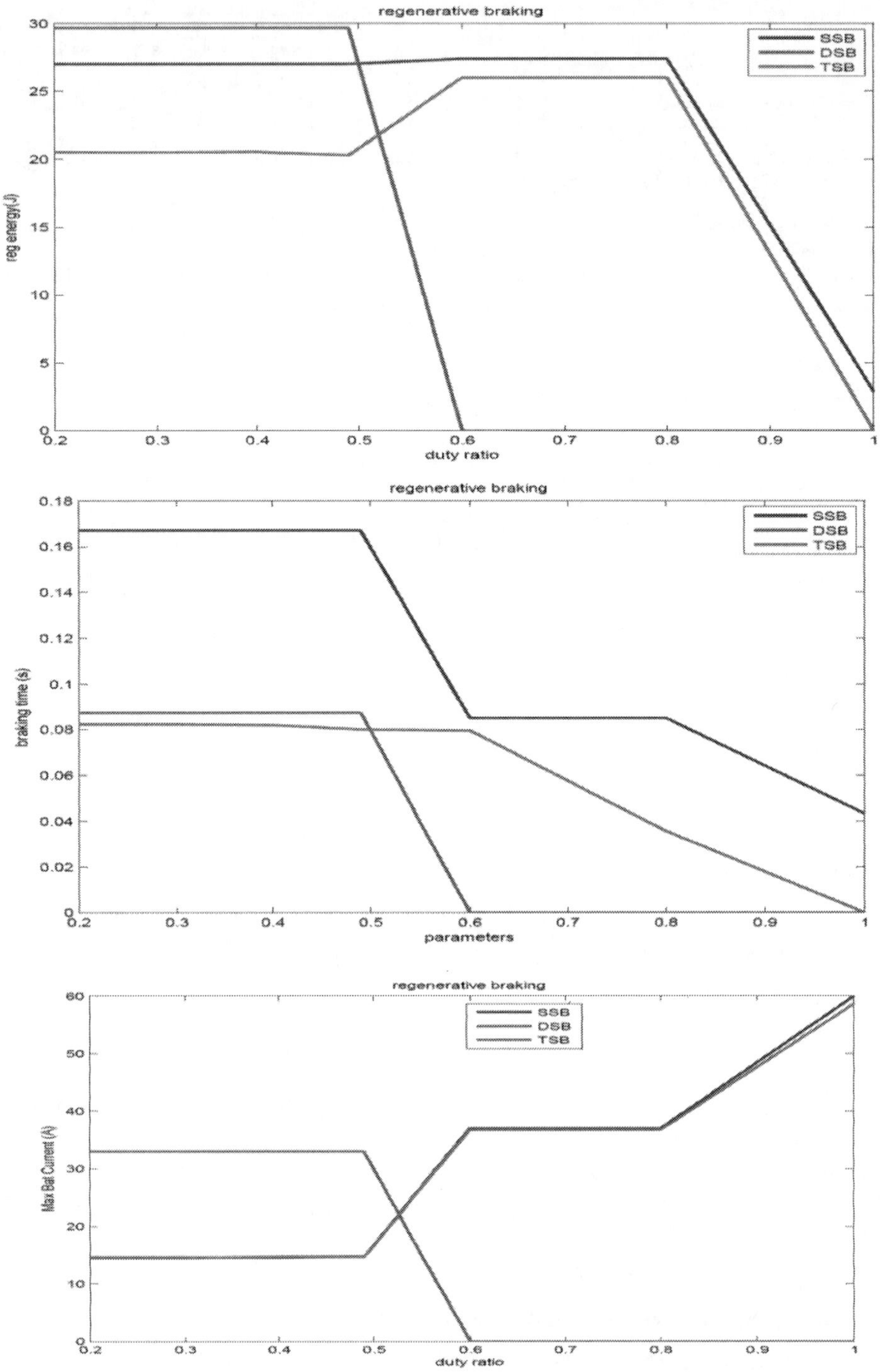

Figure 5. Comparison of three braking methods with varied duty cycle

From the above results it is observed Maximum Energy can be recovered by using double switch braking. Double switch braking method gives rise to maximum battery current during regeneration. Other two method have almost same max. battery current at low duty ratio and at higher duty ratio current in case of SSB is more than that of TSB. Braking time is least in Three Switch Braking Method. SSB method takes long braking time at low duty ratios.

https://doi.org/10.1051/e3sconf/20198701029

2.5 Performance of three braking methods when speed is varied and duty ratio kept constant

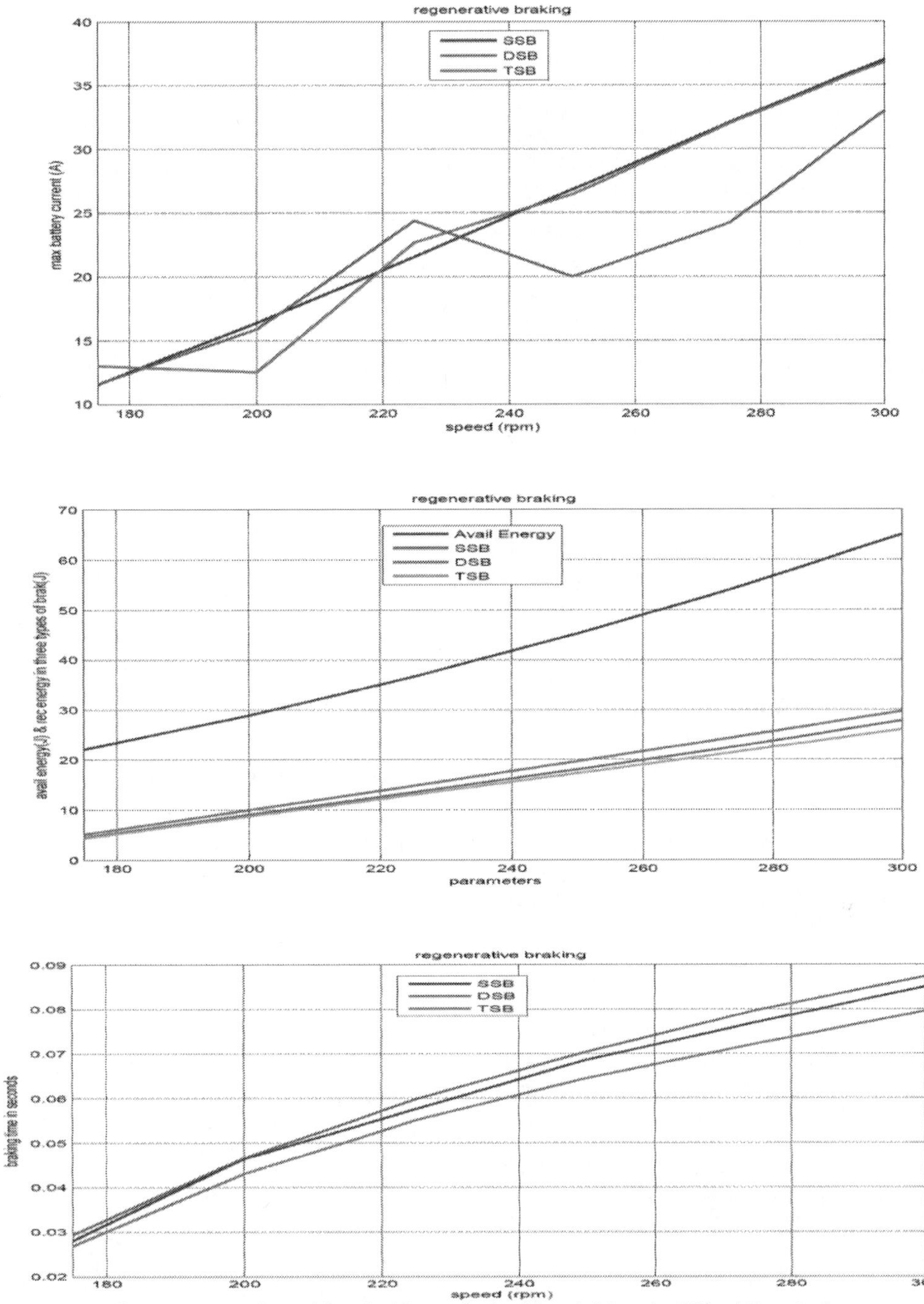

Figure 6. Comparison of three braking methods with varied Speed and Fixed Duty Cycle

From the Fig 5 and 6 it is observed that Max battery current in case of single switch braking is slightly more than that of three switch braking. and braking time is least in three switch braking method.

Conclusion

In this paper varoious regenerative braking methods are proposed for PMBLDC Motor driven electric two-wheeler. If efficiency is the major concern when the vehicle speed during the braking process is almost constant, In order to get the maximum recovered energy DSB should be used. At high speed for maximum braking torque TSB should be used, but at low speed braking torque is not sufficient in TSB, hence DSB should be used. SSB also cannot be used at low speed because of no braking torque. At high speed if energy recovery is main concern than DSB should be used.so based on the vehicle speed appropriate braking scheme is selected that means controler should works on variable switch braking scheme

References

1. C. C. Chen and K.T.Chau, *Morden Elcetric Vehicle Technology,* Oxford University Press, Inc., New York, 2001, page, 122-133

2. Bobba, P B ; Rajagopal, K R; "Compact Regenerative Braking Scheme for a PM BLDC Motor Driven Electric Two- Wheeler" *Power Electronics, Drives and Energy Systems (PEDES) & 2010 Power India, 2010 Joint International Conference.*

3. Becerra, RC; Ehsani, M; Jahns,T,"Four Quadrant Brushless ECM drive with Integrated Current Regulation" *Industry Applications Society Annual Meeting, 1989.,* Conference Record of the 1989.

4. Bobba, P B ; Rajagopal, K R; "Compact Regenerative Braking Scheme for a PM BLDC Motor Driven Electric Two- Wheeler" *Power Electronics, Drives and Energy Systems (PEDES) & 2010 Power India, 2010 Joint International Conference.*

5. Cheng-Hu Chen, Wen-Chun Chi, and Ming-Yang Cheng, "Regenerative Braking Control for Light Electric Vehicles", *IEEE PEDS 2011*, Singapore, 5 - 8 December 2011

* Corresponding author: manish.current83@gmail.com

E3S Web of Conferences **87**, 01030 (2019) https://doi.org/10.1051/e3sconf/20198701030
SeFet 2019

Artificial Neural Networks based SPWM technique for speed control of Permanent Magnet Synchronous Motor

Suresh Kumar Tummala [1,*] and *Dhasharatha* G [2]

[1]Professor, EEE Department, GRIET, Hyderabad, India
[2] Research Scholar, EEE Department, TKR Engineering College, Hyderabad

Abstract. The advancement of industry apparatuses for some methods with specific tasks to control the working of a few actuators on the field. Among these actuators, Permanent magnet synchronous motor drives are a mainly all-inclusive machine. Proficient utilization of hesitance torque, generally effectiveness, minor misfortunes and smaller size of the motor are the principle attractions of PMSM when contrasted and different drivers. Precise and rapid torque reaction is one of the parameters to determine differentiating arrangements in the ongoing past. The field-situated power perceived the likely and vigorous answer to accomplish these prerequisites to empower the figuring of streams and voltages in different parts of the inverter and motor under transient and consistent conditions. The primary objective of this paper is to investigate Artificial Neural Network based control of speed for PMSM in both open and closed loop under no-load and loaded condition. A shut circle control framework with ANN procedure in the speed circle intended to work in steady torque and transition debilitating districts. MATLAB reproduction performed in the wake of preparing the neural system (directed learning), results for reference control applications are adequate and appropriate in the process business. Speed control in shut circle at different stacking conditions talked about in detail.

1 Introduction

Permanent Magnet Synchronous Motors (PMSM) extensively utilized in low to medium power applications, for example, mechanical autonomy, outer PC flexible speed drives [1], and electric vehicles. The extension in the market of Permanent Magnet (PM) engine drives has requested the requirement for replication instruments fit for taking care of re-enactments for engine drive. Proliferations [2] including engine drives have helped the way toward growing new frameworks by decreasing expense and time. To encourage the improvement of new techniques, recreations of engine drives in a visual situation have the capacities of performing dynamic [3] Simulation.

In this article, a recreation of a field situated controlled [4,5] PM engine drive framework created including every single reasonable segment of the drive framework are introduced. A shut circle control framework with the ANN controller in the speed circle intended to work in consistent torque and motion debilitating locales [6]. Recreation results displayed for two velocities [7] of task, one underneath evaluated and another above-appraised speed.

Artificial Neural Networks are motivated by our present learning of natural sensory systems, in spite of the fact that they don't attempt [8] to be reasonable in everything about (territory of ANN isn't worried about natural demonstrating, an alternate field) [9, 10]. Some ANN models may, along these lines, be unreasonable from an environmental displaying perspective. As opposed to the ordinary advanced PC, ANN plays out their calculation utilizing numerous basic and very interconnected processors working in parallel [11].

2 Mathematical Modelling of PMSM Drive

Vector control likewise read as decoupling or field-orientated control. Vector control decouples three-stage stator current into two-stage d-q hub present, one delivering motion and other creating torque and permits coordinate control of motion and torque. In this way, by utilizing vector control, the PMSM is comparable to an independently energized dc machine. The model of PMSM is nonlinear. Along these lines, by utilizing vector control, the model of PMSM is direct.

For building a pivoting attractive field and drive the rotor, tweaked current provided to the ABC stator windings. The vector control system figured in the synchronously pivoting reference outline, ABC stator arranges, stationary α-β hub organizes and turning d-q hub facilitates equal relations of streams worked by Clarke – Park [4] and backwards changes Fig.1 demonstrates a vector chart of the PMSM. i_d slacks i_q by

* Corresponding author: sureshkumar.t@griet.ac.in

 https://doi.org/10.1051/e3sconf/20198701030

90°; in this way, id must be along the rotor motion. In the event that id is along rotor motion, the d-hub stator transition adds to the rotor motion influencing an expansion of net air hole motion [10].

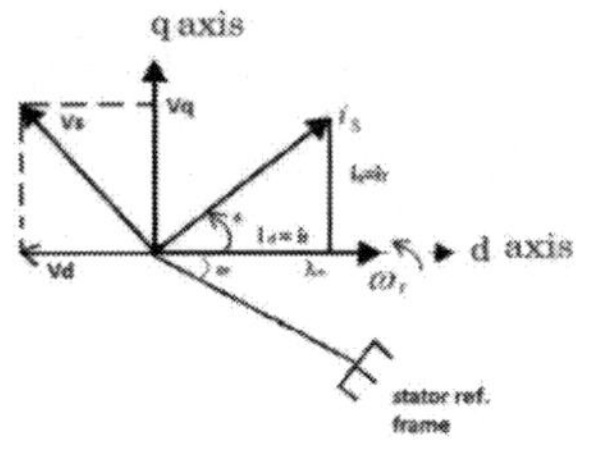

Fig. 1. Phasor representation of PMSM

$$i_a = i_s \sin(\omega_r t + \delta) \quad (1)$$

$$i_b = i_s \sin(\omega_r t + \delta - \tfrac{2\pi}{3}) \quad (2)$$

$$i_c = i_s \sin(\omega_r t + \delta + \tfrac{2\pi}{3}) \quad (3)$$

Where $\theta_r = \omega_r t$, from phasor representation we have:

$$\begin{matrix} i_q \\ i_d \end{matrix} = i_s \begin{bmatrix} \sin\delta \\ \cos\delta \end{bmatrix} \quad (4)$$

If $i_d = 0$ by $\delta = 90^o$ then the electric torque equation:

$$T_e = \left(\frac{3}{2}\left(\frac{P}{2}\lambda_m i_q\right)\right) \quad (5)$$

A constant torque is attainable if i_q is constant. Hence the electric torque is subject to only on the quadrature axis current.

d axis stator voltage:

$$V_d = Ri + L_d \frac{di_d}{dt} - L_q \omega_e i_q \quad (6)$$

q axis stator voltage:

$$V_q = Ri_q + L_q \frac{di_q}{dt} + L_d \omega_e i_d - \omega_e \lambda_{af} \quad (7)$$

d axis magnetic flux linkage:

$$\lambda_d = L_d i_d + \lambda_{af} \quad (8)$$

q axis magnetic flux linkage:

$$\lambda_q = L_q i_q \quad (9)$$

Electromagnetic torque

$$T_e = J \frac{d}{dt} \omega_r + B\omega_r + T_l \quad (10)$$

From torque equation

$$T_e = \frac{3}{2} P\left(\lambda_{af} i_q + (L_d - L_q) i_d i_q\right) \quad (11)$$

Where $\omega_r = \frac{2}{P}\omega_e$ (12)

Rotor speed in angular frequency

$$\omega_e = \int \frac{1}{\omega_r} \left[\frac{1}{J}\frac{P}{2}(T_e - T_m - B\frac{2}{P}\omega_e)\right] \quad (13)$$

3 Block Diagram Representation

In the event that the quick size of the regulation flag is higher than the bearer motion at a point in time, the yield voltage of the inverter leg should associate with the positive side of the DC connect [6]. In the event that the bearer flag is higher than the adjustment flag, the yield should identify with the negative of the DC interface.

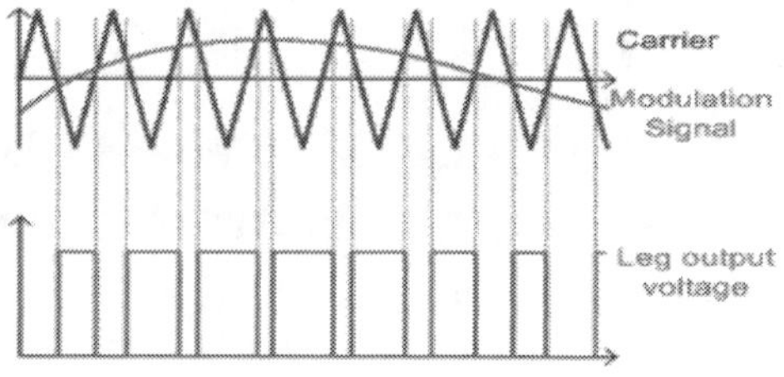

Fig. 2. PWM Principle

From the above approach, the output waveform can be made to follow any desired waveform shape. With machines, sinusoidal and trapezoidal waveform shapes are among the most common. The underlying block diagram representation of PMSM with ANN shown in figure 3. ANN has taken in controller block.

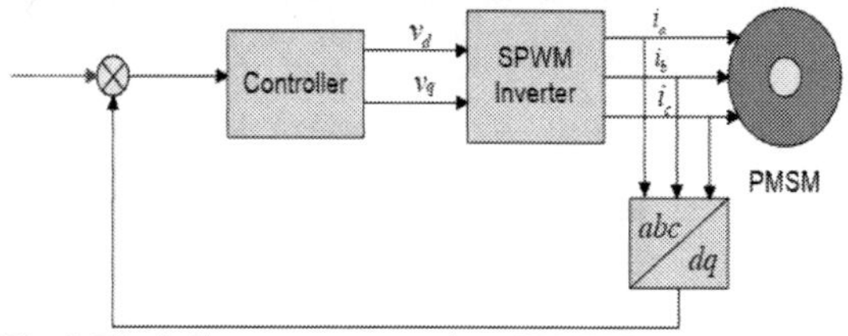

Fig. 3. Block diagram representation of PMSM

4 PMSM with ANN

Multilayer Neural Networks look like the Single Layer Neural Networks besides that despite the data and yield layer, it has no less than one disguised layer. Backpropagation Neural Network is a Multilayer Feed Forward Neural Network which utilizes the broadened inclination plummet-based delta learning standard and known as back proliferation (of mistakes) administer for preparing. The design of a Backpropagation Neural Network appeared in Fig 4.

The system has one info layer, one yield layer, and concealed layers. The information layer is associated with the thick coat and further associated with the yield layer using interconnection weights. In the figure, just a feed forward period of a system appeared, however the blunders backpropagated amid the Backpropagation (of oversights) period of learning, there might be any number of shrouded layers in a framework yet increment in the quantity of concealed layers results in the computational many-sided quality of the net. Subsequently, the time taken for intermingling and to limit the blunder might be high and an inclination accommodated both covered up and yield layer. Parameters pertaining to ANN listed in Table 1 for quick reference.

https://doi.org/10.1051/e3sconf/20198701030

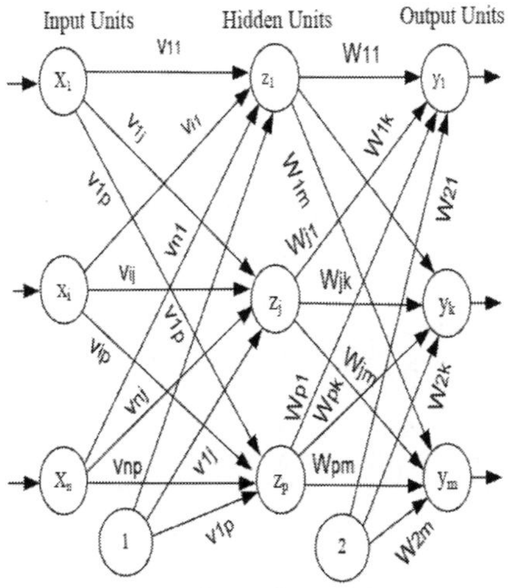

Fig. 4. Backpropagation of Neural Network

The preparation of Backpropagation Neural Network includes four phases including (an) Initialization of weights (b) Feed Forward Phase (c) Backpropagation of blunders (d) Updation of weights and predisposition. Amid the main stage, every one of the weights instated to little arbitrary qualities. In the feed forward stage, each information unit gets an information flag and communicates this message to every one of the concealed gatherings. Every obscure unit at that point processes its initiation and sends its message to each yield unit. Each yield unit figures its initiation to shape the reaction of the net for the given information design. Simulation model and experimental setup of PMSM with ANN presented in Figure 5 and 6.

Table 1. ANN Parameters

Parameter Name	Parameter Value
Hidden layer transfer function	Tansig
Output layer transfer	Purlin
Activation function	Net sum
Gradient	1.2145e-10
No. of weights	10
Regression	1
Training sets	150
Testing sets	110

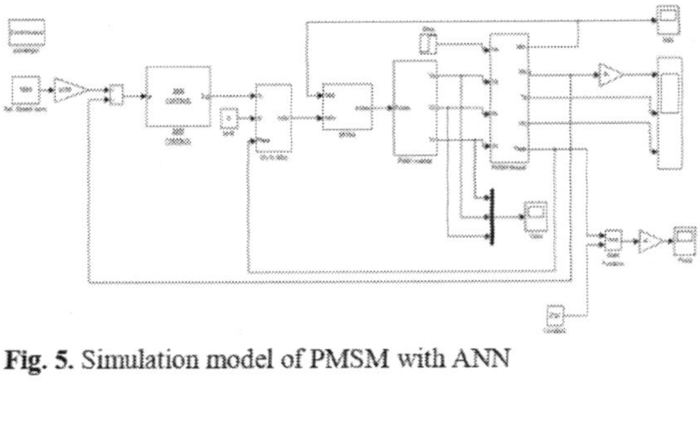

Fig. 5. Simulation model of PMSM with ANN

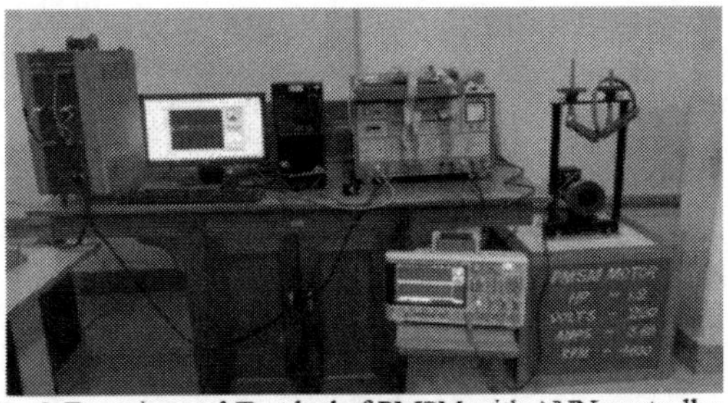

Fig. 6. Experimental Test bed of PMSM with ANN controller

5 Results and Discussion

In this segment, the reaction of different rates of PMSM under no-load and load of 1N-m connected at 0.05 sec are discussed alongside the reaction of currents of three-phases and two-phase (d, q).

5.1 Speed of 700 rpm under no-load condition

Speed representation of the ANN controller shown in Fig 7. In which speed reaches a maximum and then maintains a constant by taking 700 rpm under a no-load condition.

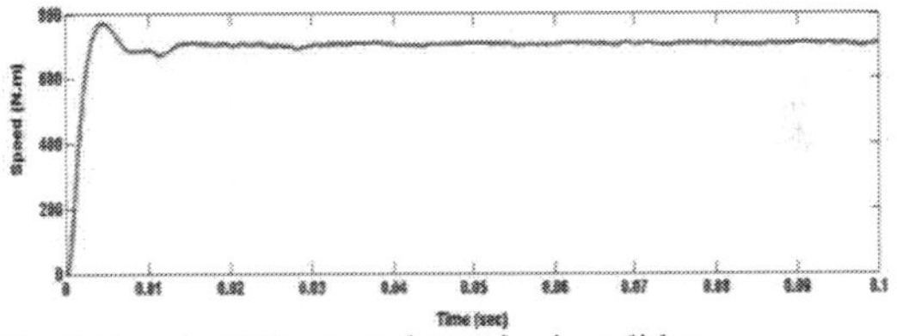

Fig. 7. Speed at 700 rpm under no-load condition

Torque representation of ANN controller shown in Fig 8. In which torque reaches to maximum torque then settled precise instant when the speed at 700 rpm under a no-load condition.

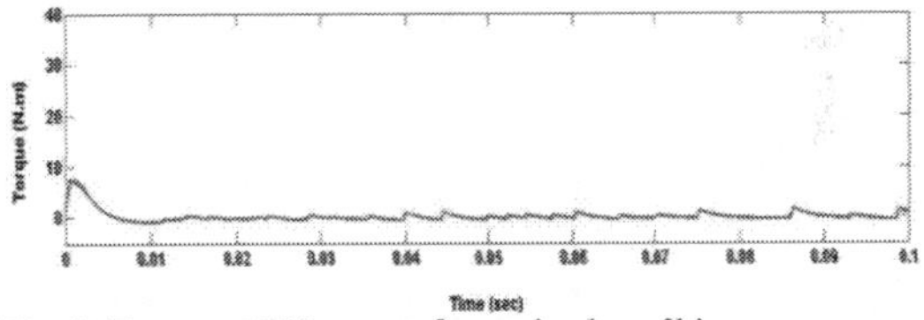

Fig. 8. Torque at 700 rpm under no-load condition

5.2 Speed at 1300 rpm under no-load condition

Speed representation of the ANN controller shown in Fig 9. In which speed reaches a maximum then maintains a constant by taking 1300 rpm under a no-load condition.

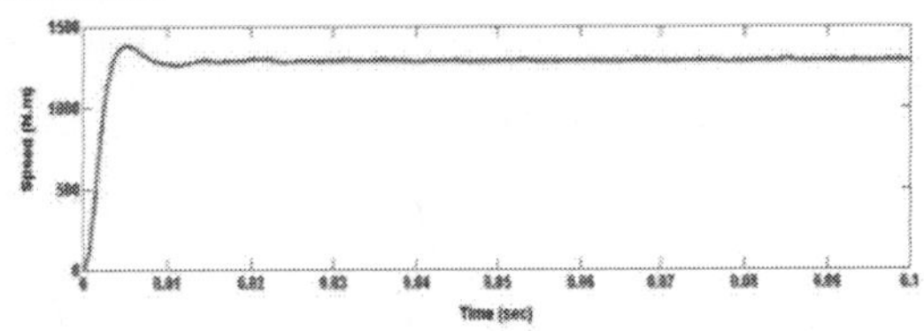

Fig. 9. Speed at 1300 rpm under no-load condition

 https://doi.org/10.1051/e3sconf/20198701030

Torque representation of ANN controller shown in Fig 10. In which torque reaches to maximum torque then settled precise instant when the speed at 1300 rpm under a no-load condition.

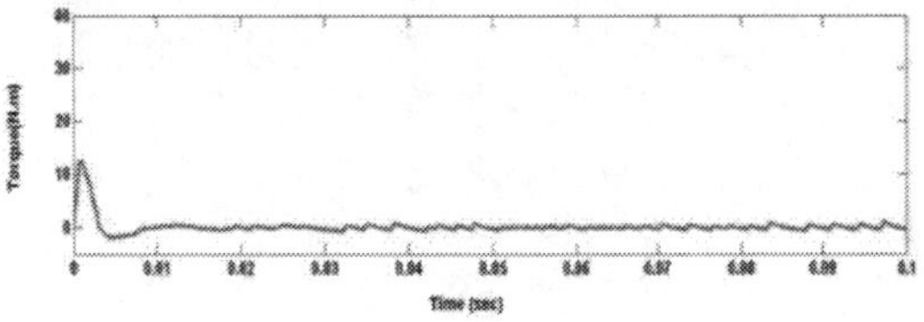

Figure 10. Torque at 1300 rpm under no-load condition

5.3 Speed at 2000 rpm under no-load condition

Speed representation of the ANN controller shown in Fig 11. In which speed reaches a maximum then maintains a constant by taking 2000 rpm under a no-load condition.

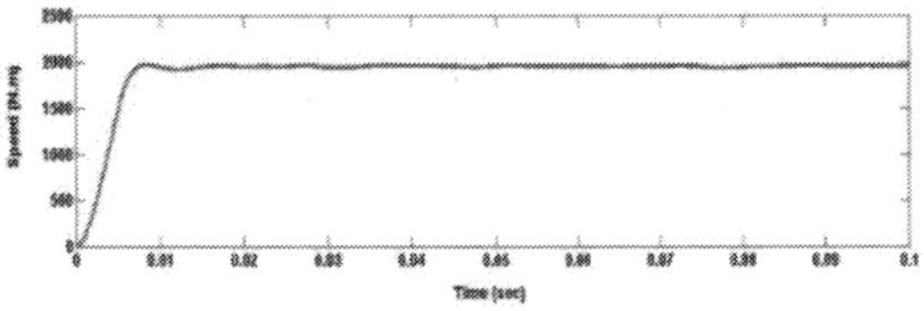

Fig. 11. Speed at 2000 rpm under no-load condition

Torque representation of the ANN controller shown in Fig 12. In which torque reaches to maximum torque then settled precise instant when the speed at 2000 rpm under a no-load condition.

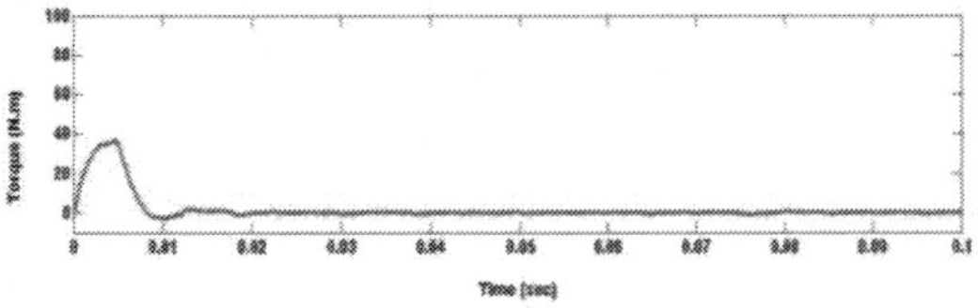

Fig. 12. Torque at 2000 rpm under no-load condition

5.4 Speed at 700 rpm under load condition

Speed representation of the ANN controller shown in Fig 13. In which speed reaches a maximum then maintains a constant by taking 700 rpm under 1 N.m load torque at 0.05 sec.

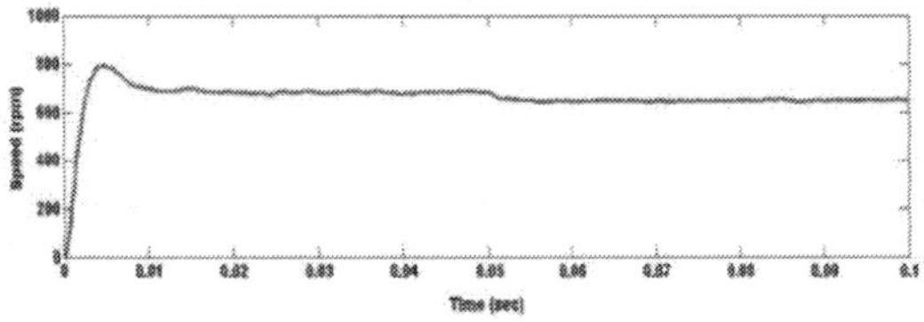

Fig. 13. Speed at 700 rpm under 1 N-m load at 0.05sec

Torque representation of the ANN controller shown in Fig 14. In which torque reaches to maximum torque then settled specific instant when speed at 700 rpm under 1 N.m load torque at 0.05 sec.

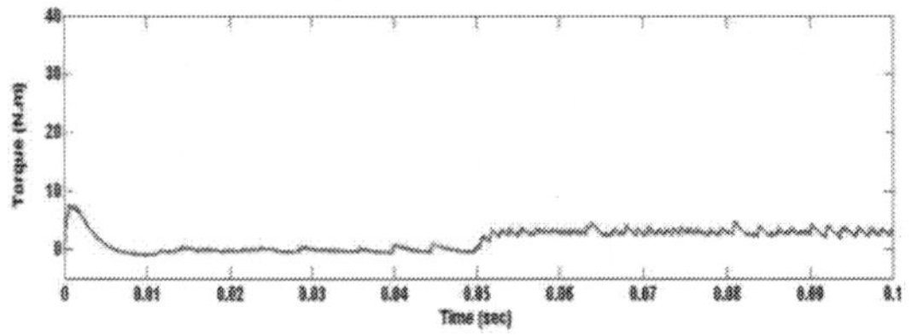

Fig. 14. Torque at 700 rpm under 1 N-m load at 0.05sec

Phase currents (Three & Two) chart is demonstrated as follows in Figure 15 & 16. In PI controller three stage streams have swells yet utilizing of ANN controller there are no swells. d-axis current is nearly equal to zero. q-axis current is away from d-axis as shown in figure 16. So, this strategy is a productive apparatus for controlling reason.

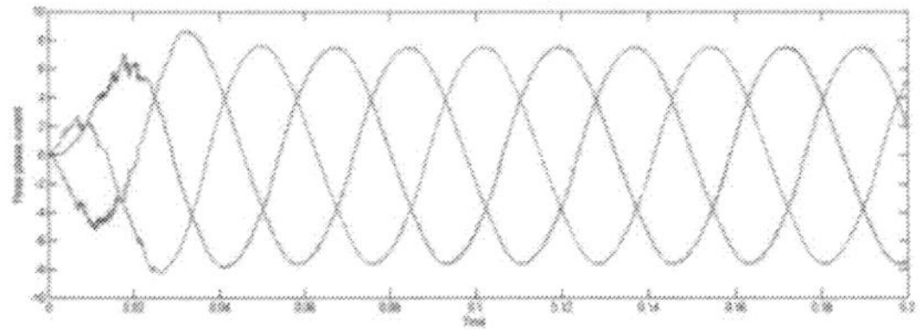

Fig. 15. Three Phase Currents of PMSM with ANN

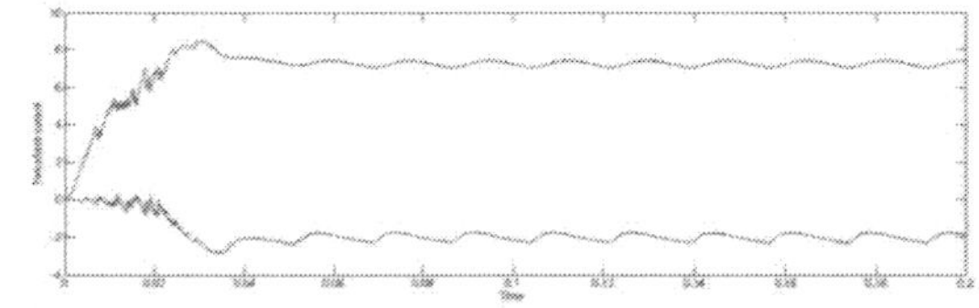

Fig. 16. Two Phase Currents (d, q) of PMSM with ANN

APPENDIX

Table 2. PMSM Specifications

Parameter Name	Parameter Value
Stator Resistance (R)	2.875 Ohm
d-axis self-inductance (L_d)	0.0066 H
q-axis self-inductance (L_q)	0.0058 H
Rotor magnetic flux linkage to the stator (Y_{af})	0.1546 Wb/m^2
Moment of Inertia (J)	0.0176 Kg-m^2
Friction coefficient (B)	0.00038818 N-m/(rad/s)
Number of poles (p)	6
Constant frequency (f)	50Hz
DC link voltage (V_{dc})	150 V
Synchronous speed (N_s)	1000 rpm

6. CONCLUSIONS

This paper talked about numerical demonstrating of executed PMSM drive framework. The concise diagram of the crucial guideline of sinusoidal heartbeat width balance changing the width of the beats in a heartbeat prepare introduced in this article. The higher the control voltage, the more extensive the subsequent vibrations move toward becoming. Steady adequacy beats describe PWM procedures which are very famous in mechanical applications. The width of these heartbeats is balanced to get inverter yield voltage control and to decrease its symphonious substance.

To enhance the framework power and the capacity of brisk recuperation of the speed from any unsettling influence and parameter variety ANN control system created. ANN-based speed controller model of PMSM drive displayed and reproduced and the outcomes spoke to for Artificial Intelligence-based control. These are intended to make an interpretation of the speed mistakes into separate driving voltage signs to the contribution of PMSM. A multilayer feed forward neural system is prepared utilizing Backpropagation learning calculation to evaluate the driving voltage contribution of PMSM. The reproduced aftereffects of shut circle speed control of ANN speed controller with PMSM have created above and underneath appraised speed at no-heap and full load conditions.

References

1. M. Arehpanahi, M. Fazil, "Position Control Improvement of Permanent Magnet Motor using Model Predictive control", ***IJE Transactions A: Basics***, Vol. 31, No. 7 (2018), 1044-1049.
2. Zheng Wang, Jian Chen, Ming-Cheng and K. T. Chau, "Field-oriented control & direct torque control for paralleled VSI fed PMSM drives with variable switching frequencies," ***IEEE Transactions on Power Electronics***, Vol. 31, No. 3 (2016), 2417-2428.
3. R. Pilla, A. S. Tummala and M. R. Chintala, "Tuning of Extended Kalman Filter using Self-Adaptive differential evolution algorithm for sensorless permanent magnet synchronous motor drive", ***IJE Transactions B: Applications***, Vol. 29, No. 11(2016), 1565-1573
4. M. Hinkkanen, T. Tuovinen, L. Hornefors and J. Luomi, "A combination position and stator-resistance observer for salient PMSM drives: Design and stability analysis," ***IEEE Transactions on Power Electronics***, Vol. 27, No. 2 (2013), 601-609.
5. S. Kouro, J. Rodriguez, B. Wu, S. Bernet and M. Perez, "Powering the future of the industry: high power adjustable speed drive topologies," ***IEEE Industry Magazine***, 2012, 26-39.
6. Z. Pan and R. A. Bharat, "Modular motor/converter system with redundancy for high speed, high power motor applications," ***IEEE Transactions on Power Electronics***, vol. 25, no. 2 (2010), 408-416.
7. Junjie, Ren, HuiFeng, HongyingRen and Yi Huang, "Simulation of PMSM vector control system based on propeller load characteristic," Proceeding of the IEEE International conference on intelligent control and information processing, Aug 13-15, 2010, pp. 735-737.
8. R. Krishnan, Permanent magnet synchronous and brushless DC motor drives, USA: Boca Raton, (2010).
9. X. Xiao, C. M. Chen, and M. Zhang, "Dynamic permanent magnet flux estimation of permanent magnet synchronous machines," ***IEEE Transactions on Applied Superconductor***. Vol. 2, No. 3 (2010), 1085-1088.
10. J. Soltani, NR Abjadi, M. Pahlavaninezhad, "An adaptive nonlinear controller for speed sensorless PMSM taking the iron loss resistance into account," ***IJE Transactions B: Applications***, vol. 21, no. 2 (2008), 151-160.
11. M. Rashed, P. F. A. Macconnell, A. F.Stronach and P. Acarnely, "Sensorless indirect rotor field orientation speed control of permanent magnet synchronous motor with stator resistance estimation," ***IEEE Transactions on Industrial Electronics***, vol. 54, no. 3 (2007), 1664-1675.

AUTHOR INDEX